Advance in Comprehensive Utilization of Tea Resource:
Proceeding of 2012 International (Hangzhou) Symposium
on Comprehensive Utilization of Tea Resource

茶资源综合利用研究进展

2012 国际(杭州)茶资源综合利用学术研讨会论文集

论文集编辑组 编

ZHEJIANG UNIVERSITY PRESS
浙江大学出版社

图书在版编目（CIP）数据

茶资源综合利用研究进展：2012 国际（杭州）茶资源综合利用学术研讨会论文集 / 论文集编辑组编. —杭州：浙江大学出版社，2013.3
ISBN 978-7-308-11241-3

Ⅰ.①茶… Ⅱ.①论… Ⅲ.①茶叶－综合利用－国际学术会议－文集 Ⅳ.①TS971－53

中国版本图书馆 CIP 数据核字（2013）第 037035 号

茶资源综合利用研究进展：
2012 国际（杭州）茶资源综合利用学术研讨会论文集

论文集编辑组 编

责任编辑	许佳颖　翁　蔚
封面设计	谢伟亮
出版发行	浙江大学出版社
	（杭州市天目山路 148 号　邮政编码 310007）
	（网址：http://www.zjupress.com）
排　　版	杭州中大图文设计有限公司
印　　刷	杭州杭新印务有限公司
开　　本	710mm×1000mm　1/16
印　　张	23.75
字　　数	414 千
版 印 次	2013 年 3 月第 1 版　2013 年 3 月第 1 次印刷
书　　号	ISBN 978-7-308-11241-3
定　　价	78.00 元

版权所有　翻印必究　印装差错　负责调换
浙江大学出版社发行部邮购电话　（0571）88925591

2012国际(杭州)茶资源综合利用学术研讨会学术委员会

主　　任：陈宗懋（兼）
副主任：梁月荣（常务）　宛晓春　江用文
成　　员：刘仲华　杨秀芳　须海荣
　　　　　翁　蔚　程启坤　俞永明
　　　　　杨贤强　陈霄雄

2012国际(杭州)茶资源综合利用学术研讨会论文集编辑组

组　　长：陈宗懋
副组长：俞永明
英　　文：陈霄雄
责任编辑：翁　蔚

科技·创新与健康

刘枫 题

茶叶深加工是实现中国茶产业跨越式发展的重要途径

陈宗懋 于二〇一三年十一月

序

中国是茶的故乡。在几千年历史进程中，茶的功能和文化内涵不断被发掘，成为国人生活和中华文化的重要组成部分，历久弥新。当代研究表明，茶不仅可以饮、可以品，而且其中提取的多种有效成分可广泛应用于食品、保健品、药品、纺织品、化妆品、装饰品等行业，精加工和综合利用的前景广阔。

杭州是中国茶文化的重要发祥地之一，明代西湖龙井茶的出现开创了我国扁形茶加工历史，径山茶宴成为日本茶道之源。近年来，杭州依托悠久的茶历史、深厚的茶文化特别是一大批"国字号"茶科研教育文化机构，积极倡导"茶为国饮"，大力推进"杭为茶都"建设，茶文化、茶产业、茶科技、茶旅游联动发展，在建设国际重要的旅游休闲中心、提升城乡居民生活品质中发挥了积极作用。杭州将继续推进"中国茶都"建设，弘扬茶文化、做强茶产业、发展茶科技、促进茶旅游，不断提高茶资源综合利用水平，让古老的茶业焕发新的光彩，更好地造福人民，为中国茶产业振兴、中华茶文化复兴作出积极贡献。

这次杭州市政府联合国内茶科研、茶教育、茶文化等单位举办的"2012国际（杭州）茶资源综合利用学术研讨会暨产品展示会"，把征集到的国内外论文汇编成册并出版，是一件很有意义的事。这些论文具有较高的学术水平，对杭州乃至全国推进茶资源综合利用、茶产业转型升级都具有重要的参考价值。我乐之为序。

中共浙江省委常委
中共杭州市委书记 黄坤明
市人大常委会主任

2012年11月8日

目 录

茶叶的多方位利用在茶产业转型中的作用 …………………………………… 陈宗懋(1)
茶及茶多酚抗糖尿病作用的分子与细胞机理研究
　… Yea-Tzy Deng　Tsai-Wen Cheng　Pay-San Tsai　Jheng-Huei Liou
　　Shen Ouyang　Chi-Li Lin　Shoei-Yin Lin-Shiau　Jen-Kun Lin(6)
儿茶素的药理应用、临床试验及其商品化 ……………………………… 原征彦(23)
茶多酚保健作用与功能食品研究开发 ………………… 王岳飞　徐　平　杨贤强(34)
茶叶深加工副产物增值利用技术研究 …………………………………… 张士康(50)
论茶叶精深加工产业发展及其路径 ………………… 陆德彪　罗列万　毛祖法(59)
中国食用植物油产销现状与茶叶籽油的优势 ……………… 程启坤　郭庆元(65)
茶染纺织品的制备及其保健功能研究
　…… 郭　丽　林　智　马亚平　吕海鹏　谭俊峰　祁尚雄　陈金择(71)
韩国爱茉莉太平洋公司含茶化妆品和保健品的研究与开发 ……… 郑镇吾(78)
孟加拉小农茶叶生产与茶叶扶贫,保障食品供给政策
　……………………………………… Mosharraf Hossain　Syed A Hasib(80)
绿茶和茶多酚对笼养蛋鸡脂类代谢的影响
　……………………… 宛晓春　周裔彬　胡经纬　邵　磊　尚岩岩(87)
遮荫栽培条件下超微茶粉制造新工艺及其应用
　………………………………………… 刘勤晋　颜泽文　席阳红(104)
茶叶在日本食品中的应用概况
　………………… Masashi Omori　Yumiko Uchiyama　Kasumi Tsukidate
　　　　Mayumi Yamashita　Yuki Okamoto　Miyuki Katoh(111)
茶资源利用与产品开发 …………………………………………………… 林志城(120)
台湾佳叶龙茶萃取物之生物活性成分、机能性及在食品加工上的应用
　……………………………………… 林圣敦　毛正伦　许清安(127)

以尼古丁受体分子抑制模式探讨儿茶素抑制抽烟诱发乳癌的分子机制
……………………………………………………………………… 何元顺(146)
茶树花新资源的活性成分研究现状与市场前景
………… 屠幼英 徐元骏 杨子银 李 博 金玉霞 汤 雯(158)
高新技术在茶叶多酚提取分离中的应用及发展趋势
………………… 王洪新 娄在祥 马朝阳 尔朝娟 秦晓娟(166)
中国茶饮料市场未来发展趋势 ………… 罗之纲 李松原 杨干辉(176)
福鼎白茶爽保健糖开发与研究 …… 张星海 虞赔力 许金伟 王岳飞(180)
茶色素染纸技术研究 ………………… 吴媛媛 王 洁 屠幼英(186)
茶饼干的配方优化研究 ………… 陈贞纯 金恩惠 孟 庆 屠幼英(197)
茶多酚在护肤品中的应用 ……… 江和源 龙 丹 张建勇 王伟伟(206)
苦瓜绿茶含片的生产工艺研究 ………… 梁 进 侯如燕 宛晓春(214)
径山抹茶及其在茶食品中的应用 … 吴茂棋 庞英华 吴步畅 施海根(224)
茶-粮复合发酵体系茶叶发酵行为研究
………… 李大伟 施海根 朱跃进 张海华 张士康(231)
脱咖啡因花草袋泡绿茶配方及其抗肠癌活性研究初探
………… 涂云飞 杨秀芳 张士康 朱跃进 王盈峰 孔俊豪(239)
夏秋茶鲜叶发酵茶酒的品质形成规律
………… 黄友谊 肖 平 张嫣嫣 倪 超 潘 欣 乔如颖
黄莹捷 常银凤(251)
十两茶水提物降糖作用及机制研究
………… 黄春桃 杜万红 刘仲华 吴浩人 刘小阳 施兆鹏(262)
茶多酚和茶黄素通过 MAPK 途径诱导癌细胞凋亡
……………………………………………………… 李 伟 屠幼英(270)
十两茶提取物对小鼠免疫功能的影响
………… 杜万红 彭世喜 刘仲华 段家怀 吴浩人 施 玲
周重旺 霍 治 施兆鹏(276)
茶叶籽油中反式脂肪酸和苯并(α)芘含量的测定
………… 朱晋萱 朱跃进 张士康 刘国艳 金青哲(283)
花卷茶提取物对高胆固醇血症大鼠血脂和内皮功能的影响 …………
………… 杜万红 刘仲华 施 玲 周重旺 郭立新 刘小阳
姜德建 施兆鹏(291)
茶皂甙结构修饰型引气剂及其引气混凝土性能

………………………… 朱伯荣　叶　勇　杨　杨　瞿　佳(299)
关于茶树修剪枝再利用的探讨 …… 郑生宏　柴红玲　李　阳　何卫中(307)
钙离子对绿茶茶汤沉淀形成的影响
………… 许勇泉　钟小玉　陈根生　邓余良　袁海波　尹军峰(315)
基于响应曲面法的茶黄素发酵工艺优化
………………………… 孔俊豪　杨秀芳　张士康　涂云飞(324)
光照对PET瓶装绿茶饮料品质稳定性影响初探
………… 刘　平　许勇泉　汪　芳　袁海波　刘盼盼　尹军峰(334)
茶籽多糖的表征和生物活性研究 … 王元凤　姜慧仙　毛芳芳　魏新林(344)
松阳茶叶精深加工现状及今后发展方向 ……………………… 叶火香(346)
茶氨酸的保健功能研究进展 …………………… 沈晓玲　龚正礼(350)

Contents

Role of Multi-directional Utilization of Tea in the Development of Tea
　　Industry ··· CHEN Zong-mao(1)
Studies on the Molecular and Cellular Mechanisms of Anti-diabetic
　　Effects of Tea and Tea Polyphenols
　　　··· Yea-Tzy Deng　Tsai-Wen Cheng　Pay-San Tsai　Jheng-Huei Liou
　　　　　Shen Ouyang　Chi-Li Lin　Shoei-Yin Lin-Shiau　Jen-Kun Lin(6)
Pharmacological Applications, Trials and Commercialization of Tea
　　Catechins ··· YUKIHIKO Hara(23)
Health Protective Function and Development of Functional Foods
　　Derived from Tea Polyphenols
　　　···················· WANG Yue-fei　XU Ping　YANG Xian-qiang(34)
Study on the Technology for Increase Additional Value of By-products
　　of Refined Tea Processing　···················· ZHANG Shi-kang(50)
Discussion on the Way to Develop Multipurpose Utilization of Tea
　　···················· LU De-biao　LUO Lie-wan　MAO Zu-fa(59)
Current Situation of Production and Marketing of Edible Oil in China and
　　Superiority to Develop Chinese Tea-seed Oil
　　　···················· CHEN Qi-kun　GUO Qing-yuan(65)
Study on the Tea Dyeing Textiles and Their Health Functions
　　···················· GUO Li　LIN Zhi　MA Ya-ping　LV Hai-peng
　　　　　　TAN Jun-feng　QI Shang-xiong　CHEN Jin-ze(71)
Introduction of Products Used Green Tea in Amore Pcific, Research and
　　Development of Various Products Containing Tea ······ Jinoh Chung(78)
Food Security and Poverty Alleviation through Smallholding Tea Plantation

in Bangladesh ·················· Mosharraf Hossain Syed A Hasib(80)

Effect of Green Tea and Tea Catechins on the Lipid Metabolism of Caged Laying Hens
·················· WAN Xiao-chun ZHOU Yi-bin HU Jing-wei
SHAO Lei SHANG Yan-yan(87)

New Producing Technology and Application of Ultra-micro Tea Powder Under the Condition of Shading Cultivation
·················· LIU Qin-jin YAN Ze-wen XI Yang-hong(104)

All Purpose Utilization and Application of Japanese Tea
·········· Masashi Omori Yumiko Uchiyama Kasumi Tsukidate
Mayumi Yamashita Yuki Okamoto Miyuki Katoh(111)

Comprehensive Utilization of Tea Resource and Product Development
·································· Chih-Cheng LIN(120)

Bioactive Components and Functional Properties of Taiwan γ-aminobutyric Acid (GABA) Tea Extract, and Their Application in Processed Food
·················· Sheng-Dun Lin Jeng-Leun Mau Ching-An Hsu(127)

Tea-polyphenol (−)-Epigallocatechin-3-gallate as a Good Candidate to Prevent Smoking-induced Human Breast Cancer Cells Proliferation through Inhibition of 9-nicotinic Acetylcholine Receptor
·································· HO Yuan-Soon(146)

Studies and Utilization in the Future Market of the Bioactive Compounds in Flowers of Tea (*Camellia Sinensis*): A New Resources
·················· TU You-ying XU Yuan-jun YANG Zi-ying
LI Bo JIN Yu-xia TANG Wen(158)

The Application and Development Tendency of Hi-tech in the Extraction and Separation of Tea Polyphenols
·················· WANG Hong-Xin LOU Zai-xiang MA Zhao-yang
ER Chao-juan QIN Xiao-juan(166)

The Trend of Tea Beverage Market in China
·················· LUO Zhi-gang LI Song-yuan YANG Qian-hui(176)

Cool Health Sugar of Fuding White Tea Development and Research
······ ZHANG Xing-hai YU Pei-li XU Jin-wei WANG Yue-fei(180)

Research on the Technology of Paper-dyeing with Tea Pigments

.... WU Yuan-yuan　WANG Jie　TU You-ying(186)

Research on Optimizing the Formula of Tea Cookies
　　　... CHEN Zhen-chun　JIN En-hui　MENG Qing　TU You-ying(197)

Application of Tea Polyphenols on the Skin Care Products
　　　........................... JIANG He-yuan　LONG Dan
　　　　　　　　　　　　ZHANG Jian-yong　WANG Wei-wei(206)

Research on Processing Technology of Bitter Gourd-green Tea Tablet
　　　.................. LIANG Jin　HOU Ru-yan　WAN Xiao-chun(214)

Study and Development of Jingshan Mattea for Foods
　　　......... WU Mao-qi　PANG Ying-hua　WU Bu-chang　SHI Hai-geng(224)

The Behavior of Tea Fermentation in the Tea-cereals Composite Fermentation System
　　　.................... LI Da-wei　SHI Hai-gen　ZHU Yue-jin
　　　　　　　　　　　　ZHANG Hai-hua　ZHANG Shi-kang(231)

Optimization the Formula of Herbal Green Tea Bags and Preliminary Evaluation the Anti-colon Cancer Activity In-vitro
　　　............... TU Yun-fei　YANG Xiu-fang　ZHANG Shi-kang
　　　　　　　ZHU Yue-jin　WANG Ying-feng　KONG Jun-hao(239)

Quality Formation of Law in Tea Wine Fermented by Fresh Leaves (*Camellia sinensis*)
　　　......... HUANG You-yi　XIAO Ping　ZHANG Yan-yan　NI Chao
　　　　　PAN Xin　QIAO Ru-ying　HUANG Ying-jie　CHANG Ying-feng(251)

Research on the Anti-diabetic Effects of Shi-liang Tea Extract
　　　............... HUANG Chun-tao　DU Wan-hong　LIU Zhong-hua
　　　　　　　　WU Hao-ren　LIU Xiao-yang　SHI Zhao-peng(262)

Tea Polyphenols and Theaflavins Induce Apoptosis of Cancer Cells through MAPK Signaling Pathways .... LI Wei　TU You-ying(270)

Effects of the Extract of Shi-liang Tea on Immune Function in Mice
　　　...... DU Wan-hong　PENG Shi-xi　LIU Zhong-hua　DUAN Jia-huai
　　　　　　　WU Hao-ren　SHI Ling　ZHOU Zhong-wang　HUO Zhi
　　　　　　　　　　　　　　　　　　SHI Zhao-peng(276)

Detection of Trans Fatty Acids and Benzo(α)pyrene in Tea Seeds Oil
　　　............... ZHU Jin-xuan　ZHU Yue-jin　ZHANG Shi-kang

Effects of the Extract of Hua-juan Tea on Blood Lipids and Endothelial Function in Hypercholesterolemic Rats
…… DU Wan-hong　LIU Zhong-hua　SHI Ling　ZHOU Zhong-wang
　　GUO Li-xin　LIU Xiao-yang　JIANG De-jiang　SHI Zhao-peng(291)

Structural Modified Tea Saponin as an Air Entraining Agent and Its Performance in Concrete
………………… ZHU Bo-rong　YE Yong　YANG Yang　QU Jia(299)

Study on Reutilization for Tea Branches Produced by Pruning
…… ZHENG Sheng-hong　CHAI Hong-ling　LI Yang　HE Wei-zhong(307)

Effect of Calcium on Tea Sediment Formation in Green Tea Infusion
………… XU Yong-quan　ZHONG Xiao-yu　CHEN Gen-sheng
　　DENG Yu-liang　YUAN Hai-bo　YIN Jun-feng(315)

Fermentation Technique Optimization of Theaflavins Based on Response Surface Methodology
… KONG Jun-hao　YANG Xiu-fang　ZHANG Shi-kang　TU Yun-fei(324)

Effect of Illumination on the Quality Stability of PET-bottled Green Tea Beverage
………………… LIU Ping　XU Yong-quan　WANG Fang
　　YUAN Hai-bo　LIU Pan-pan　YIN Jun-feng(334)

Characterization and Biological Activities of Purified Polysaccharides from Crude Extract of Tea Seeds
………………… WANG Yuan-feng　JIANG Hui-xian
　　MAO Fang-fang　WEI Xin-lin(344)

Multipurpose Utilization of Tea in Songyang and its Developing Trend
………………………………………… YE Huo-xiang(346)

Research Development of Health Function of Theanine
………………………… SHEN Xiaoling　GONG Zheng-li(350)

LIU Guo-yan　JIN Qing-zhe(283)

茶叶的多方位利用在茶产业转型中的作用

陈宗懋

中国农业科学院茶叶研究所,中国 浙江

摘　要：新中国成立60年来中国茶产业呈现可持续发展态势。2011年,我国茶园面积是1950年的12.1倍,茶叶产量是1950年的23.9倍。21世纪以来中国茶园面积有较大增长。本文提出,茶叶的多方位利用对茶产业发展具有有效解决中低档茶叶的出路、延伸产业链、提升产业附加值和拓展茶的应用领域等作用,可实现可持续发展。在茶叶多方位利用的内涵上,本文从茶饮料、茶食品、饲料添加剂、日常生活中的应用、防辐射、保健品、食品保鲜和化妆品中的应用等8个方面讨论了茶叶的综合利用。

关键词：茶叶；茶产业；多方位利用

Role of Multi-directional Utilization of Tea in the Development of Tea Industry

CHEN Zong-mao

Tea Research Institute, Chinese Academy of Agricultural Sciences, Zhejiang, China

Abstract: The China tea industry showed a tendency of sustainable development in the past 60 years. The tea acreage in 2011 was 12.1 times higher than that in 1950, and the tea production in 2011 was 23.9 times higher than that in 1950. The tea acreage in China in the new century showed a relatively significant increase. How to bring about the further sustainable development in China tea industry, the multi-directional utilization of tea showed the following four roles in the development of tea industry in this paper, i.e., to solve the utilization of middle-low grade tea, to extend the tea industrial chain, to elevate the extra-value of industry and to open up the utilization field of tea. With regarding the intention of multi-directional utilization of tea, the following aspects on the multi-utilization of tea were discussed in this paper: ready-to drink tea, foodstuff processed by tea, feed additives, application of tea in every-day use articles, anti-radiation, food anti-staling and application in the cosmetics etc.

Keywords: tea; tea industry; multi-directional utilization

1 我国茶产业的现状

新中国成立后的 60 年,我国茶产业经历的是一条可持续发展的道路。目前,我国的茶园面积和茶叶产量均居世界首位,茶园面积 2011 年已达 3300 万亩,占世界茶园总面积的 52%;茶叶产量 2011 年为 155.2 万 t,比世界第二名的印度的茶叶产量几乎高 1/3。我国的茶叶出口 2011 年达 32.2 万 t,居世界第二。特别是进入 21 世纪以来,我国茶产业稳定持续发展,产业链延长,第一产业持续增长;第二产业发展迅速,已接近或超过第一产业;第三产业方兴未艾。

如果将我国茶产业的发展按每 10 年一个周期进行统计,可以发现,各个周期的面积和产量的增长呈现一种不平衡的特点。茶园面积的增长除了 20 世纪 70 年代 10 年增加 5.5 万 hm^2 外,一般 10 年间的增加量为 1 万多 hm^2,2001—2011 年茶园面积年均增长达 11 万 hm^2。与此相关联的是茶叶产量的大幅度上升。由于 20 世纪 70 年代茶园面积大增,80 年代出现茶产量的大增,平均每年产量增加 2.2 万 t。2001—2011 年平均每年产量的增加达到空前的 8.5 万 t。可以预计,2011—2020 年茶叶产量还会出现更大的增幅。

产出和销售间的协调和平衡情况如下。根据 2011 年的数据,茶叶出口量 2011 年是 32.2 万 t,内销量是 115 万 t,两者合计为 147.2 万 t。与生产量 155.4 万 t 相比,产销差额为 8 万 t。这个剩余数字是可以容纳的。未来的 3 年中估计会有 700 万亩茶园陆续投产,预计每年产量会增加 28 万～30 万 t,但内销量如果按每年 5% 增加,出口量估计增加不大,这样出口加内销的量和产量的差额就会比 8 万 t 这个数字大得多。这将会造成产大于销的局面,需要从扩大第二产业的消费量来缩小这个差额。

2 茶叶多方位利用的内涵和对产业发展的作用

茶叶多方位利用是 21 世纪我国茶产业转型升级的核心内容。茶产业转型升级就是从现有的茶叶经营模式向更新、更科学、更高效的经营组织模式的转变,是茶产业运行与组织模式的根本性变化,是一种制度性变迁,意味着茶叶增长方式的转变,是产业结构的优化和茶叶组织模式的重新调整。

在当前茶叶产大于销的特殊背景下,充分发挥茶叶的多方位利用是增加消费的重要手段。以下八个方面的开发和应用可供思考。

2.1 茶饮料

利用夏秋季的茶叶原料进行茶饮料生产是一个促进消费、增加产值的有效措施。我国从1997年起步生产,到2011年,茶饮料的产量从1997年的20万t增加到2011年的近1600万t。用6%的中低档鲜叶原料产生了占茶叶总产值50%以上的产值,这就是深加工的巨大潜力。随着我国旅游业的进一步发展,以及功能性茶饮料的开发,茶饮料的生产预期还会有更大的发展前景。

2.2 茶食品

应用夏秋季茶鲜叶原料提取茶多酚类化合物,并添加到各种食品中发挥其抗氧化功能。利用鲜叶加工成不同细度的茶粉,翠绿的色泽和抗氧化功能使得各种食品具有不同的保健功能,而且产生明显的经济效益。

2.3 饲料添加剂

将茶多酚添加到鸡饲料中可以提高鸡肉中的维生素和肌酸含量,同时可以显著降低鸡蛋中的胆固醇,这种低"胆固醇"鸡蛋可为胆固醇患者和老年人提供更健康的鸡蛋,同时提高了经济效益。将茶多酚添加到猪饲料中可使猪肉的肥肉率降低、瘦肉率提高,同时改善了养猪场的卫生环境。将茶多酚添加到宠物的饲料中可以减轻宠物饲养的异味,改善卫生质量。在以动物性饲料为食料的肉食性动物饲料中添加茶多酚后,由于多酚类化合物的抗氧化作用可以减缓体内的脂质过氧化,可以提高肉食性动物的寿命和健康水平。

2.4 人们日常生活中的应用

在内衣、袜、鞋加工的棉织物中添加茶多酚类化合物可以杀灭鞋、袜中的真菌、消除异味。在汽车和空调器的出气和进气处放上茶多酚可以起到吸收异味和污染物质的功能。40%的日本产空调机中有茶多酚。

2.5 防辐射

茶多酚具有防辐射的功效。电视机使用时会有微量的放射线释放。茶多酚片一方面可作为休闲食品,另一方面具有消除这种微量辐射线对人体的不利影响。手机使用时也会释放微量的辐射性物质,国外生产有一种可以黏附在手机表面的小型物品,可以起吸附辐射的作用。

2.6 其他保健品

茶多酚是一种抗氧化剂,具有多种保健功效。茶多酚含片具有增进口腔健康、减少口腔中龋齿细菌和预防口腔癌的功效。茶多酚片具有减肥、降血脂等功效。茶黄素是红茶中的一种茶多酚类化合物。茶黄素片也具有减肥、降血脂的功效。茶氨酸是茶叶中特有的一种氨基酸,具有松弛人体紧张,解除疲劳的功效。茶氨酸片具有消除神经紧张和增强人体免疫力的功能。在香烟的过滤嘴中加入茶多酚可以分解尼古丁,降低吸烟的危害。含茶多酚的牙膏可以增强防龋效果。

2.7 食品保鲜剂

基于茶多酚化合物的抗氧化活性,在食品中添加茶多酚可以发挥其对食品的保鲜功能。如新鲜鱼和水产品用茶多酚处理后可以保持新鲜的色泽。在火腿和肉类加工时加入茶多酚可以延长肉制品的保鲜期,减轻和延缓肉类因脂质过氧化造成的"饸"味。月饼和糕饼加工中加入茶多酚可以减轻和延迟产生"饸"味的时间,使食品保持清净卫生。食用油脂放置时间较长会发生油的过氧化,降低食用品质,加入茶多酚后可以延迟油脂过氧化的进程,提高油脂的品质,延长商品的上架期。

2.8 化妆品的添加剂

茶多酚具有抗氧化功能,在防晒霜中加入茶多酚,可以保护皮肤免受紫外线的辐射伤害,减少引起皮肤癌的可能性。在沐浴露中加入茶多酚可以减轻皮肤的瘙痒感。在化妆品中加入茶多酚还可以减少和延缓皮肤上老年斑的产生。

茶叶的多方位利用是用化学工程的原理和技术,基于医学、食品科学和轻工业的视野,从茶树鲜叶、成茶和茶叶加工过程中产生的副产品中提取、分离其中的茶叶次生代谢物和有效活性成分,并应用于农业、医药业、食品业、畜牧业、轻工业、服装业等方面,由此获得高附加值的深度加工终端产品。

茶的多方位利用对中国产业的发展具有如下作用。

(1)有效解决中低档茶出路。我国茶园面积已有 3300 万亩,到 2015 年,会有几百万亩茶园面积陆续投入生产。如何进一步开发利用大量的夏秋茶是未来若干年茶产业的重要目标。

(2)提升茶叶附加值。茶产业除了依靠第一产业外,要从第二产业的开发利用上获取更大的附加值。实践表明,通过茶叶的多方位利用,可以获得 10～100 倍的效益。

(3)延伸茶产业链。茶产业通过第一产业(加工成各种茶类)、第二产业(通过茶叶多方位利用,包括茶叶中有效成分的提取和应用、茶饮料的开发等)和第三产业(茶文化、茶旅游等)延伸,获得更大的经济效益。

(4)进一步拓展茶叶应用领域。

可以预期,在当前茶园面积大量增加的情况下,茶叶多方位利用是实现可持续发展的有效途径。本次会议邀请了国内外专家介绍在茶叶多方位利用方面的成就和贡献。

茶及茶多酚抗糖尿病作用的分子与细胞机理研究

Yea-Tzy Deng　Tsai-Wen Cheng　Pay-San Tsai　Jheng-Huei Liou
Shen Ouyang　Chi-Li Lin　Shoei-Yin Lin-Shiau　Jen-Kun Lin
台湾大学药学院 生化与分子生物研究所，中国 台湾

摘　要：糖尿病患者最显著的特征就是代谢混乱，每年占全球死亡率的5%。2型糖尿病占糖尿病患者的95%。胰岛素抗性是2型糖尿病患者的主要特征，引发胰岛素信号缺陷。研究表明，茶多酚（EGCG）有抗糖尿病作用，但是其对胰岛素增强机理作用仍未被人们了解。本文研究了EGCG通过人体HepG2细胞、大白鼠胰腺β细胞、C2C12小鼠骨骼肌细胞对胰岛素的反应。研究结果表明，EGCG是预防2型糖尿病的潜在化合物。

关键词：2型糖尿病；EGCG；胰岛素受体底物-1

Studies on the Molecular and Cellular Mechanisms of Anti-diabetic Effects of Tea and Tea Polyphenols

Yea-Tzy Deng　Tsai-Wen Cheng　Pay-San Tsai　Jheng-Huei Liou
Shen Ouyang　Chi-Li Lin　Shoei-Yin Lin-Shiau　Jen-Kun Lin
Institute of Biochemistry and Molecular Biology, College of Medicine,
National Taiwan University Taiwan, China

Abstract: Diabetes mellitus is the most common metabolic disorder that causes about 5% of all death globally each year. Type 2 diabetes comprises 95% of diabetic individuals. Insulin resistance is the primary characteristic of type 2 diabetes which as a result of insulin signaling defects. It has been suggested that the tea polyphenol (－)-epigallocatechin-3-gallate (EGCG) displays some antidiabetic effects, but the mechanism for EGCG insulin-enhancing effects is incompletely understood. In the present study, the investigations of EGCG on insulin signaling are performed in insulin-responsive human

HepG2 cells, rat pancreatic β cells and C2C12 mouse skeletal muscle cells. We found that the high glucose condition causes significant increasing Ser307 phosphorylation of insulin receptor substrate-1 (IRS-1), leading to reduce insulin-stimulated phosphorylation of Akt. As the results, the insulin metabolic effects of glycogen synthesis and glucose uptake are inhibited by high glucose. However, the treatment of EGCG improves insulin-stimulated down signaling by reducing IRS-1 Ser307 phosphorylation. In addition, EGCG could preserve the insulin secretory machinery and stimulate insulin receptor substrate 2 (IRS-2) signaling in rat pancreatic β cells, RIN-m5F and the reduced glucolipotoxic effects of EGCG through activating AMP-activated protein kinase (AMPK) signaling to inhibit the activities of lipogenic enzymes and ameliorating mitochondrial function. Furthermore, EGCG could reduce IRS-1 Ser307 phosphorylation in C2C12 mouse skeletal muscle cells and inhibit lipid accumulation in the intracellular site. Taken together, our results indicated that EGCG might be a potential compound to prevent type 2 diabetes.

Keywords: type 2 diabetes; EGCG; IRS-1

1 Introduction

High levels of circulating glucose, or hyperglycemia, is a serious problem with type 2 diabetes. Once chronic hyperglycemia becomes apparent, it in turn damages insulin target tissues and aggravates insulin resistance, forming a vicious circle that is collectively called glucotoxicity[1]. Recently studies have found that the lacking functions of insulin receptor substrate (IRS), a family of docking molecules connecting insulin receptor (IR) activation to essential downstream kinase cascades, may be the key molecular lesion signature of insulin resistance[2]. This defect appears to be a result of insulin-stimulated IRS-1 tyrosine phosphorylation resulting in reduced IRS-1-associated phosphatidyl inositol 3 kinase (PI3K) activities[3]. Previous result demonstrates that hepatic IRS-1 and IRS-2 have complementary roles in hepatic metabolism control; particularly, IRS-1 is more closely related to glucose homeostasis[4]. Furthermore, disruption of mice IRS-2, not IRS-1, would lead to the spontaneous apoptosis of β cells and to evolve into individuals with the similarities of diabetes[5] IRS-2 acts critically in β cells, which could activate downstream signaling, Akt, FoxO1 (Forkhead-O transcription factor 1), and PDX-1 (Pancreas-duodenum homeobox-1), and preserve the integrity of β cells during the high metabolic demand.

Diabetes mellitus (DM) is the most common metabolic disorder that count about 5% all deaths globally each year. The etiology of DM could be genetic or environmental origins in which type 2 diabetes comprise 90% diabetic individuals two major characteristics of type 2 diabetes are pancreatic beta cell dysfunction and peripheral insulin resistance which both could from an imbalance energy metabolism. Pancreatic beta cell failure induce critical metabolic disorder in the development of type 2 diabetes. Decreased viability and dysfunction of beta cells would accelerate the diabetic pathogenesis associated with higher mortality. In this study, the tea polyphenol EGCG and the buckwheat flavonoid rutin were investigated to attenuate the induced glucotoxicity in beta cells. EGCG and rutin could present insulin secretory machinery and stimulate insulin receptor substrate 2 (IRS-2) signaling in rat pancreas beta cells. (R1N. msF). These findings further demonstrated the reduced glucolipotoxic effects of EGCG and Rutin. The intrinsic protective effects of EGCG and Rutin is preserving the insulin signaling and regulating lipogenesis, manipulating cell cycling and maintaining mitochondrial function to achieve the integrity of beta cells which highlight the possibilities of EGCG and Rutin as novel strategies for the strategy of type 2 diabetes (14).

Tea (Camellia sinensis) is one of the most popular beverages in the world. Of our interest, many folk remedies such as traditional Chinese medicine have mentioned the antidiabetic properties of tea for diabetes management[6]. Epidemiological study has shown a strong correlation between the consumption of tea and the prevention of diabetes[7]. In particular, the bioactivities of (−)-epigallocatechin-3-gallate (EGCG), the major polyphenol isolated from green tea, have been investigated intensively in several studies. It has been demonstrated that EGCG displays some antidiabetic activities in tea. For example, EGCG was verified to be the predominant active compound in green tea and have insulin-enhancing activity[8]. Dietary supplementation with

EGCG markedly enhances glucose tolerance in diabetic rodents[9]. In addition, EGCG mimics insulin actions by inducing PI3K-sensitive phosphorylation of transcription factor FOXO1a (Forkhead box O1a) which is sensitive to scavengers of free radicals[9].

TNF-α is a major player mediating the activation of signaling cascades in adipocytes that are central to inflammation and insulin resistance. In the regulation, TNF-α triggers adipocyte activation of the mitogen activated kinase (AMPKs) inhance cellular signal-regulated kinase (ERK1/2), c-jun N-terminal kinase (JNK) and p38 and of transcription factor AP-1 and nuclear factor κB. These signaling cascades are in part redox regulated, given binding of TNF-alpha to its receptor results in the activation of NAPDH oxidase leading to an increase in oxidant production. Activation of these signaling pathways leads to oxidant production. And leads to pro-inflammatory IL-6, IL-8, IL-1 and MCP-1, NF-κB also drives an tyrosine phorylation 1B (P1P1B), a negative regulator of insulin signaling of which dephosphorylation tyrosine induces of the insulin receptor (IR) and insulin receptor substrate-1 (IRS-1). Furtherance proliferation-activator receptor gamma (PPARγ) that plays a major role in the regulation of both glucose and lipid metabolism.

Flavonoids and phytopolyphenolic compounds that are widely present in human diets. Epidemiological studies have shown can inverse relationship between consumption of flavonoid-rich foods and insulin of chronic diseases such as cancer and diabetes. Health effects of flavonoids can be in part of a inhibited to their capacity to regulate oxidant production and proinflammatory signals. Recent reports support an anti-inflammatory drug signal. Addition of a polyphenol-rich grape powder extract to human adipocytes, inhibits TNF-α trigger activation of MAPKs and NF-κB and the expression of proteins involved in inflammation and insulin resistance. High concentration of the phenolic compounds p-curcurmic acid, quercetin and resveratrol prevent TNF-α-induced increase in several parameters of inflammation and oxidative stress and of decrease insulin sensitivity in 3T3L1 and human adipocytes. (−)Epicatechin is one of the most abundant flavonoids in human diets being present in high concentration in grapes, cocoa tea and many other fruits and vegetables. A recent report describes (−)Epicatechin prevents TNF-α-induced activation of signaling cascade involved in inflammation and insulin sensitivity in 3T3L1 adipocytes. (15) Baicalein (3,6,7-trihydroxyflavone) is a polyphenol compound and major bioactive flavonoid isolated from the root of scutellavia baicalensis Georqi which is widely used in traditional Chinese medicine. Many report have shown that baicalein could activate AMPK in hunman tumor cells, a recent report demonstrated that baicalein can activate AMPK in vivo and in vitro and it might be an ideal could treat metabolic syndrome. Erivdictyal is an catechein derivatives that can increase glucose uptake and improve insulin resistance, suggesting that it may process antidiabetic properties. Our studies have demonstrated that (−)epicatechein-gallate(EGCG) could attenuated the insulin resistance in human hepatoma cells (13) and in rat pancreatic β cell (14). Our recent investigation has demonstrated that high glucose induced insulin blockade can be attenuated by EGCG in HepG2 cell (13).

2 Specific Aims

In this study, we used human HepG2 cells, rat pancreatic β cells and C2C12 mouse skeletal muscle cells to investigate the effects of EGCG on signaling pathways to prevent type 2 diabetes.

3 Methods

3.1 Cell Culture

HepG2 cells were maintained in DMEM, supplemented with 10% fetal bovine serum, 100units/mL penicillin, 100μg/mL streptomycin, 2mmol/L L-glutamine (Invitrogen, CA, USA), and kept at 37℃ in a humidified atmosphere of 5% CO_2 in air. RIN-m5F rat insulinoma pancreatic β cells were obtained from the NHRI cell bank (National Health Research Institutes, Taiwan, China) and maintained in RPMI 1640 containing 11mmol/L glucose supplemented with 10mmol/L HEPES, 1 mmol/L sodium pyruvate, 10% (V/V) fetal bovine serum, 100units/mL penicillin, and 100μg/mL streptomycin (HyClone, Logan, UT) in a humidified 5% CO_2 incubator at 37℃. Mouse skeletal muscle cell lines, C2C12 myoblasts, were maintained in Dulbecco's modified Eagle's medium (DMEM) supplemented with 10% (V/V) fetal bovine serum, 100units/mL penicillin, and 100 μg/mL streptomycin (HyClone, Logan, UT) at 37°C in a humidified 5% CO_2 atmosphere. After 3 days of plating, the cells had reached 80%~90% confluence (day 0). Differentiation was then induced by replacing the growth medium with DMEM supplemented with 2% horse serum instead of 10% fetal bovine serum (differentiation medium). Myotubes formation was achieved after 4~5d of incubation, and the cells were used for subsequent experiments.

3.2 Western Blot Analysis

Cells were lysed with lysis buffer [10% glycerol, 1% Triton X-100, 0.1% sodium dodecyl sulfate (SDS), 10mmol/L NaF, 50mmol/L Tris-HCl, pH 8.0, 5mmol/L ethylenediaminetetraacetic acid, 150mmol/L NaCl, 0.5mmol/L phenylmethylsulfonyl fluoride, 10μg/mL aprotinin, 10μg/mL leupeptin, and 0.5mmol/L dithiothreitol], and the cell lysates were centrifuged at 12000r/min for 30min at 4°C, and then, the supernatants were collected as whole cell extracts. For Western blotting, equal amounts of total cellular protein (50μg) were subjected to 8% SDS-polyacrylamide gel electrophoresis and then transferred to a polyvinylidene difluoride membrane (Millipore), and the membranes were then blocked for 1h at room temperature with 1% BSA in PBS. The membranes were next immunoblotted with primary antibodies at dilution of 1∶1000 followed by secondary antibodies with a 1∶5000 dilution of anti-mouse or rabbit IgG-conjugated with horseradish peroxidase. The immunocomplexes were visualized with enhanced chemiluminescence system (Perkin-Elmer Life Sciences, Boston, MA).

3.3 Glucose Uptake

The glucose uptake rate was measured by adding fluorescent D-glucose analogue 2-[N-(7-nitrobenz-2-oxa-1,3-diazol-4-yl) amino]-2-deoxy-D-glucose (2-NBDG) as a tracer to the culture medium. After treatments, cells plating in the 24-well were incubated with or without insulin (100 nM) for 15 min, and then, 2-NDBG was added at a 50μmol/L final concentration for another

20min. The medium was then removed, and the cells were washed three times with cold PBS. The cells in each well were suspended with PBS after trypsinization, and the fluorescence intensity was read by fluorescent spectrometry at an excitation of 485nm and an emission of 535nm.

4 Results

4.1 Effects of EGCG Treatment on Insulin-induced Akt Phosphorylation in HepG2, Rat-cells and C2C12 Mouse Skeletal Muscle Cells

To determine whether the increase in insulin sensitivity by EGCG is due to the alternation of IRS-1 phosphorylation state, we found that the EGCG treatment recovers tyrosine phosphorylation of IRS-1 in a dose-dependent manner. Furthermore, cotreated EGCG also decreased Ser307 phosphorylation of IRS-1 in a dose-dependent manner (Fig. 1(b)). These data indicated that high glucose promotes the activation of serine phosphorylation and deactivation of tyrosine phosphorylation of IRS-1 in HepG2 cells, whereas EGCG reverses the state of IRS-1 phosphorylation and enhances insulin-stimulated Akt Ser473 phosphorylation (Fig. 1(a)).

Rutin, a flavonoid from buckwheat (Fagopyrum esculentum), has been shown the ability to reduce blood glucose concentration in diabetic model rats[10], yet the underlying molecular mechanism has not been elucidated. To determine the metabolic effects of EGCG and Rutin in β cells by protecting the activity of IRS-2 signaling, cells were incubated in 33mmol/L glucose condition with or without the EGCG and Rutin treatment. The results showed that EGCG and Rutin stimulated the IRS-2 signaling through enhancing tyrosine phosphorylation after 2h or 24h under high glucose condition (Fig. 2(a)), and then the stimulated effects could be sustained until 48 h (Fig. 2(b)). EGCG and Rutin could effectively reduce the level of serine phosphorylation of IRS-2 after 48h under high glucose incubation (Fig. 2(b)). To assess the action of EGCG and Rutin in the activation of Akt, the treated cells under high glucose condition were evaluated by Western blotting, which revealed that EGCG and Rutin both could preserve and extend the activation of Akt under high glucose condition (Fig. 2(c)).

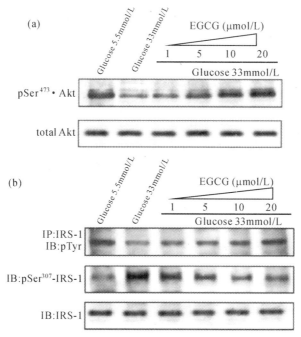

Fig. 1 Consequences of insulin: stimulated Akt phosphorylation by EGCG

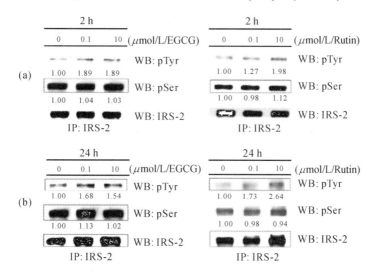

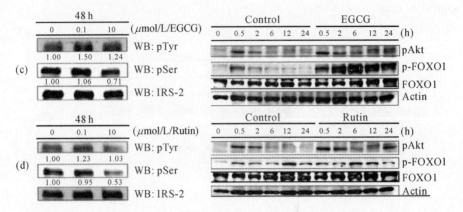

Fig. 2 Effects on the stimulation of IRS-2 signaling

C2C12 myotubes were stimulated by 0.5μmol/L TPA for 30min and then treated with EGCG in different doses. The result showed a decreased content of phospho-IRS-1 Ser307 and an increase in Akt phosphorylation only in a high concentration of EGCG (40μmol/L) treatment (Fig. 3).

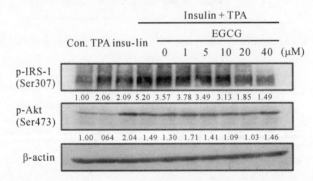

Fig. 3 Effects of EGCG on TPA: induced IRS-1 signaling down-regulation

4.2 EGCG Increases Insulin-stimulated Glucose Uptake in HepG2 and C2C12 Cells

In response to insulin, glucose uptake increased three-fold over basal levels (Fig. 4(a)). In contrast to the insulin-stimulated control group, the high glucose treatment in HepG2 cells inhibited insulin-stimulated glucose uptake by over 50%. As expected, EGCG-stimulated glucose uptake demonstrated in a dose-dependent manner under high glucose, with 20μmol/L stimulating glucose uptake by ~80% compared with the high glucose group.

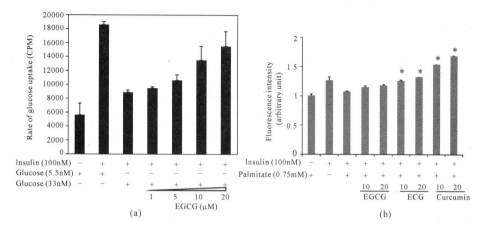

Fig. 4 EGCG increases insulin-stimulated glucose uptake in HepG2
* $P<0.05$

To evaluate whether EGCG, ECG, and curcumin enhance glucose uptake in the state of FFA-induced insulin resistance, C2C12 myotubes were incubated with 0.75mmol/L palmitate for 16h and then treated with polyphenols for 3h. As shown in Fig. 4(b), cells were exposed to 2-NBDG for 20min and then washed and displayed a significant rescue of fluorescence intensity after treatment with these three polyphenols, especially curcumin.

The search for the modulation of IRS activities has been suggested to focus on its phosphorylation state. Most notably, serine/threonine phosphorylation has been shown to modulate both positive and negative signaling transmission via IRS. Based on the fact, some high IRS-1 serine phosphorylation (i.e., Ser307) leads to an insulin-signaling blockade by inhibiting insulin-induced IRS-1 tyrosine phosphorylation. This provides the indication that increasing Ser307 IRS-1 phosphorylation could affect insulin action. By this reason, we suggest that EGCG may reduce IRS-1 serine phosphorylation. Although the understanding of high glucoseinduced insulin resistance ultimately leads to metabolic disorders has advanced considerably in recent years, effective therapeutic strategies to prevent or to delay the development of this damage remain limited. Further experiments directed at the determining the mechanism of EGCG or other novel phytochemicals action may lead to the identificationof molecular targets for the generation of therapeutic agents useful in the management of insulin resistance disease like diabetes.

5 Discussion

Defects in several insulin signaling pathways, such as reduced IRS tyrosine phosphorylation and decreased PI3K activity, are reported in human and several animal models of diabetes[3]. However, EGCG has previously been shown to display some anitdiabetic properties. For instance, the intake of green tea extract ameliorates fructose-induced insulin resistance[11]. Besides, EGCG also has been reported to lower glucose production and decreases the expression of genes that control gluconeogenesis[12]. These results indicate that EGCG indeed has insulin-potentiating activity; however, a clear molecular mechanism underlying these effects has not been established. In this study, we demonstrate that the exposure to high glucose results in Ser307 phosphorylation of IRS-1. Moreover, we also display that the EGCG treatments significantly reverse these high glucose effects. Therefore, EGCG appears to stimulate hepatic insulin sensitivity through a signaling pathway distinct from the insulin signaling itself. In this study we have confirmed that EGCG is indeed inhibitory to Ser307 phosphorylation in IRS-1. EGCG activates AMPK at Ser789 which in turn attenuate phosphorylation of IRS-1 at Ser307. In addition EGCG attenuates PKC/JNK signalings then phosphorylation of IRS-1 at Ser307 is blocked then PI3K/Akt is activated (Chat 1).

This study also highlights the novel function of EGCG in rat pancreatic β cells, integrating the molecular effects of glucose and lipid metabolism. EGCG activates AMPK and Tyr phosphorylation on IRS-2 then Akt is activated. Akt phosphorylates FOXO1 can promote synthesis of insulin. Consequently, these data suggested that EGCG might be potential agents for the attenuation of type 2 diabetes through the protection of pancreatic β cells. In C2C12 mouse skeletal muscle cells, an increased FFA level induces serine kinase PKC activation, resulting in Ser307 phosphorylation of IRS-1 and inhibition of its Tyr phosphorylation, reducing the binding affinity of IRS-1 to IR. These results

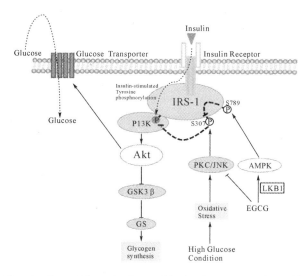

Chat 1 Potential mechanisms of high glucoseinduced insulin signaling blockade in HepG2 cells

down-regulate the PI3K-Akt signaling pathway and reduce insulin-stimulated glucose uptake.

A potential mechanism of this attenuation by EGCG was summarized in Chat 1. The high glucose induced oxidative stress triggers serine kinase cascades by activating PKC and JNK. This ultimately induces Ser 307 phosphorylation of IRS-1 site. This serine phosphorylation reducing IRS-1 tyrosine phosphorylation and these by inhibiting IRS-1 binding to insulin receptor (IR). These results down regulate PI3K/Akt signaling pathway and reduce insulin-stimulated glucose uptake and glycogen synthesis. However IRS-1 tyrosine phosphorylation has been demonstrated to be positively regulated by AMPK activation. Our results indicates that EGCG is shown to both reduce PKC/JNK activation and activate AMPK activity leading to block IRS-1 serine phosphorylation at Ser 307 and attenuate high glucose induced insulin signaling blockade (Chat 1). Further experiments directed at the determine the mechanisms of EGCG or other phytochemical actions may lead to the identification of molecular targets for the generation of therapeutic agents useful for the management of insulin resistance disorder like diabetes (13).

17

A possible molecular mechanisms dealing with the attenuations of glucotoxicity through modulating insulin receptor substrate 2 (IRS-2) signalings have been summarized in Chat 3. Chronic high glucose exposure could directly increase intracellular ROS generation and deteriorate mitochondrial function to uncouple with ATP generation, impairing the glucose stimulated insulin resistance. Suppression of IRS-2 signaling could be induced by serine phosphorylation cased from TNK and GSK3 activation which would disrupt the tyrosine phosphorylation and the recruitment of P13K to actuate downstream signaling Akt, FoxO1 and PDX-1. Akt display multiple roles in pancreatic beta cells regulating cell proliferation through cyclin D1 and BCL2 and suppression cell apoptosis through p21 and BAX inhibition. Otherwise, Akt phosphorylation FoxO1 to request it from the nucleus, facilitative the of PDX-1 (important for insulin and glucose function). Which in for insulin synthesis and cell survival chronic high glucose exposure suppressed AMPK activity and increased lipogenesis through activition lipogenic enzymes acetyl-CoA carboxylase (ACC) and FAS (Fatty acid synthetase). Lipid accumulation in beta cells would with the cellular function SREBP-1a conserved transcription factor in lipid metabolism would suppress IRS-2 and PPX-1 expression and increase with cellular lipogensis, EGCG and Rutin could increase Akt IRS-2 and AMPK suppression from high glucose exposure the of pancreatic beta cells to deal with glucotoxicity. The dotted line the of EGCG and Rutin could the IRS-2 degradation through Chat 2 (see Ref. 14). Chronic glucose exposure could stress pancreatic beta cells to compensate the over loaded metabolism and then that could induce the development of glucotoxicity and lipotoxicity situations.

AMPK activity is considered to not only but to be linked with insulin resistance and cellular lipogenesis. Developed antidiabetic drugs, metformin and TZDs (Thiazolidinedione) could activate AMPK, meating for insulin resistance and disorders associated with the metabolic syndrome. EGCG and Rutin could increase AMPK activity through NF-κB and then suppress ROS generation (Chat 3). This the function of EGCG and Rutin in rat pancreatic beta cells with effects of glucose and lipid metabolism, consequences, these data suggested that EGCG and Rutin might be agents for the attenuation of type 2 diabetes the of pancreatic beta cells (14).

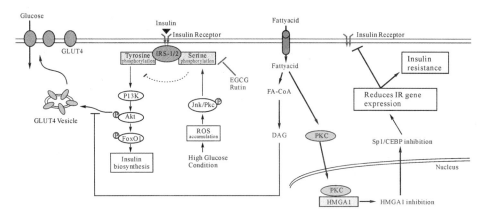

Chat 2　Fatty acid and high glucose induced insulin resistance

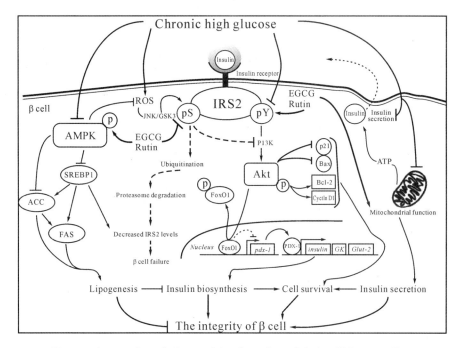

Chat 3　Attenuation of glucotoxicity through modulating IRS-2 signalings

References

[1] Rossetti L. Glucose toxicity: the implications of hyperglycemia in the pathophysiology of diabetes mellitus[J]. Clin Invest Med, 1995, 18(4): 255−260.

[2] Thirone A C, Huang C, Klip A. Tissue-specific roles of IRS proteins in insulin signaling and glucose transport[J]. Trends Endocrinol Metab, 2006, 17(2): 72−78.

[3] Petersen K F, Shulman G I. Etiology of insulin resistance[J]. Am J Med, 2006, 119(Suppl 1): 10−16.

[4] Taniguchi C M, Ueki K, Kahn R. Complementary roles of IRS-1 and IRS-2 in the hepatic regulation of metabolism[J]. J Clin Invest, 2005, 115(3): 718−727.

[5] Withers D J, et al. Disruption of IRS-2 causes type 2 diabetes in mice[J]. Nature, 1998, 391(6670): 900−904.

[6] Hale P J, et al. Xiaoke tea, a Chinese herbal treatment for diabetes mellitus[J]. Diabet Med, 1989, 6(8): 675−676.

[7] Wolfram S, Wang Y, Thielecke F. Anti-obesity effects of green tea: from bedside to bench[J]. Mol Nutr Food Res, 2006, 50(2): 176−187.

[8] Anderson R A, Polansky M M. Tea enhances insulin activity[J]. J Agric Food Chem, 2002, 50(24): 7182−7186.

[9] Wolfram S, et al. Epigallocatechin gallate supplementation alleviates diabetes in rodents[J]. J Nutr, 2006, 136(10): 2512−2518.

[10] Kawa J M, Taylor C G, Przybylski R. Buckwheat concentrate reduces serum glucose in streptozotocin-diabetic rats[J]. J Agric Food Chem, 2003, 51(25): 7287−7291.

[11] Li R W, et al. Green tea leaf extract improves lipid and glucose homeostasis in a fructose-fed insulin-resistant hamster model[J]. J Ethnopharmacol, 2006, 104(1/2): 24−31.

[12] Koyama Y, et al. Effects of green tea on gene expression of hepatic gluconeogenic enzymes in vivo[J]. Planta Med, 2004, 70(11): 1100−1102.

[13] Lin C-L, Huang H-C, Lin J-K. Theaflavins attenuate hepatic lipid accumulation through activating AMPK in human HepG2 cells[J]. J Lipid

Res, 2007, 48(11): 2334—2343.

[14] Cai E P, Lin (J-K). Epigallocatechin Gallate (EGCG) and Rutin suppress the glucotoxicity through activating IRS-2 and AMPK signaling in rat pancreatic β cells[J]. J Agric Food Chem, 2009, 57: 9817—9827.

[15] Marcela A, Vazquez-P, et al. (−)Epicatechein prevnts TNF alpha-induced activation of signaling cascade involved in inflammation and insulin sensitivity in 3T3L1 adipocytes (2012) Arch[J]. Biochem Biophys, 101016: 2012—2019.

[16] Pu P, et al. Baicalein, a natural product, selectively activating AMPKα(2) and ameliorates metabolic disorder in diet-induced mice[J]. Mol Cell Endocrinol, 2012, 362(1/2): 128—138.

[17] Zhang W Y, et al. Effect of eriodictyol on glucose uptake and insulin resistance in vitro[J]. J Agric Food Chem, 2012, 60: 7652—7658.

[18] Kuhajda F P, Landree L E, Ronnett G V. The connections between C75 and obesity drug-target pathways[J]. Trends Pharmacol Sci, 2005, 26(11): 541—544.

[19] Lin Y L, Lin J K. (−)-Epigallocatechin-3-gallate blocks the induction of nitric oxide synthase by down-regulating lipopolysaccharide-induced activity of transcription factor nuclear factor-kappaB[J]. Mol Pharmacol, 1997, 52(3): 465—472.

[20] Liang Y C, Lin-shiau S Y, Chen C F, et al. Suppression of extracellular signals and cell proliferation through EGF receptor binding by (−)-epigallocatechin gallate in human A431 epidermoid carcinoma cells[J]. J Cell Biochem, 1997, 67(1): 55—65.

[21] Lin Y L, Cheng C Y, Lin Y P, et al. Hypolipidemic effect of green tea teaves through induction of antioxidant and Phase II enzymes including superoxide dismutase, catalase and glutathione-S-Transferase in rats[J]. J Agric Food Chem, 1998, 46: 1893—1899.

[22] Liang L-S, Cheng L. EGCG arrests cell cycle at G1 phase in human breast carcinoma cells[J]. J Cell Biochem, 1999, 75: 1—12.

[23] Lin J K, Liang Y C, Lin-Shiau S Y. Cancer chemoprevention by tea polyphenols through mitotic signal transduction blockade[J]. Biochem Pharmacol, 1999, 58(6): 911—915.

[24] Lee H H, Ho C T, Lin J K. Theaflavin-3,3'-digallate and penta-O-galloyl-beta-D-glucose inhibit rat liver microsomal 5alpha-reductase activity and the expression of androgen receptor in LNCaP prostate cancer cells [J]. Carcinogenesis, 2004, 25(7): 1109—1118.

[25] Lin C L, Chen T F, Chiu M J, et al. Epigallocatechin gallate (EGCG) suppresses beta-amyloid-induced neurotoxicity through inhibiting c-Abl/FE65 nuclear translocation and GSK3 beta activation[J]. Neurobiol Aging, 2009, 30(1): 81—92.

[26] Lin C L, Lin J K. Epigallocatechin gallate (EGCG) attenuates high glucose-induced insulin signaling blockade in human hepG2 hepatoma cells [J]. Mol Nutr Food Res, 2008, 52(8): 930—939.

[27] Lin J K, Lin-Shiau S Y. Mechanisms of hypolipidemic and anti-obesity effects of tea and tea polyphenols[J]. Mol Nutr Food Res, 2006, 50(2): 211—217.

[28] Pan M H, Liang Y C, Lin-Shiau S Y, et al. Induction of apoptosis by the oolong tea polyphenol theasinensin A through cytochrome c release and activation of caspase-9 and caspase-3 in human U937 cells[J]. J Agric Food Chem, 2000, 48(12): 6337—6346.

[29] Way T D, Lee H H, Kao M C, et al. Black tea polyphenol theaflavins inhibit aromatase activity and attenuate tamoxifen resistance in HER2/neu-transfected human breast cancer cells through tyrosine kinase suppression[J]. Eur J Cancer, 2004, 40(14): 2165—2174.

[30] Deng Y-T, Chang T W, Lee M S, et al. Suppression of free fatty acid-induced insulin resistance by phytopolyphenols in C2C12 mouse skeletal muscle cells[J]. J Agric Food Chem, 2012, 60: 1059—1066.

[31] Yeh C W, Chen W J, Chiang C T, et al. Suppression of fatty acid synthase in MCF-7 breast cancer cells by tea and tea polyphenols: a possible mechanism for their hypolipidemic effects [J]. Pharmacogenomics J, 2003, 3(5): 267—276.

儿茶素的药理应用、临床试验及其商品化

原征彦

Tea Solutions, Hara Office Inc., 日本 东京

摘 要：本文将回顾本公司研究儿茶素的情况。Tsuneo Kada 博士 1980 年研究报告，绿茶溶液对培养基中的芽孢杆菌 NIG1125 具有抑制作用。我们与 Kada 博士合作研究发现，产生这种抗突变生物作用取决于茶多酚。20 世纪 80 年代初，我们大量地从绿茶中提取 EGCG 和其他儿茶素。从 1996 年起，我们在美国国家癌症研究所（NCI）20 个病例用"多酚 E"进行临床试验。此外，1990 年，我们与北京癌症中心合作研究发现，对因感染乳头瘤病毒而罹患生殖器疣的试验患者局部施用儿茶素可减少疣的数量。德国一家制药公司基于我们的研究结果在欧盟和美国开展 2～3 期临床试验。2006 年，美国食品和药品管理局（FDA）批准"多酚 E"软膏上市销售。它成为美国食品和药品管理局首次批准的"植物性药物"，也是迄今为止世界上唯一以茶提取物为原料制作的药品。

关键词：儿茶素；EGCG；多酚 E；茶酚 E；植物性药物

Pharmacological Applications, Trials and Commercialization of Tea Catechins

YUKIHIKO Hara

Tea Solutions, Hara Office Inc., Tokyo, Japan

Abstract: Though a series of new clinical trials with tea catechins are planned in my Tea Solutions, Hara Office Inc. let me report in this review what had been done in the past with tea catechins where I was involed in. In 1980, the late Dr. Tsuneo Kada reported that the mutability of Bacillus subtilis NIG1125 was suppressed when the brewed green tea was added to the medium. We worked with Dr. Kada and found that the responsible compound that shows the bio-antimutagenecity was (—)-Epigallocatechin gallate (EGCG). Taking advantage of these findings we started in early 1980's to isolate EGCG and other catechins in large amounts from green tea. We had made use of these samples, for ourselves and for our collaborators around the world, for the elucidation of physiological functions of tea catechins, under the assumption that the various so far claimed health benefits of tea drinking should have derived from tea catechins. From 1996 onward, we have been supplying the defined tea catechin mixture, Polyphenon® E to the US National Cancer Institute (NCI) for their chemoprevention clinical trials in over 20 instances. Separately, we found in 1990 that the topical application of tea catechins on genital warts caused by HPV eliminates the warts in a number of trials in Beijing Cancer Center. The results prompted a German pharmaceutical company to do the Phase 2/3 trials in EU and US. In 2006, the US FDA (Food and Drug Administration) approved the marketing of Polyphenon® E ointment, Veregen®, as the first "Botanical Drug" under FDA regulation. Currently, new and various developmental works with Theaphenon® E by Tea Solutions, Hara Office Inc. is underway.

Keywords: tea catechins; EGCG; Polyphenon® E; Theaphenon® E; Botanical drug

1 Introduction

Over the last 30 years, various beneficial health functions of tea catechins have been investigated. Those representative properties and beneficial health effects were reviewed in two books by Hara[1] and Kuroda and Hara[2]. The topics include the history of tea, tea catechins and their extraction methods, anti-oxidative action[3], anti-bacterial action[4], anti-viral action[5], prevention of cancer[6], hypolipidemic action[7], hypoglycemic action[8], hypotensive action[9], effects on intestinal flora [10] and practical applications of those functions.

Recently, very extensive reviews on "*Tea and Health*" were edited by Yang C S et al. [11][12], in which the habitual intake of tea catechins, with their very potent anti-oxidative, radical scavenging actions, is quoted to have protective effects against such life-style-related, age-related diseases as cancer, cerebro, coronary diseases as well as the effects on body weight or on bone health are discussed. Tea catechins also show very potent affinity with specific molecules, compounds or microbes, which impart tea catechins with properties such as anti-viral and anti-bacterial actions against life threatening infectious diseases and deodorizing actions against oral or environmental odors. Recently, an entirely new function of EGCG against cancerous cells has been proposed by Tachibana, H. where the binding of EGCG to 67kD Laminin Receptor, specific to cancerous cells, triggers the apoptotic process[13].

In this review, recent developments in the application of tea catechins in the prevention and treatment of diseases and a trial for the wider use of tea are discussed.

2 Prevention of Cancer

From the point of view of safety and public acceptability, chemoprevention by food or food ingredients seems to be a most welcome idea but no food or food ingredients have ever been approved by Health Authorities, such as the US or EU FDA, to be marketed as a chemoprevention drug.

In attempts to prove the chemopreventive efficacy of tea catechins, the US NCI and we (Mitsui Norin Co. Ltd.) worked for more than 10 years from 1996, mainly on the safety of the intended agent[14]. We manufactured "Polyphenon® E", by extraction from green tea leaves (*Camellia sinensis*), which is composed of about 70% EGCG and the remainder of other tea catechins[15]. In 2004 the US FDA issued a guideline for "Botanical Drug Products" wherein active components (Active Pharmaceutical Ingredient, API) of a botanical drug should be the crude extract of plants without being purified. The critical feature of a botanical drug is that no particular effective component is assumed[16]. The agent will be approved as a botanical compound if the defined crude extract in the drug formula shows efficacy in clinical phase trials. As a crude extract of green tea, Polyphenon® E is composed of more than 10 different catechins as well as other miscellaneous components including trace of unknowns. In the manufacturing of Polyphenon® E under cGMP (current Good Manufacturing Practice) on the Botanical Drug system, certain criteria must be fulfilled in the specification, such as "consistency", "stability", "absence of adulterants" and "traceability". In other words, as an FDA official put it, "process is the product". The huge document which supports the above specification is termed DMF (Drug Master File) and is filed with the FDA. In the last 10 years, NCI/DCP (Dept. of Chemoprevention) have been supporting more than 20 clinical phase 1 and 2 trials of chemoprevention, making use of Polyphenon® E capsules/placebos, in the States in collaboration with investigators as well as Mitsui Norin Co.[17]. See the following List. The difficulty in these trials is to identify biomarkers that accurately predict an agent's clinical benefit or cancer-incidence reducing effect. Another difficulty is the recruiting of the subjects since those subjects are essentially healthy individuals with little motivation to join in the trials.

3 List: Clinical Trials with Polyphenon® E[①]

(No.) (Study)

1 (R) Study of Polyphenon E in Men with High-grade Prostatic Intraepitherial Neoplasia (Phase 2)

Sponsor: H. Lee Moffit Cancer Center and Research Institute

2 (C) Erlotinib and Green Tea Extract (Polyphenon E) in Preventing Cancer Recurrence in Former Smokers Who Have Undergone Surgery for Bladder Cancer (Phase 2)

Sponsor: University of California, Los Angeles

3 (R) Safety and Neuroprotective Effects of Polyphenon E in Multiple Sclerosis (Phase 2)

Sponsor: National Center for Complementary and Alternative Medicine (NCCAM)

4 (R) Study to Evaluate Safety and Toxicity of Polyphenon E in HIV-1-infected Individuals (Phase 1)

Sponsor: Baylon College of Medicine

---

① (A): Active, not recruiting, (C): Completed, (N): Not yet recruiting, (R): Recruiting, (S): Suspended, (T): Terminated, (U): Unknown

5 (U) Safety of Polyphnon E in Multiple Sclerosis Pilot Study (Phase 1)

Sponsor: National Center for Complementary and Alternative Medicine (NCCAM)

6 (A) Safety of Polyphenon E in Addition to Erlotinib in Advanced Non Small Cell Lung Cancer (Phase 1/2)

Sponsor: Louisiana State University Health Science Center, Shreveport

7 (C) Efficacy and Safety Study of Polyphenon E to Treat External Genital Warts (Phase 3)

Sponsor: MediGene

8 (R) Phase 1 Chemoprevention Trial with Green Tea Polyphenon E and Erlotinib in Patients with Premalignant Lesions of the Head and Neck (Phase 1)

Sponsor: Emory University

9 (A) Green Tea Extract in Preventing Esophageal Cancer in Patients with Barrett's Esophagus (Phase 1)

Sponsor: M.D. Anderson Cancer Center

10 (R) Treatment of Epidermolysis Bullosa Dystrophica by Polyphenon E (Phase 2)

Sponsor: Centre Hospitalier Universitair de Nice

11 (U) Pilot Study of Green Tea Extract in Ulcerative Colitis (Phase 2)

Sponsor: University of Louisville

12 (A) Green Tea or Polyphenon E in Preventing Lung Cancer in Former Smokers with Chronic Obstructive Pulmonary Diseases (Phase 2)

Sponsor: University of Arizona

13 (A) A Study of the Effect of Polyphenon E on Breast Cancer Progression (Phase2)

Sponsor: Louisiana State University Health Science Center, Shreveport

14 (A) Green Tea Extract in Treating Women with Hormone Receptor-Negative Stage 1,2,3 Breast Cancer (Phase 1)

Sponsor: M.D. Anderson Cancer Center

15 (C) Green Tea Extract in Preventing Cancer in Healthy Participants (Phase 1)

Sponsor: University of Arizona

16 (C) Pharmacokinetic Study of Topically Applied Veregen 15% Com-

pound with Oral Intake of Green Tea Beverage (Phase 4)

Sponsor: MediGene

17 (A) Green Tea Extract in Treating Patients with Stage 0, 1, 2 Chronic Lymphocytic Leukemia (Phase 1/2)

Sponsor: Mayo Clinic

18 (S) Defined Green Tea Catechin Extract in Treating Patients with Localized Prostate Cancer Under-going Surgery (Phase 2)

Sponsor: Case Comprehensive Cancer Center

19 (N) Treatment of the Recessive Nonbullous Congenital Ichthyosis by the Epigallocatechin Cutaneous (Phase 4)

Sponsor: Centre Hospitalier Universitaire de Nice

20 (U) Green Tea Extract in Treating Patients with Nonmetastatic Bladder Cancer (Phase 2)

Sponsor: University of Wisconsin, Madison

21 (T) Green Tea Extract in Treating Patients with Actinic Karatosis (Phase 2)

Sponsor: University of California, Irvine

22 (R) Green Tea Extract in Treating Current or Former Smokers with Bronchial Dysplasia (phase 2)

Sponsor: British Columbia Cancer Agency

23 (C) Green Tea Extract and Prostate Cancer (Phase 2)

Sponsor: Louisiana State University Health Science Center, Shreveport

24 (T) Green Tea Extract in Preventing Cervical Cancer in Patients with Human Papillomavirus and Low-grade Cervical Intraepithelial Neoplasia (Phase 2)

Sponsor: National Cancer Institute (NCI)

25 (R) Green Tea Extract in Treating Patients with Monoclonal Gammopathy of Undermined Significance and/or Smoldering Multiple Myeloma (Phase 2)

Sponsor: Barbara Ann Karmanos Cancer Institute

26 (R) A Pilot Study of Chemo-prevention of Green Tea in Women with Ductal Carcinoma in Situ (Phase 2)

Sponsor: University of Chicago

27 (A) Green Tea Extract in Preventing Cancer in Former and Current Heavy Smokers with Abnormal Sputum (Phase 2)

Sponsor: British Columbia Cancer Agency

28 (C) Defined Green Tea Catechins in Treating Patients with Prostate Cancer Undergoing Surgery to Remove the Prostate (Phase 1)

Sponsor: National Cancer Institute (NCI)

For more details go to http://clinicaltrials.gov/ and search with the key word: "Polyphenon E".

Since chemoprevention trials are very laborious and time/money consuming affairs with no assured results, it will be better to carry out a thorough feasibility study before launching them. In particular the business, commercial outlook and the risk/benefits thereof should be studied in depth.

On the other hand, if we continue with our present food grade status, with no clinical claims on the label by doing no clinical phase trials, we have to remain within the same regulatory framework as before in spite of all the evidence for possible clinical chemopreventive efficacy.

In order to overcome the above stalemate, I would like to propose the following measures to the health agencies which have no "Botanical Drug Guidelines".

First of all, to enact the regulation of "Botanical Drugs". The active pharmaceutical ingredient should be the crude extract derived from plants. Secondly the agency should support the realization of Botanical Drugs and related processes financially, anticipating a world in which chemoprevention drugs will be available with firm scientific confirmation of the specific biomarkers.

4 An FDA-Approved Botanical Drug

Here is a brief overview of our successful registration of a tea extract product with the US FDA. Along with the NCI projects, the development of a botanical drug was successfully completed in a certain clinical setting. Application of Polyphenon® E ointment on genital warts (Condyloma acuminata) caused by HPV removed the warts effectively in Beijing Cancer Center (Hospital) in 1990 in a fact finding trial. This is a therapeutic effect of tea catechins on benign tumors and the results were patented[18]. On this basis, a German pharmaceutical company spent time and money on Phase 2/3 clinical trials internationally. In 2006 the US FDA approved the marketing of the ointment as a prescribed Botanical Drug in the States under the trademark of "Veregen®". This is the first ever Botanical Drug approved under the FDA Guideline, with no subsequent approvals up until now[19].

On my retirement from Mitsui Norin Co., in 2008, I handed over to my successors the whole Polyphenon® world, which I had built up with my colleagues and collaborators over 30 years of efforts. Recently, I registered a new trade mark of "Theaphenon®" and started operations at my new venture, "Tea Solutions, Hara Office Inc."[20]) aiming to supply Theaphenon® E as well as EGCG and other catechins to the concerned parties.

5 Ultimate RTD Teas

Currently, I am directing a national project in Japan to manage R&D for the Ultimate RTD (Ready-to-Drink) Teas of the world. I am of the belief that, in the years to come, the carbonated sugar drinks world-wide should be replaced, at least in part, by tea drinks of various efficacies with regards to health benefits. This idea, which anticipates several entirely new RTD tea drinks, was supported and was funded in 2009 by the National Science and Technology Agency (Japan) with 10 million USD over a five year period. The basic concept is for science to shed light on the traditional way of tea manufacturing, which is why I term it "*Tea Renaissance*"[21]. Teas manufactured and consumed today, green tea, black tea, oolong tea and others are the outcome of historical trial and error by so many people in various regions. If we could control the actions of enzymes in tea by the radiation of certain wavelength or by addition of enzymes from external sources, we would be able to manufacture very different RTD teas that are more flavorful than present ones with certain marked health benefits.

Within the next year, I am determined to convince several big beverage companies to take up the idea and go forward with commercialization of the Ultimate RTD Teas for the benefits of consumers as well as for that of beverage companies.

6   ICOS 2013 in Shizuoka, Japan

The 5th International Conference on O-Cha (Tea) Culture and Science(ICOS) is to be held in Shizuoka, Japan on 6-8th, November 2013, hosted by myself[22]. The plenary session was planned focusing on the clinical trials of tea catechins. At the same time the audience will learn that tea catechins are extracted and purified in large amount in China, and not elsewhere in the world.

Prof. Saverio Bettuzzi from Italy will talk on the prevention of prostate cancer by oral intake of tea catechins. He is the first person ever to have shown clearly the chemopreventive efficacy of tea catechins in human clinical settings. Prof. Michael Quon from the USA will talk on the anti-metabolic, anti-diabetic efficacy of EGCG and on planned human trials on those functions with an NIH group. Prof. Liu Zonghua from China will talk on the large scale extraction, isolation and purification systems of such tea components as soluble teas (Instant Teas), catechins, caffeine, theanine and others for the Chinese and world market.

7   Conclusions

From my 40 years experience in tea and its R&D, as beverages or as a medicinal plant, I would say that there will be a huge possibility for tea to be a health promoting material for every human body and much research has to be done to prove those efficacies and to elucidate the mechanisms thereof. A recent finding that there is a specific receptor protein of EGCG on the membrane of cancerous cells is very promising and warrants future clinical trials[13].

As regards botanicals other than tea, there should be many other natural plants or their ingredients with the potential to be a botanical drug. "Botanical Drug" is a nice idea of the US-FDA. If "crude extract" of tea is approved as a cancer preventive/curative agent, a great number of consumers would become heavy tea drinkers or buy tea capsules and that will reduce the incidence of cancer to a certain extent. The government could reduce the huge medical expenditures through the pocket money of consumers being used on tea products. In the case of the tea catechin drug against genital warts, it unfortunately could not drive people into drinking more tea. I am sure that if a crude extract of a certain medicinal plant or a certain food ingredient works on a certain human disorder on a folklore basis or in animal experiments, there could be chances of this crude extract being approved as a "Botanical Drug" if only chemically well defined, processed under cGMP with proven efficacy/safety in clinical phase trials. Separately, RTD teas are good carriers of various other herbs or medicinal plant extracts. In addition to the new RTD teas, combination of tea (*Camellia sinensis*) and those medicinal plants in RTD beverages should be pursued more extensively, aiming at their commercial production for people's good health. In the coming ICOS 2013, 6-8th November, I wish many people concerned will attend, present papers and discuss various possibilities of teas on R&D basis.

References

[1] Hara Y. Green Tea-Health Benefits and Applications. Taylor & Francis, 2001.

[2] Kuroda Y, Hara Y. Health Effects of Tea and its Catechins. Kluwer Academic/Plenum Publishers, 2004.

[3] Matsuzaki T, Hara Y. Antioxidative Activity of tea leaf catechins. J Agric Chem Soc Japan, 1985(59): 129—134.

[4] Ishigami T, Hara Y. Antibacterial activities of tea polyphenols against foodborne pathogenic bacteria(Studies on Antibacterial Effects of Tea Polyphenols Part 3). Jpn Soc Food Sci Technol, 1989(36): 996—999.

[5] Nakayama M, Suzuki K, Toda M, Okubo S, Hara Y, Shimamura T. Inhibition of the infectivity of influenza virus by tea polyphenols. Antiviral Res, 1993(21): 289—299.

[6] Kuroda Y, Hara Y. Antimutagenic and anticarcinogenic activity of tea polyphenols. Mutation Res, 1999(436): 69—97.

[7] Fukuyo M, Muramatsu K, Hara Y. Effect of green tea catechins on plasma cholesterol level in cholesterol-fed rats. J Nutr Sci Vitaminol, 1986(32): 613—622.

[8] Hara Y, Honda M. The inhibition of α-amylase by tea polyphenols. Agric Biol Chem,1990(54):1939－1945.

[9] Hara Y, Tono-oka F. Effect of tea catechins on blood pressure of rats. J Jpn Soc Nutr Food Sci,1990(43):345－348.

[10] Goto K, Kanaya S, Nishikawa T, Hara H, Terada A, Ishigami T, Hara Y. The influence of tea catechins on fecal flora of elderly residents in long-term care facilities. Annu Long-Term Care,1998(6):43－48.

[11] Yang C S, Lambert J D. Special Issue:Tea and Health. Pharmacological Research. Elsevier,2011.

[12] Hara Y. Tea Catechins and their applications as supplements and pharmaceutics. Pharmacological Research,2011(64):100－104.

[13] Tachibana H, Koga K, Fujimura Y, Yamada K. A receptor for green tea polyphenol EGCG. Nature Structural & Molecular Biology,2004(11):380－381.

[14] Chow H H S, Cai Y, Hakim I A, Crowell J A, Shahi F, Brooks C A, Dorr R T, Hara Y, Alberts D S. Pharmacokinetics and Safety of Green Tea Polyphenols after Multiple-Dose Administration of Epigallocatechin Gallate and Polyphenon E in Healthy Individuals, Clinical Cance Rersearch,2003(9):3312－3319.

[15] U. S. Patent 2010/0069429 A1. Tea Polyphenol Composition and the Method for Producing the Same,2010.

[16] U. S. Food and Drug Administration. Center for Drug Evaluation and Research. 2004. Guidance for Industry, Botanical Drug Products.

[17] U. S. Food and Drug Administration. Clinical Trial. A Service of the U. S. National Institute of Health as of 2011.

[18] U. S. Patent 5,795,911. Composition for Treating Condyloma Acuminata,1998.

[19] Chen S T, Dou J, Temple R, Agarwal R, Wu, K M, Walker S. New therapies from old medicines. Nature Biotech,2008(26):1077－1083.

[20] http://www.theaphenon.com.

[21] http://www.shizuoka-tiikikesshu.jp/.

[22] http://icos2013.jp/main/.

茶多酚保健作用与功能食品研究开发

王岳飞 徐 平 杨贤强

浙江大学茶学系,中国 浙江

摘 要:茶多酚是茶叶中最主要的活性物质,被证明对人体健康具有广泛的促进作用。当前茶多酚应用范围不断拓展,其中,在保健品领域中的应用被认为对茶产业发展和茶多酚价值的充分发挥具有重要意义。本文介绍了我国保健食品开发受理范围,茶多酚作为功效成分的契合度,以及我们在茶多酚保健食品(包括辅助降血脂、辅助降血糖、增强免疫力、抗氧化和延缓衰老、美容祛斑、减肥、缓解体力疲劳、抗辐射等八方面功能)开发所做的研究工作,以期为茶多酚在保健食品应用领域的更广泛研究提供参考。

关键词:茶多酚;保健功能;保健食品

Health Protective Function and Development of Functional Foods Derived from Tea Polyphenols

WANG Yue-fei XU Ping YANG Xian-qiang

Department of Tea Science, Zhejiang University, Zhejiang, China

Abstract: Tea polyphenols are the main active components in tea, and have been shown great benefits for human health. Currently, tea polyphenols have been applied in broad fields, among which tea polyphenols derived-functional foods are supposed to be important not only for valuing tea polyphenols, but also for the whole tea industry. In the present paper, the work done regarding tea polyphenols derived-functional foods, including functions of antihyperlipidemic, antihyperglycemic, immunopotentiating, antioxidantive and antiaging, beauty-improving, antiobesity, relieving physical fatigue and resistradiation, in our lab was reviewed to provide some related examples for the further development.

Keywords: tea polyphenols; health protective function; functional foods

1 我国保健食品开发受理范围

国家药品食品监督管理局(FDA)的保健食品功能受理审批和范围有如下27项(2003年5月1日起实施),且规定同一配方保健食品申报和审批功能不超过两个:①增强免疫力;②辅助降血脂;③辅助降血糖;④抗氧化;⑤辅助改善记忆;⑥缓解视疲劳;⑦促进排铅;⑧清咽;⑨辅助降血压;⑩改善睡眠;⑪促进泌乳;⑫缓解体力疲劳;⑬提高缺氧耐受力;⑭对辐射危害的辅助保护;⑮减肥;⑯改善生长发育;⑰增加骨密度;⑱改善营养性贫血;⑲对化学性肝损伤的辅助保护;⑳祛痤疮;㉑祛黄褐斑;㉒改善皮肤水分;㉓改善皮肤油分;㉔调节肠道菌群;㉕促进消化;㉖通便;㉗对胃黏膜有辅助保护作用。

2 保健食品所选功效成分要求与茶多酚的契合度

当前,保健食品开发所选功效成分需满足五大要求。

2.1 功效成分明确

功效成分是保健食品机体调节功能的物质基础,保健食品不仅需明确产生功能作用的功效成分的名称,而且需要明确该成分的化学结构和稳定的形态。一定的化学物质在机体内是通过一定的生化反应或生理作用产生生理调节功能的,只有明确了参与调节功能的物质,才能科学地评价食品对人体作用的机理,也才有可能分析或预见可能对人体产生的其他影响或不良作用。因此,明确了功效成分,才有可能更科学、更有针对性地选食保健食品,真正起到功能保健作用。目前,对于茶多酚化学结构及性质已研究得非常透彻。

2.2 含量应可以测定

功效成分在体内发挥调节作用需要一定的量来保证,过少过多都不好。若摄入量过少,起不到功能调节效果,过多则可能产生不利影响。因此,保健食品要求其中的功效成分可以测定,只有有了量化指标,才能使保健食品保持产品的稳定,才能使消费者更科学、更合理地食用,避免摄入过多或不足。含量的测定方法应方便、可靠、稳定,尽可能采用国家有关部门或行业推荐的测定方法。茶多酚总量的测定(福林酚测定)及其儿茶素组成(HPLC)均已经有很完整的检测方法。

2.3 作用机理明确

保健食品的功效成分是按一定的作用机理达到调节生理功能目的的。主要功效成分在体内的作用机理，包括作用部分、反应方式、作用路线、代谢过程等一般都应清楚，而且应该科学合理。当前，茶多酚对辐射损伤的防护、抗突变、抗肿瘤、延缓衰老、抗氧化、调节免疫功能、抑菌抗病毒、减缓香烟毒害等功能的机理研究已经达到了分子水平。

2.4 研究资料充分

保健食品开发需要在充分的实验和研究基础上才能完成。研究资料不仅包括在研制开发过程中自己的研究和实验资料，而且包括国内外其他人的相关文献资料，资料要求充分、真实、可靠。当前，茶多酚作为最热门的天然产物之一，国内外对其健康功能的研究极其翔实并且已非常深入。

2.5 临床效果肯定

保健食品必须具有稳定而有效的临床效果。目前对于茶多酚临床研究也在不断开展。我们在浙江大学附属一院、二院，湖南省肿瘤医院，杭州市第三医院，宁波市第二医院，上海市杨浦区肿瘤防治院、北站医院和市北医院等开展了关于茶多酚抗氧化、辅助降血糖、辅助降血脂、缓解疲劳、祛斑等一系列的临床研究，结果表明茶多酚具有临床意义上的广谱药理功效。

因此，茶多酚完全契合保健食品开发要求，是茶类保健食品开发的首选成分。

3 茶多酚在保健食品开发中的应用范围

从目前国内外关于茶多酚研究进展及我国关于保健食品受理范围来看，我们认为茶多酚作为保健食品功效成分可以开发的保健食品至少有 16 项：①增强免疫力；②辅助降血脂；③辅助降血糖；④抗氧化；⑤促进排铅；⑥清咽；⑦辅助降血压；⑧缓解体力疲劳；⑨提高缺氧耐受力；⑩对辐射危害的辅助保护；⑪减肥；⑫对化学性肝损伤的辅助保护；⑬祛黄褐斑；⑭调节肠道菌群；⑮促进消化；⑯通便。

4 茶多酚保健食品研究开发

近年来,我们围绕上述潜在开发范围,并按照《保健食品检验与评价技术规范》的要求进行了一系列茶多酚保健食品的研究开发。

4.1 茶多酚辅助降血脂保健食品研究开发

脂质是人体内的重要物质,包括各种脂肪酸、甘油酯、磷脂、类固醇等,其生理作用是能量利用及能量储存,也是细胞膜的主要成分,并在维持体温、神经传导等方面起重要作用。脂质是非水溶性的,必须与能和脂质结合的蛋白质(称为载脂蛋白,apo)结合,形成脂蛋白形式在体内运输。脂蛋白中的脂质主要是胆固醇(TC)、低密度脂蛋白胆固醇(LDL-C)、极低密度脂蛋白胆固醇(VLDL-C)、高密度脂蛋白胆固醇(HDI-C)和甘油三酯(TC)及脂蛋白(a)等,上述各项指标中有一种或一种以上不正常称为血脂异常。其中低密度脂蛋白胆固醇易沉积于血管壁上,并产生斑块,使动脉发生粥样硬化,从而阻碍血液向身体各部位运送营养物质,进而导致冠心病、心肌缺血,以及中风,甚至死亡,因此是一种"坏"胆固醇。高密度脂蛋白胆固醇能清除血管内沉积的胆固醇,因此是一种"有益"胆固醇。当人体进食大量饱和脂肪(甘油三酯)及高胆固醇食物,或身体产生过量的胆固醇,就会使血液中胆固醇和甘油三酯的含量过高,形成高胆固醇血症和高血脂症,从而引发一系列心脑血管等病症。

保健食品是否具有调节血脂作用,需要进行两方面的实验研究:①建立高血脂症动物模型,检测其对动物高血脂症各项指标的影响;②人体试验,只有通过动物试验的受试物才可进行人体试食或临床试验。调节血脂作用动物实验的原理是用高胆固醇和脂类饲料喂养动物,使其形成脂代谢紊乱动物模型,再给予动物或同时给予受试样品,可检测受试样品对高血脂症的影响,并可判定受试样品对脂质的吸收、脂蛋白的形成、脂质的降解或排泄产生的影响。

研究结果见图1。茶多酚保健食品处理组与模型对照组及各剂量组相比,总胆固醇TC明显下降,且呈剂量效应关系,其中高剂量组下降了28.0%,达到显著性差异($P<0.05$),同时茶多酚保健食品处理组甘油三酯TG明显下降,高剂量组下降了27.4%,达到显著性差异($P<0.05$),已接近空白对照。这些结果表明茶多酚保健食品具有降低血清总胆固醇TC和动物甘油三酯TG的作用。此外,整个实验过程中不同处理组的老鼠体重无显著变化。

选择单纯血脂异常者(指 TC、TG、LDL 3 项中 1 项或 1 项以上增高,或

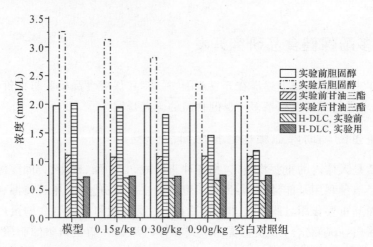

图1 茶多酚保健食品对动物胆固醇、甘油三酯和高密度脂蛋白的影响

HDL减少者)30例为受试者。年龄40~70岁(平均56岁),男24例、女26例;其中TC异常者35例,TG异常者45例,HDL-C异常者30例,LDL异常者39例。血脂正常值:TC＜5.72mmol/L、TG＜1.69mmol/L、HDL-C男＞1.05mmol/L,女＞1.15mmol/L。按照临床伦理相关要求开展了临床试验,由实验结果(图2)可知茶多酚保健食品制剂对50例患者3种血脂的影响(包括正常及异常者)为:TC下降12.9%($P>0.05$),TG下降21.9%($P<0.05$),LDL-C下降10.8%($P<0.05$),HDL-C上升8.6%($P<0.05$),TC-HDL-C/HDL-C下降21.7%($P<0.05$)。

根据《保健食品检验与评价技术规范》标准,由上述实验结果可判定所试茶多酚保健食品具有辅助降血脂功能。

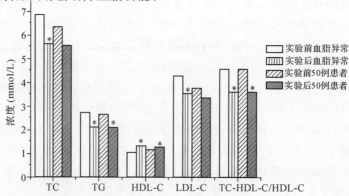

图2 茶多酚保健食品制剂对血脂异常和50例患者血脂的改善幅度

4.2 茶多酚辅助降血糖保健食品研究开发

采用腹腔注射四氧嘧啶生理盐水溶液制备高血糖小鼠模型,研究茶多酚保健食品制剂的降血糖功效,实验设 900mg/kg、300mg/kg、150mg/kg 3 个样品剂量组和模型对照组(双蒸水),研究周期 30d。结果表明,茶多酚保健食品制剂对于糖尿病组小鼠的体重增长没有显著影响($P>0.05$),对正常小鼠的体重增长也没有显著影响($P>0.05$)。而糖尿病小鼠在实验结束时,血糖都有所下降。对照组(D组)、低剂量组(C组)、中剂量组(B组)、高剂量组(A组)小鼠的空腹血糖下降率分别是 11.38%、13.25%、16.68% 和 22.00%。其中高剂量组(A组)小鼠的空腹血糖下降值与对照组(D组)相比,达到极显著水平($P<0.01$)(表1)。

表 1　糖尿病小鼠实验始末空腹血糖浓度比较

组　别	动物数(只)	实验前血糖浓度(mmol/L)	实验后血糖浓度(mmol/L)
D(对照组)	10	19.1±1.6	17.0±1.5
C(低剂量)	10	19.7±2.5	17.1±2.5
B(中剂量)	10	19.2±2.1	16.1±2.6
A(高剂量)	10	19.6±1.8	15.3±2.2

此外,在葡萄糖耐量实验中各组的血糖曲线下面积分别是:对照组 45.11±2.74、低剂量组 41.31±3.97、中剂量组 41.43±3.00、高剂量组 41.32±4.01。与对照组相比,各个剂量组均有显著差异($P<0.05$),表明茶多酚保健食品制剂有显著改善糖尿病小鼠的葡萄糖耐量的作用(图3)。

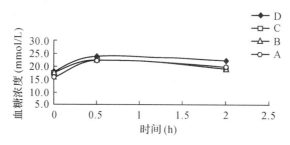

图 3　茶多酚保健食品制剂对小鼠葡萄糖耐量的影响

4.3 茶多酚增强免疫力保健食品研究开发

机体的免疫功能由免疫系统完成。免疫系统是脊椎动物和人类的防御系统,它与神经系统、内分泌系统、呼吸系统等一样,是机体的一个重要系统。它是在系统发育过程中长期适应外界环境而形成的,主要包括:中枢免疫器官、外周免疫器官、免疫活性细胞(T、B淋巴细胞等)和免疫活性介质(抗体、细胞因子和补体等)。免疫细胞是参与免疫应答或与免疫应答有关的细胞群的统称,包括淋巴细胞、单核细胞、巨噬细胞、多形核细胞、肥大细胞和辅佐细胞等。其中能接受抗原刺激而活化、增殖、分化发生特异性免疫应答的淋巴细胞,称免疫活性细胞,即T淋巴细胞和B淋巴细胞(简称T细胞、B细胞)。有免疫功能的细胞除T细胞、B细胞外,还有一些淋巴细胞:杀伤细胞(K细胞)和自然杀伤细胞(NK细胞)。

我们通过ConA诱导的小鼠淋巴细胞转化试验,二硝基氟苯小鼠迟发型变态反应试验,溶血空斑数和血清溶血素抗体积数影响试验、小鼠碳廓清试验、小鼠腹腔巨噬细胞、吞噬鸡红细胞试验、NK细胞活性试验等研究了茶多酚保健食品制剂增强免疫力功能。结果表明(图4~6),茶多酚保健食品制剂在300mg/kg·d和600mg/kg·d(相当于人体推荐摄入量的10倍和20倍)剂量时,具有增强细胞免疫功能的作用;在150mg/kg·d、300mg/kg·d和600mg/kg·d(相当于人体推荐摄入量的5倍、10倍和20倍)剂量时,对抗体生成细胞试验结果未见有影响;在300mg/kg·d和600mg/kg·d(相当于人体推荐摄入量的10

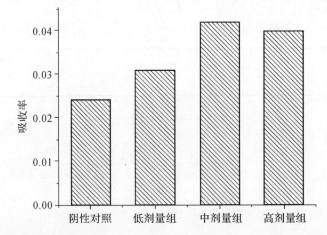

图4　茶多酚保健食品制剂对ConA诱导的小鼠淋巴细胞转化试验

倍和 20 倍)剂量时,具有增强自身免疫功能的作用;在 600mg/kg·d(相当于人体推荐摄入量的 20 倍)剂量时,具有增强小鼠单核吞噬功能和小鼠单核巨噬细胞吞噬功能的作用。

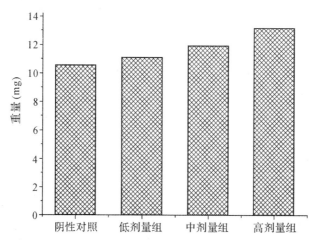

图 5 茶多酚保健食品制剂对二硝基氟苯小鼠迟发型变态反应的影响

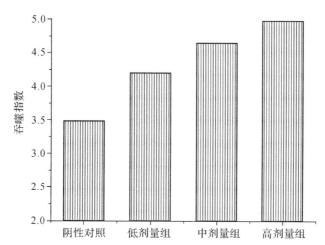

图 6 茶多酚保健食品制剂对小鼠巨噬细胞吞噬指数的影响

根据《保健食品检验与评价技术规范》中结果判定的原则:在细胞免疫功能、体液免疫功能、单核巨噬细胞吞噬功能、NK 细胞活性 4 个方面任两个方面结果阳性,可判定该茶多酚保健食品具有增强免疫力功能。

4.4 茶多酚抗氧化和延缓衰老保健食品研究开发

衰老的自由基理论由 Harman 于 1955 年在美国原子能委员会上提出。该理论有两个基本要点：①衰老是由氧自由基对细胞成分的有害攻击造成的；②维持体内适当水平的抗氧化剂和自由基清除剂可以延长寿命和推迟衰老。众所周知,茶多酚是一类出色的天然抗氧化剂,为此,我们开展了茶多酚抗氧化保健食品的研究。

我们以 SD 雄性大白鼠为研究对象,茶多酚保健食品制剂设低、中、高三剂量组,分别为 0.15g/kg、0.30g/kg、0.90g/kg,相当于人体剂量的 5 倍、10 倍和 30 倍,并设老龄动物对照组及青年动物对照组(以蒸馏水灌胃)。各剂量组给以样品,对照组以蒸馏水灌胃,连续 30d。第 31d,5 组动物摘眼球取血,离心,取血清。测定如下生化指标:血中过氧化脂质降解产物丙二醛(MDA)含量和血中超氧化物歧化酶(SOD)活力。研究结果见图 7 和图 8。茶多酚制剂 0.15g/kg、0.30g/kg、0.90g/kg 剂量组与老年对照组相比,MDA 含量分别下降了 13.6%、24.8% 和 39.9%,其中茶多酚保健食品制剂 0.30g/kg、0.90g/kg 剂量组与老年对照组相比,差异分别达显著($P<0.05$)和极显著($P<0.01$)水平。同时,茶多酚制剂 0.15g/kg、0.30g/kg、0.90g/kg 剂量组与老年对照组相比,SOD 含量分别上升了 12.26%、28.4% 和 48.6%,其中茶多酚保健食品制剂 0.30g/kg、0.90g/kg 剂量组与老年对照组相比,差异分别达显著($P<0.05$)和极显著($P<0.01$)水平。

与此同时,我们采用 Oregon K 野生型黑腹果蝇研究了茶多酚保健食品制剂延缓衰老的作用。结果表明(表 2),茶儿茶素 0.06%、0.18% 剂量组雄性果蝇和 0.18% 剂量组雌性果蝇的半数死亡时间与对照组相比均延长 4d 以上;0.06%、0.18% 剂量组雄性果蝇和 0.18% 剂量组雌性果蝇的平均寿命及最高寿命与对照组相比,均有显著性差异($P<0.05$)。

表 2 茶多酚保健食品制剂对果蝇生存试验的影响($\bar{x}\pm s$)

浓度(%)	样本数(只)		平均体重(μg)		半数死亡时间(d)		平均寿命(d)		平均最高寿命(d)	
	雄	雌	雄	雌	雄	雌	雄	雌	雄	雌
普通对照	200	200	640	1060	39	46	40±12	46±13	62±8	70±4
0.01	200	200	650	990	41	46	42±13	47±12	65±5	71±5
0.02	200	200	700	1000	41	48	42±12	48±13	65±4	71±2
0.06	200	200	710	1060	43	48	44±14*	49±12	70±4*	72±2
0.18	200	200	690	1010	43	55	45±12*	56±10*	70±4*	76±2*

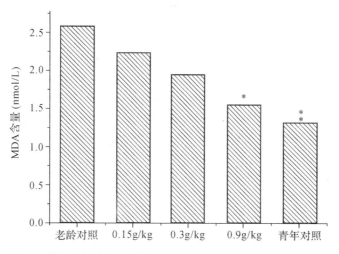

图 7 茶多酚保健食品制剂对 SD 大鼠血中丙二醛含量的影响

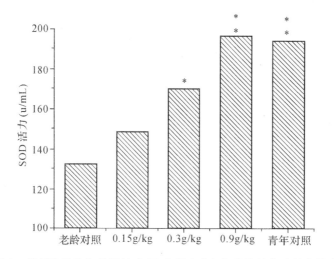

图 8 茶多酚保健食品制剂对 SD 大鼠血中超氧化物歧化酶活力影响

根据《保健食品功能学评价程序与检验方法规范》准则，这些实验结果表明，茶多酚保健食品制剂具有抗氧化和延缓衰老的功效。

4.5 茶多酚美容祛斑保健食品研究开发

按中国国家药品食品监督管理局（2003）目前的保健食品功能受理审批规

定,仅限于祛痤疮功能、祛黄褐斑功能、改善皮肤水分功能、改善皮肤油分功能4项。本研究仅考察茶多酚保健食品制剂祛黄褐斑功能。黄褐斑由黑色素沉着于皮肤中形成,黑色素由黑色素细胞制造。在表皮,每个黑色素细胞被一群角蛋白细胞围绕,进入角蛋白细胞的黑色素可被溶酶体降解后随角化细胞而脱落。当机体内的肌脂质含量升高时,溶酶体的活力就下降,黑色素的降解速度随之变慢,就会在皮肤表面形成黄褐斑。

茶多酚美容祛斑保健食品采用临床人体试验,包括组间对照设计和自身对照设计。黄褐斑受试者均为女性,年龄18～65岁,无严重心、肝、肾等并发症,按黄褐斑颜色、面积情况随机分为受试组和对照组,并尽可能考虑影响结果的主要因素如户外活动情况、性别、年龄等,进行均衡性检验。根据纳入标准和排除标准筛选100例黄褐斑受试者,随机分为受试组和对照组。受试组食用茶多酚保健食品制剂胶囊,每日2次,每次3粒,每粒0.3g,连续观察30d。对照组食用安慰剂,剂量与受试组相同。两组受试者在试验期间停止使用其他口服及外用有关养颜祛斑的用品。试验期间不改变原来的饮食习惯。通过比较试验前后黄褐斑颜色深浅和面积大小来判定功效。

人体试验结果见图9(黄褐斑颜色)和图10(面积大小)。发现人体试食后茶多酚保健食品制剂组黄褐斑面积平均减少3.75cm^2,与试食前比较差异达极显著水平($P<0.01$),与对照组比较差异也有显著性($P<0.05$);人体试食后祛黄褐斑作用功效50例中有28例有效,总有效率达56%,而对照组仅6例,相比较差异达极显著水平($P<0.01$)。并且食用茶多酚保健食品1个月,两组血相、肝肾功能各项检测指标均在正常范围内,说明茶多酚保健食品制剂对机体健康无损害。

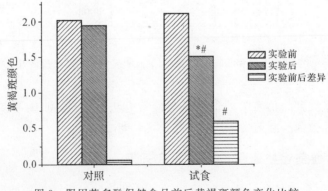

图9 服用茶多酚保健食品前后黄褐斑颜色变化比较

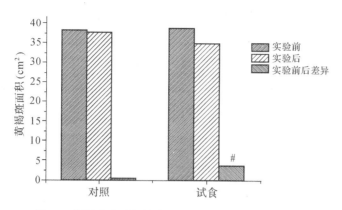

图10　服用茶多酚保健食品前后黄褐斑面积变化比较

4.6 茶多酚减肥保健食品研究开发

采用预防肥胖模型法进行试验,雄性 SD 大鼠 50 只,按体重随机分为 5 组,每组 10 只。设模型对照组:蒸馏水灌胃;茶多酚保健食品制剂低、中、高 3 个剂量组,分别灌胃给予茶多酚保健食品制剂 175mg/kg·d、350mg/kg·d、700mg/kg·d(3 个剂量组分别相当于人体推荐摄入量的 5 倍、10 倍、20 倍),以上 4 组每只动物每日给予等量的营养饲料,饲料给予量以多数动物吃完为原则。同时设空白对照组,蒸馏水灌胃,给予基础饲料,饲料给予量与营养饲料相同。灌胃体积为 10.0mL/kg,连续灌胃 35d。

试验结束时,大鼠体重增加量和体内脂肪重量及脂/体比见图 11 和图 12。模型对照组体重增重均值与空白对照组比较差异有显著性($P<0.05$),且茶多酚保健食品制剂大鼠高剂量组体重增重均值显著低于模型对照组($P<0.05$)。模型对照组动物脂/体比均值与空白对照组比较差异有显著性($P<0.05$),且茶多酚保健食品制剂高剂量组动物脂/体比均值明显小于模型对照组($P<0.05$),说明茶多酚保健食品制剂胶囊内容物在 700mg/kg·d(相当于人体推荐摄入量的 20 倍)时,具有降低实验动物脂/体比的作用。在此基础上,我们进一步开展了人体试食试验,选择年龄在 18～65 岁的单纯性肥胖者为受试者,要求成人 BMI≥30,或总脂百分率达到男≥25%,女≥30%。本次试验共选 100 例受试者,采用双盲对照法,将 100 例单纯性肥胖者按年龄、性别、体重指数、体内脂肪百分率等,随机分为对照组和试食组两组,其中试食组 50 例,对照组 50 例,分别食用茶儿茶素制剂和安慰剂,每日 2 次,每次 3 粒。观察期间保持饮食习惯不变。30d 后,结果表明:茶儿茶素制剂对单纯性肥胖者的体重、体重指数、超

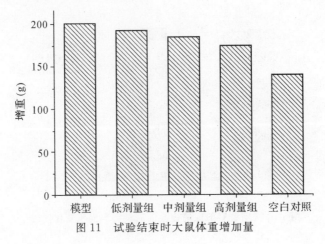

图 11 试验结束时大鼠体重增加量

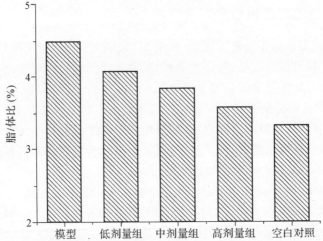

图 12 茶多酚保健食品制剂对大鼠体内脂肪重量及脂/体比的影响

重度、体内脂肪重量、脂肪百分率有降低作用($P<0.01$),能减少右三角肌下缘、右肩胛下角、右脐旁 3cm 和右髂前上棘皮下脂肪厚度及腰围、臀围($P<0.05$),而对运动耐力无影响。试食前后血象及肝肾功能各项检测指标均在正常范围,血尿酮体阴性,对机体健康无明显损害。根据发布的结果判定的原则,已充分证明茶多酚保健食品制剂对人体具有减肥功能。

4.7 茶多酚缓解体力疲劳保健食品研究开发

疲劳是现代生活中常见的问题,过度疲劳是导致"亚健康"的主要原因。本研究将小鼠按体重随机分为空白对照组(给予蒸馏水)和 3 个对照组,灌胃给予试验样品135mg/kg·d、270mg/kg·d、540mg/kg·d,分别是人体推荐摄入量的5倍、10倍、20倍,灌胃容量为20mL/kg,每天1次,连续30d。通过负重游泳试验和血清尿素含量、肝糖原含量和血乳酸含量测定来评价缓解体力疲劳功效。

结果表明,茶多酚保健食品制剂提高了小鼠的负重游泳时间,增加了小鼠肝糖原的含量,减少运动后血清尿素含量,减小了血乳酸含量及其曲线下面积。这些试验结果直接反映了制剂的缓解体力疲劳作用。运动耐力是反映机体疲劳最直接的指标。从表3可知,3个剂量组小鼠的负重游泳时间均显著高于对照组($P<0.05$)。

表3　茶多酚西洋参复方对小鼠游泳时间的影响($\bar{x}\pm s$)

剂量(mg/kg)	动物数(n)	负重游泳时间(min)
0	10	11.02±2.84
135	10	13.14±2.10*
270	10	13.75±3.02*
540	10	15.17±2.69*

肝糖原作为能量物质,它的大量消耗直接导致机体的活动能力下降,进一步引起全身性疲劳。本实验中,小鼠灌胃茶多酚保健食品制剂30d,肝糖原储备量显著增加(表4),说明预防疲劳产生及缓解体力疲劳能力提高。

表4　茶多酚西洋参复方对小鼠肝糖原的影响($\bar{x}\pm s$)

剂量(mg/kg)	动物数(n)	肝糖原(mmol/L)
0	10	12.28±1.76
135	10	13.60±2.03
270	10	14.44±2.47
540	10	15.62±1.92*

同时,当剧烈运动导致肝糖原的大量消耗时,蛋白质和含氮化合物的分解代谢加速,直接引起机体内血清尿素含量增加,使机体对运动负荷能力的适应力降低,引起疲劳。而本试验中3个剂量组小鼠血清尿素没有增加,反而有显著性地下降。根据《保健食品功能学评价程序与检验方法规范》准则,通过这些实验结果,可以判定所用茶多酚制剂具有缓解体力疲劳功效。

4.8 茶多酚抗辐射保健食品研究开发

当前,电脑、手机、微波炉等的普及使得日常生活中有害的辐射也不断增加,同时,臭氧层不断变薄,更多来自太空的辐射粒子进入了我们生活的大气层,对人类已经造成的辐射危害日益严重。紫外辐射对人类健康产生了极大影响,能引起白内障和皮肤癌等。因此,我们研究了茶多酚抗辐射保健食品功效。

我们建立γ射线辐射损伤小鼠模型,通过骨髓有核细胞计数、小鼠骨髓细胞微核试验、外周血白细胞计数、血/组织中超氧化物歧化酶(SOD)活性检测、小鼠血清溶血素测定等试验评价茶多酚制剂抗辐射作用。观察辐照后第3d,各剂量组骨髓有核细胞数与辐照模型对照组相比,中剂量组小鼠骨髓有核细胞数明显增加,差异显著($P<0.05$)。其余两组与辐照模型对照组相比无显著差异,中剂量的茶多酚抗辐射制剂会促进骨髓有核细胞数的显著增加,微核细胞率显著降低(表5)。

表5 茶多酚抗辐射制剂对辐照后小鼠骨髓细胞微核率和骨髓有核细胞数的影响($\pm SD, n=10$)

组别	PCE计数（只）	含微核PCE数（只）	微核率（‰）	抑制率（%）	骨髓有核细胞数（$\times 10^9$ 个/mL）
辐照模型对照	10000	92	9.2±2.0	—	66.0±28.8
阴性对照	10000	12	1.2±0.9	—	—
低	10000	83	8.3±2.2	11.3	76.0±47.8
中	10000	70	6.6±1.6**	32.5	110.0±29.9**
高	10000	73	7.3±1.8	23.8	68.0±30.7

茶多酚抗辐射制剂对外周血白细胞数的影响见表6。各剂量组在辐照前、辐照后第3d、辐照后第14d,与辐照模型对照组相比,白细胞数均无显著差异。辐照后,白细胞数先减少,后回升。减少和回升的幅度,4个组无显著差异。在辐照后第14d,中剂量组白细胞数为3.4×10^9个左右,其他3组白细胞数均为$2.6\times10^9 \sim 2.8\times10^9$个,中剂量茶多酚对于促进辐照后白细胞数的恢复有更好的作用。在辐照前到辐照后14d这段时间里,4个组没有明显差异。剂量组与辐照模型对照组的SOD活性,在7d实验后没有显著差异。但各剂量组SOD活性均比辐照模型对照组略高,说明茶多酚抗辐射制剂还是具有轻微的保持SOD活力的功效,但效果不明显。此次试验的数据显示,3个剂量组中,高剂量

组 SOD 活力数值最高,中剂量组最低。根据《保健食品检验与评价技术规范》,有两项结果呈阳性,可判定茶多酚样品对辐射危害有辅助保护的功能。

表6 茶多酚抗辐射制剂对小鼠辐照前后外周血白细胞、血中超氧化物歧化酶活性和血清溶血素抗体积数值的影响

组 别	动物数	白细胞数($\times 10^9$ 个/L)		红细胞 SOD 活力 (u/gHb)	血清溶血素抗体积数值
		辐照前	辐照后第 14 天		
辐照模型对照	10	10.6±3.0	2.7±0.6	118223.94±36519.91	161.7±21.6
低剂量	10	10.5±1.8	2.6±0.6	122866.64±17319.29	179.9±30.1
中剂量	10	10.7±2.4	3.4±1.4	120126.49±25471.65	172.9±32.1
高剂量	10	11.0±2.2	2.8±0.7	140949.06±31119.84	171.0±23.6

以上是我们这些年在茶多酚保健食品研究中获得的主要成果。当然,在该领域还有很多研究可以开展,比如其他未涉及的功能开发,还有茶多酚与不同中草药配伍的作用等,这些工作需要我们茶叶科技工作者共同努力完成。

总之,在我国人民生活水平不断提升,健康问题越来越受到重视,以及茶多酚研究不断深入的大背景下,茶多酚保健食品研究开发正当时。我们认为,中国茶行业中的传统茶产业、茶饮料业、茶叶提取物以及茶馆业虽然在未来还会有较大的发展,但这种发展在较长一段时期内相对有限,因为会受到土地、气候等众多条件的制约,只能算是一种量变;而茶多酚保健食品的发展将是一种质变,对整个茶产业将带来巨大的影响。茶多酚保健食品业将来会出现重量级的企业,其单个企业或某一种产品的产值会超过目前整个茶产业,而这种变化是革命性的。茶多酚所具有的综合保健作用及几无副作用的特点,使利用其为原料的某一产品将成为国人保健首选品种。而且,由于其最初原料还是茶叶,所以又会促进传统茶叶行业的发展,会刺激茶叶种植面积扩大,刺激产量和茶价上升。所以,发展茶叶保健食品是整个茶叶行业自身拓展和向其他行业渗透的需要。

茶叶深加工副产物增值利用技术研究

张士康

中华全国供销合作总社杭州茶叶研究院，中国 浙江

摘　要：茶叶是我国优势特产资源。茶叶深加工发展迅猛，由此产生的茶叶深加工副产物产量剧增且未得循环利用，给环境保护造成很大压力。因此，本文就茶叶深加工副产物增值加工技术进行综述，概括了目前国内外关于茶叶深加工副产物的转化技术研究报道与应用状况，结合笔者所提出的茶产业发展"全价利用、跨界开发"理念，切合实际地引导本单位研究团队对茶叶深加工副产物进行转化增值加工技术相关研究。

关键词：茶叶；深加工；副产品

Study on the Technology for Increase Additional Value of By-products of Refined Tea Processing

ZHANG Shi-kang

Hangzhou Tea Research Institute, CHINA COOP, Zhejiang, China

Abstract: Tea is a special local product in China. Recent years, tea deep processing has been sharply developed. Herein, a large mount of tea deep processing by-products come out and strengthened the environment. Therefore, reprocessing techniques of tea deep processing by-products is urgently needed and has been developed. In this paper, the reported or used reprocessing techniques were concluded. Meanwhile, the works has been done by our team has been stated, which is obeyed to the thought of "Complete utilization, Cross-border development" for directing tea industry developing. That the tea deep processing by-products are value-added and reutilized is warmly expected.

Keywords: tea; deep processing; by-products

茶叶是我国主要的传统优势特产资源，其产量位居世界首位，是我国具有

较强市场竞争力的特色农产品资源,其相关产业已被十几个省列为农业支柱产业或优先发展产业。尤其是近年来,茶叶资源在我国发展迅猛,与十年前相比,种植面积增加了70%,产量增加了近1倍。2011年,我国茶叶产量162万t,增产9.9%[1]。除去日常消费,大部分用于茶叶深加工行业。茶叶中已鉴定的功效成分百余种,从茶叶深加工角度来看,目前茶饮料、速溶茶、茶多酚和茶油作为几种主要的茶深加工产品进入市场流通,这几种茶产品已经具有成熟的深加工工艺且产量可观。这种以单一功能成分为主要目标的加工工艺不可避免地产生大量茶叶废料,这些废料中含有的大量茶膳食纤维、茶蛋白质等功能成分未得到有效利用。以茶多酚生产为例,绿茶中茶多酚的含量一般占干茶的25%～35%,即便是加工中茶多酚能够100%获得,生产30g茶多酚将会产生70g废料,几乎是茶多酚产品的两倍。红茶、乌龙茶中茶多酚含量比绿茶更低。实际上,按2011年我国茶叶产量的1/3用于茶叶深加工计算,仅2011年茶叶深加工副产物就超过30万t。这些大量的茶叶深加工副产物几乎都没有被利用,仅有少部分作为普通肥料,绝大部分都被当作废物扔掉,这样不但造成环境的污染,而且造成茶资源的巨大浪费。

1 国内外茶叶深加工副产物增值技术与新产品研究开发

1.1 茶叶蛋白制备技术与产品开发

茶叶中蛋白质占茶叶干重的20%～30%,而70%～80%的茶叶蛋白为水不溶性蛋白,这部分蛋白大量存在于茶叶深加工产物——茶渣中。国内外对于水不溶性茶蛋白的研究仍旧停留在提取层面,对其保健功能略有研究。水不溶性茶蛋白多与多酚类、纤维类以共价或非共价方式联接,因此提取制备难度较大,尤其是多酚类的存在,使得水不溶性茶蛋白的提纯变得非常困难。在水不溶性茶蛋白的提取方面,目前主要通过传统提取蛋白质方法——碱提酸沉法、酶法、复合法进行提取[2-3],其中复合法提取率最高为81.03%。几种提取方法各有利弊,碱提酸沉法是提取蛋白最常用的方法,工艺简单,适合大规模生产,但水不溶性茶蛋白溶液是一个复杂的混合体系,所以溶液的等电沉淀点不是一个值,而是一个范围,所以要提高蛋白质得率,必须反复改变溶液体系中的pH值,这样在反复提取中会造成蛋白质的损失和品质变化,因此该方法的提取率偏低,不足60%[4];酶解法提取茶蛋白条件比较温和,但提取茶蛋白被水解后提取出,蛋白被酶解成低分子量,原有的长链蛋白将被水解成断链,原蛋白固有的功能品质将受到影响;复合法主要是碱法和酶法复合,尽管提取率有所提高,但是碱法和酶法的弊端仍无法避免,茶蛋白原有的功能品质受到提取条件的影响。

对茶渣蛋白的保健功能研究主要集中在茶渣蛋白对抗癌、防癌作用上。李燕等[5]对茶渣蛋白酶解液预防辐射引起的突变效应进行研究,经 Ames 试验、V79 细胞染色体畸变试验、小鼠嗜多染红细胞(PCE)微核试验和显性致死试验,发现茶渣蛋白酶解液可抵抗^{60}Co 辐射对沙门氏菌的诱变作用,降低中国仓鼠肺 V79 细胞的染色体畸变数,使小鼠 PCE 微核数明显减少,对^{60}Co 辐射引起小鼠的显性致死损害有保护作用,综合以上结果可以认为茶蛋白对预防放射治疗时引起的致突变效应有保护作用。华炜[6]通过研究不同剂量的茶渣蛋白对高脂大鼠血清 TG、TC 及衰老模型小鼠血液 SOD、MDA、GSH.PX 的影响,发现高脂大鼠灌胃 7 周后,与高脂对照组比较,茶蛋白 10g/kg、20g/kg 剂量组能显著降低血液中 TG、TC 含量;老龄小鼠与衰老模型组比较,茶蛋白 10g/kg、20g/kg 剂量组显著降低血液中 MDA 含量,且茶蛋白 10g/kg 剂量组能明显升高血液中 SOD 的活性;血液中 GSH.PX 含量基本无变化,综合以上指标认为蛋白具有一定的降血脂及抗氧化作用。

1.2 茶渣饲料

迄今为止,已有为数不少的报告介绍茶叶及其提取物用作饲料添加剂以改善饲料的特性,或将茶叶、茶渣或茶叶提取物按一定比例加入日粮中,以提高饲料的转化率。但总的来说,这类报道零星而分散,缺乏系统性,缺乏明确的研究目标,尤其是缺少理论上的探讨,这大大限制了茶叶在畜牧饲料中的应用。浙江农业大学动物科学院研究证明,日粮添加 0.05%～0.1%的茶多酚可抑制肉鸡法式囊病的发生,日粗粮加入 0.25%茶多酚可显著提高鸡血液 C36 受体免疫复合物的含量,并能保护肝脏正常的免疫功能,而对鸡的授食量没有影响。经分析,茶多酚能有效地提高鸡的胸腺淋巴细胞的增殖,茶多酚的抑菌能力与鸡的免疫力之间有显著的相关性[7-8]。湖南农大研究证明,家猪日粮中加入 5%的茶渣,替代 5%的麦麸,结果对肉质无影响,猪血液生化成分与对照组相比无显著差异,因此猪日粮饲料中搭配茶渣是可行的。经对茶渣成分进行分析,其中含粗蛋白 18.7%,氨基酸组成合理(赖氨酸 1.75%、蛋氨酸 0.54%、粗纤维 16.3%)。茶渣表现消化能和表现代谢能分别为 8.478mg/kg 和 8.0mg/kg,可消化蛋白 1168mg/kg,氮的表现代谢率 44.92%,在 5%添加量下可完全替代麦麸,而产量品质不减,成本至少节约 10%[9]。

最近,中日合作开发成功一项将茶饮料生产中的副产物——茶叶渣转化加工成青贮饲料的技术。在茶饮料生产中,经提取处理后的茶叶废渣几乎不含有可转化为青贮饲料所必需的乳酸菌和葡萄糖等糖类,因此无法像牧草和玉米等饲料作物那样借助于乳酸菌发酵的作用转化成青贮饲料。该技术在茶叶渣中适量添加乳酸菌及可以将植物纤维分解为糖的酶制剂,经过4个月,茶渣发酵物即转化为可以利用的青贮饲料。这种转化饲料的特点是pH值低,含有蛋白质、胡萝卜素、维生素E和茶独有的有效成分儿茶素等,营养价值高,乳酸含量高,是高品质的茶渣青贮饲料。经喂养实验证明,食用这种茶叶渣饲料的家畜,其排泄物中有益的乳酸菌量增加,有害的大肠菌和葡萄球菌等数量减少,由此证实了茶叶渣青贮饲料有利于改善喂养动物肠道菌群的功效[10]。

1.3 食用菌培养基质

近年来,随着茶产业和食用菌产业两大产业的发展与扩大,关于茶资源与食用菌栽培的结合已成为解决茶资源利用度不足和食用菌栽培基料供需矛盾的有效途径。以茶资源为食用菌基料的研究也有部分报道及专利成果,如利用茶籽壳为主料栽培茶新菇、利用废茶产业次级产物栽培猴头菇、用废弃茶叶为基质培养灵芝等,这些研究证实了以茶资源为基料栽培的食用菌产品色泽好、产量高[11-12]。

1.4 茶渣肥料

从茶叶中提取可溶性组分后所剩的茶渣含有18%～31%的粗蛋白、16%～22%的粗纤维、1%～2%的茶多酚、0.1%～0.3%的咖啡因,还含有少量维生素、矿质元素及茶皂素等成分,具有较高的潜在利用价值。目前,世界上大多数国家施用的氮肥多为碳酸氢铵、磷酸氢铵和尿素等铵态氮肥,这些铵态氮肥一旦进入土壤,很快硝化转为易溶的硝酸根,硝酸根易被雨水冲刷而流失,造成土壤长效肥力降低。而茶渣中的多酚类物质、蛋白类物质具有供氢能力,能够固化铵离子,减缓硝化作用,进而增强肥力[13]。

茶渣中有机质含量较高,土壤施用以茶渣为基料的肥料后,可明显增加土壤有机质含量,促进土壤微生物菌群的良好生长。对此,夏会龙[14]对施用茶渣有机-无机复合肥的茶园土壤进行了观察研究,其结果表明茶园土壤各土层的有机质含量比施用尿素、菜饼明显增多,土壤pH值稳定,土层中细菌、放线菌和真菌的总量均较高,因此茶渣有机-无机复合肥能够改善茶园土壤的生态特性。孙志栋、邱富林等[15-17]关于茶渣肥料对作物生产的肥效试验还表明茶渣有机－无机复合肥能够显著促进作物新梢生长、产量和品质的提高。

1.5 茶渣树脂、纸浆

日本"环保产品展2011"在东京国际会展中心举行,在日本最大的茶饮料生产企业伊藤园公司的展厅里,摆放着多款利用茶渣生产的环保产品。原本仅仅用来生产肥料的茶渣,被用于树脂和纸浆的生产。据伊藤园公司的员工介绍:"由于茶渣中含有儿茶素等有益成分,用含有茶渣的树脂生产的鞋垫、座椅、水桶等产品,具有除臭、杀菌等多种功效。用含有茶渣的纸浆制作的纸张,也有一种清新的茶香味。这些产品都深受消费者的欢迎。"[18]

1.6 保健品

以产茶出名的日本静冈县,把脑筋动到了泡过的茶叶渣上,磨粉再制成了浓缩保健补给丸,其中含有胡萝卜素、维生素A、维生素E等。维生素E能抗氧化、保护肌肤。有日本业者说:"喝煎茶,只能摄取3成左右的养分,茶叶渣里含有更多的养分。"[19]

1.7 茶渣蚊香、茶渣枕头

将晒干的茶渣在夏季的黄昏点燃起来,可以驱除蚊虫,和蚊香的效果相同。因茶性属凉,故茶渣枕可以清神醒脑,增进思维能力。

尽管国内外对于茶叶深加工行业的副产物包括固体废弃物的利用途径研究报道很多,但是真正在实际中形成规模的还非常少见,尤其在国内。而且,伴随着茶产业的转型快速发展,茶叶深加工产生的固体废弃物的综合利用问题越来越突出,因此需突破传统方式,开辟新型利用途径,加速实现茶叶固体废弃物再利用转化,促进生态绿色发展,提升环境质量。

2 "全价利用、跨界开发"理念推动茶叶深加工副产物增值转化

基于前人的研究报道和国内外茶产业发展现状,作者一直认为"全价利用、跨界开发"才是中国茶产业优化突破有效路径过程中均衡、高效的发展方向[20]。基于"全价利用、跨界开发"理念,破除茶产业领域界限,全面推动以茶产业与其他领域交叉互用为最终目的的茶产业发展。基于"全价利用、跨界开发"这一理

念,作者领导研究院研究团队以茶叶深加工副产物增值转化为一个研究方向,针对目前我国茶叶深加工副产物利用度低、无附加值的现状,研究设计开发系列茶叶深加工副产物增值转化技术,并倾力推动技术成果孵化。

2.1 茶渣蛋白组织化技术与仿生肉产品开发

传统的碱提酸沉法制备茶渣蛋白技术不仅得率不高,而且伴随产生大量废水,增加环境污染,也因此该技术一直停留于实验研究层面,并未能实际转化。对此,中华全国供销合作总社杭州茶叶研究院研究团队针对茶渣蛋白的超声波辅助碱法提取工艺[21]、理化性质[22]、功能特性以及茶渣蛋白的挤压组织化制备仿生肉进行了系统研究,获得了茶渣蛋白分子量(图1)、茶渣蛋白的超声波辅助提取最佳工艺参数,以及茶渣蛋白的乳化性(图2)、起泡性(图3)等信息和挤压组织化产品(图4)。

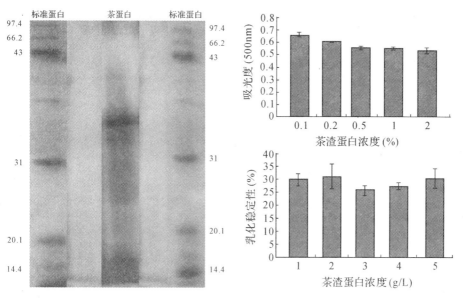

图1 茶渣蛋白的 SDS-PAGE 电泳图　　图2 茶渣蛋白浓度对其乳化性和乳化稳定性的影响

这些茶渣蛋白的信息对于指导茶渣蛋白的应用具有重要的理论意义,以此为基础,中华全国供销合作总社杭州茶叶研究院正在开展茶渣蛋白在肉制品中应用的研究,如承担的浙江省农业成果转化工程"茶及功能性成分在食品加工中的产业化应用技术与示范(2012T202－06)"项目。

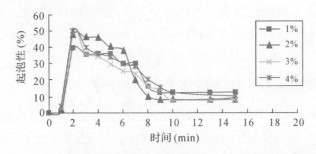

图3 茶渣蛋白浓度对其起泡性和泡沫稳定性的影响

5%茶蛋白添加量　　　　　　　25%茶蛋白添加量

图4 茶渣蛋白挤压组织化产品

2.2 茶蛋白和茶叶膳食纤维联产

茶渣中不仅有高含量的茶渣蛋白,还含有约30%的茶渣纤维。在茶渣蛋白的研究开发基础上,针对茶渣蛋白和茶渣纤维,中华全国供销合作总社杭州茶叶研究院设计一套茶渣蛋白和茶渣纤维联产制备技术,能够同时回收茶渣蛋白和茶渣纤维,做到茶渣全价利用。

2.3 与食用菌产业联合

以茶渣作为食用菌基料,国内已经有所研究,甚至已经申请专利保护,但并未实现转化,对食用菌种的筛选和菌料的循环利用也未有系统研究。针对此种现状,我院研究团队以茶渣为食用菌基料,按食用菌中木腐菌和草腐菌进行分类,已成功从金针菇、秀珍菇、凤尾菇、平菇、茶树菇、香菇、灵芝等十几种食药用菌中筛选出适宜在茶渣基料生长的品种,对这些适宜品种的生物学特性、生物学效率进行了深入系统的研究,并已对其中的金针菇进行了实际工厂化扩大栽培。头潮菇生物学效率达 30%（图5）。

图 5　茶渣基料生长的 3 种食用菌（从左到右：平菇 650、秀珍菇、金针菇 F_{21}）

3　结语与展望

如何实现茶叶深加工副产物高效转化，国内外专家、学者们做了很多工作。然而，茶叶深加工副产物增值转化仅仅是茶业发展道路上的一块顽石，在开发利用茶叶深加工副产物时，还要意识到整个茶产业发展道路上的其他顽石的存在，需以高瞻远瞩的目光来综合考虑茶产业发展。因此，作者建议在对待茶产业发展上，不妨借鉴"两手抓，两手都要硬"的战略发展方针，推行"全价利用、跨界开发"的茶产业发展理念，在茶产业内部抓名优茶、品牌茶开发，优化茶产业体制，规范茶产业管理，实现茶叶多元化，满足消费者更加追求饮茶的生理和心理双重需求；在茶产业之外抓交叉跨界利用，带动相关产业进步与增效，开展茶食品、茶药品等交联产品与技术研发。在茶产业周围形成以茶产业为核心的具有一定发展力、带动力、推动力的经济带，实现茶资源全价开发、物尽其用，方能实现茶产业高效、快速、稳步发展。

参考文献

[1] 中华人民共和国 2011 年国民经济和社会发展统计公报. http://www. stats. gov. cntjgbndtjgb/qgndtjgb/t20120222_402786440. htm.

[2] 陆晨，邹雨虹，张士康等. 响应面法优化超声辅助提取茶渣蛋白的工艺条件[J]. 食品与生物技术学报，2012，31(3)：319－325.

[3] 陆晨，张士康，朱科学等. 碱提酸沉法提取茶叶蛋白质的研究[J]. 现代食品科技，2011，27(6)：673－677.

[4] 王忠英. 茶叶中蛋白的提取及理化性质的研究[D]. 杭州：浙江工商大

学, 2006.
[5] 李燕, 蔡东联, 夏雪君等. 茶蛋白液预防辐射引起的突变效应[J]. 癌变. 畸变. 突变, 2001, 13(1): 32—36.
[6] 华炜. 茶蛋白降血脂及抗氧化作用的实验研究[C]. 达能营养中心第九次学术年会论文集, 2006: 296—301.
[7] 吴树良. 在养殖业中巧用茶饲料效果好[J]. 茶叶科学技术, 2001(3): 29—30.
[8] 倪星虹. 茶渣提取蛋白及饲料化利用的初步研究[D]. 南京: 南京农业大学, 2012.
[9] 高凤仙, 田仲雄, 王继成. 速溶茶渣饲用价值研究(2)饲喂生长育肥猪的消化代谢试验[J]. 湖南农业大学学报, 1998, 24(6): 465—467.
[10] http://info.china.alibaba.comnewsdetail/v0-d1024587786.html.
[11] 高鹏. 茶渣培养灵芝菌丝及提取灵芝腺苷技术研究[D]. 福州: 福建农林大学, 2012.
[12] 一种栽培食用菌的培养料: 中国, 00410040038[P].
[13] 仇方方, 杨晓萍. 茶渣肥料的应用[J]. 福建茶叶, 2011(6): 21—23.
[14] 夏会龙. 茶渣复合肥对茶园土壤的生态效应[J]. 污染防治技术, 2003, 16(4): 76—78.
[15] 孙志栋, 梁月荣, 戴国辉等. 茶渣有机无机复混肥在克服小菘菜连作障碍上的应用研究[J]. 土壤通报, 2008(1): 2000—2002.
[16] 孙志栋, 张松强, 陈惠云等. 茶渣有机无机活性肥改良大棚葡萄土壤初步研究[J]. 中国农学通报, 2010, 26(4): 178—181.
[17] 邱富林, 曹炎成, 江义鸿. 茶渣有机无机——复混肥对柑生长结果影响的试验初报[J]. 浙江农业科学, 2008(3): 276—277.
[18] http://info.tjkx.com/detail/875446.htm.
[19] http://news.t0001.com/2009/0417/article_93785.html.
[20] 张士康. 全价利用, 跨界开发——中国茶产业优化突破有效路径探索[J]. 中国茶叶加工, 2010(2): 3—5.
[21] 陆晨, 邹雨虹, 张士康等. 响应面法优化超声辅助提取茶渣蛋白的工艺条件[J]. 食品与生物技术学报, 2012, 31(3): 319—325.
[22] 张士康, 陆晨, 朱科学等. 茶渣蛋白理化特性研究[J]. 中国茶叶加工, 2011(2): 38—40.

论茶叶精深加工产业发展及其路径

陆德彪　罗列万　毛祖法

浙江省农业厅经济作物管理局,中国 浙江

摘　要：本文论述了发展茶叶精深加工的意义和迫切性,分析浙江茶叶精深加工产业现状与问题,提出当前茶叶精深加工的重点和促进产业发展的建议。

关键词：茶叶；精深加工；产业

Discussion on the Way to Develop Multipurpose Utilization of Tea

LU De-biao　LUO Lie-wan　MAO Zu-fa

Department of Agriculture, Zhejiang Provincial People's Government, Zhejiang, China

Abstract: Both significance and urgency of development of the industry of multipurpose utilization of tea is discussed in this paper. It also analyzed the situation of the industry in Zhejiang province and issues to be solved. Key-points for development of the industry of multipurpose utilization of tea and promotion steps are put forwarded.

Keywords: tea; multipurpose utilization of tea; industry

1　全价利用——发展茶叶精深加工的意义

茶叶精深加工是指以茶鲜叶、毛茶及加工副产品、茶树花果等为原料,利用现代高新技术及加工工艺将其加工成终端产品。它主要包括两大方面：①将传统工艺加工的产品进行更深层次的加工,形成新型茶饮料品种；②提取茶叶中的功能性成分,将其应用于食品、化工、医药、环保等行业,形成各种新的终端产品。

据专家分析,茶叶精深加工是一个蕴藏巨大商机的朝阳产业,茶叶精深加工产品在21世纪有广阔的市场前景。①茶饮料因其天然、方便、健康等特点而得到消费者特别是年轻消费者的青睐。保健茶和果味茶在国际市场上发展很快,欧洲市场上已占饮料总销量的50%;国内茶饮料特别是调味茶和固体奶茶市场呈迅猛发展势头,近年来液体茶饮料以每年35%的速度快速增长,目前年销量已超过800万t,产值超过400亿元。②茶叶精深加工产品尤其是茶叶有效功能性成分提取物的开发将有较快发展。国外研究表明,茶多酚在医学上的用途非常广泛,有望开发成清除自由基、抗癌、抗突变、抗辐射等方面的药物;茶叶中含有的咖啡因、茶氨酸、茶皂素、茶色素、茶多酚等成分也可广泛用于医药和食品行业。近30年来,茶叶的用途已经从其冲泡饮用的单一方式扩展到饮茶、吃茶、用茶、赏茶等人们生活的方方面面,茶叶的功能成分已经被分离,并已转化成高科技产品。例如,日本的茶叶精深加工产品已渗透到医药保健、食品、日用化工、养殖等行业;年产值已达1400亿元人民币;美国仅开发以茶多酚为原料的终端产品的产值已达100亿美元。发展茶叶精深加工,不仅有益夏秋茶资源利用和茶农增收,而且可以培育出年产值数百亿元的新产业,潜力巨大,意义深远。

(1)可以将资源优势转化为经济优势。

目前,浙江的茶叶资源利用率不高,一是不少茶区只采摘春茶,而占茶叶理论产量50%左右的夏秋季鲜叶通常未被利用;二是在茶叶精制加工中产生的20%左右低档茶和副产品,常常因为没有市场而被废弃。因此,发展茶叶精深加工,可以充分利用夏秋茶资源和副产品,为茶农和企业增收。

(2)可以丰富茶叶产品,满足人们对茶叶产品的多元化需求。

通过精深加工,可以创造出如速溶茶、茶饮料、茶食品、茶保健品,以及茶日用化工品等更多、更新、既方便又健康的茶制品,极大地满足人们对茶叶产品个性化消费的需求,大大提高茶叶的附加值。

(3)可以拓展茶的功能与用途,延长产业链。

通过茶叶精深加工,将茶多酚、茶皂素、茶氨酸、咖啡因等天然产物分离、提纯与应用,使其成为食品、日用化工和医药工业的重要原料,开发出一系列基于茶叶功能成分的终端产品,并形成新的产业。

此外,茶叶精深加工广泛采用高新技术,引入先进的运行机制和管理理念,对促进整个茶产业的转型升级具有重要意义。

2 时不我待——发展茶叶精深加工的迫切性

茶业是浙江省最具竞争力的特色优势产业,在全省农业主导产业中占有重要地位。近年来,在浙江省委省政府的正确领导下,各地坚持茶产业发展、茶品牌培育、茶文化弘扬并举,积极调整品种结构,大力开发名优茶,全省茶产业取得长足发展。2011年浙江省茶园面积273万亩,茶叶产量17万t,茶叶产值101亿元,茶叶出口17.6万t,茶叶出口额4.9亿美元。浙江茶叶以全国8%的面积,创造了全国11%的产量和全国14%的产值,绿茶出口占全国绿茶出口的63%,茶产业已成为浙江农业的一张"金名片"。

但是,在茶叶产业繁荣的背后,一些深层次的问题正暴露出来。①茶叶产能快速扩张与传统茶叶产品市场容量相对有限之间的矛盾日益突出;②以茶叶精深加工为重点的第二、第三产业严重滞后,使茶叶"一、二、三"产业之间的发展不协调,造成产业层次不高,产品附加值低,产业链短,产业潜力没有充分发挥。尤其是近年来,贵州、四川、湖北等中西部地区茶园种植规模快速扩张,产能不断释放,"供过于求"、"丰产不丰收"矛盾将日益凸显。为此,改变目前过分依靠第一产业扩张的发展模式,解决制约茶叶精深加工产业发展的瓶颈,把茶叶精深加工产业培育成新的增长点,实现产业转型升级已成为当务之急。

科学研究已经证明,茶叶(树)全身都是宝。如果我们进一步研究茶叶中的有效成分,充分发挥茶叶有效成分对人体健康的作用,加快茶叶精深加工终端产品的开发进程,开发新用途,激发新需求,开辟新市场,就可以解决大量中低档茶和副产品的出路,充分利用茶叶资源,丰富茶产品的种类和档次,实现茶产品的形态与功能多样化,扩大茶叶产品的市场。因此,加快发展茶叶精深加工不仅是茶产业转型升级的重要内容,也是继续保持浙江茶产业全国领先地位的重要举措。可以说,浙江在名优茶战略取得成功之后,精深加工应当成为下一阶段浙江茶产业转型升级的突破口。

3 忧深思远——浙江茶叶精深加工现状与问题

3.1 浙江茶叶精深加工产业现状

2010年浙江省茶叶产业协会曾对浙江省茶叶精深加工产业现状做过深入调研。调查结果表明,浙江省茶叶精深加工研究始于20世纪80年代,80年代中期率先从茶叶中提取出以茶多酚为主体的天然抗氧化剂,并成功应用于食品行业,有效地延长了食品的保质期。20世纪90年代,随着研究的不断深入,应用茶叶精深加工研究成果而形成的产业开始起步。目前浙江茶叶精深加工企业总数在20余家,产品主要有3类:①加工速溶茶、茶饮料和固体奶茶;②萃取茶多酚原料和加工茶多酚保健品;③开发各种袋泡茶。另外,有几家从事抹茶生产、低咖啡因茶加工和茶籽油榨取的企业。这3类中产销量最大的是速溶茶和茶饮料,浙江茗皇天然食品开发有限公司、易晓食品(衢州)有限公司、杭州茗宝食品有限公司、浙江东方茶业科技有限公司绍兴分公司、浙江塔塔茶业科技有限公司等以生产速溶茶为主,全省速溶茶企业生产能力达到1万t左右,满足生产能力可消化干茶5万t,实际年产量约5000t,占全国速溶茶产量的一半左右;杭州娃哈哈集团有限公司生产的调味型茶饮料,年产量约150万t,占全国茶饮料总量的15%左右;临安九诚茶业有限公司和长兴茶乾坤食品有限公司主要生产袋泡茶,年产量3000t,大多销往日本市场。浙江萃取茶多酚原本处于全国领先地位,但现在企业基本上处于停产状态,只有单体儿茶素如EGCG等有少量萃取。茶多酚单体成分萃取目前还没有实现产业化的主要原因是高纯度单体儿茶素市场需求量不大,一般只用作实验室进行化学分析和小规模实验;茶多酚片剂、抹茶、低咖啡因茶和茶籽油生产企业规模比较小。

3.2 茶叶精深加工发展"瓶颈"分析

虽然浙江茶叶精深加工起步早,也初具规模,但除少数企业发展比较快以外,大多数企业举步维艰。阻碍企业发展的"瓶颈",主要表现在以下方面。

(1)精深加工产品单一,应用途径少,制约了产业发展。目前茶叶精深加工产品以茶多酚和速溶茶为主。原先浙江的茶多酚大多出口美国代替麻黄碱作为减肥药的组成原料,后来因新的减肥药配方未获美国FDA批准,茶多酚出口陷入低谷,茶多酚销售价格也从最初的每千克1000多元下降到现在的每千克100多元。茶多酚价格暴跌,除了有技术进步、萃取率提高和成本降低的因素外,主要是供求关系变化。速溶茶以前生产总量的70%左右用于出口,由于金融危机导致国际需求的减少,以及国内茶饮料和固体奶茶的发展,速溶茶在国内的销量超过出口,但是由于应用范围的限制,速溶茶的实际生产量仅为生产能力的一半左右。

(2)缺少产品标准,企业仍处于低层次竞争。目前绝大多数茶叶精深加工产品没有国家标准、行业标准和地方标准,只有部分企业根据自身需要制订了企业标准,使消费者对产品的评判缺乏统一的尺度,企业的生产过程也存在一定的随意性。

(3)企业资金实力不强,产品研发能力弱,限制了新产品开发。如茶多酚作为食品添加剂使用,只需通过卫生学检测,而生产保健品则要完成急慢性毒理试验和功能试验,很多功能还要进行人体临床试验。由于茶叶精深加工产业规模小,企业研发资金不足,很难维持这些长周期和高额的研发、试验投入,致使企业放弃新产品研发而生产技术工艺简单的产品,导致产品趋同,产能过剩,市场竞争激烈。

(4)茶叶精深加工产业化程度低,上下游产业链没有形成,缺乏循环利用基础。精深加工原料对茶叶外形要求不高,如果简化加工工艺完全可以达到要求,而当前茶叶精深加工产品的原料来自于经过全套加工程序的各类中低档茶叶和副产品,无谓浪费了大量的能源;经过精深加工的茶叶残渣只有少部分再利用,其他被废弃,不仅浪费了可利用资源,而且极易造成环境污染。

4 终端开发——茶叶精深加工的重中之重

茶叶精深加工就生产过程来讲,涉及毛茶原料的生产、茶叶功能成分的提取、终端产品的开发等。目前,原料生产和成分提取条件成熟,今后工作重点应是茶叶精深加工终端产品的产业化开发,做精产品,做大品牌,做强产业。根据现有技术条件与市场需求,当前茶叶终端产品开发可着重于以下几个方面。

(1)茶饮料:包括罐装茶水、速溶茶等。

(2)茶食品:包括茶糕点、茶料理、茶面食等,如以抹茶、超细绿茶粉为配料的各种食品等。

(3)保健品:如心脑血管病药、降糖降脂产品、降血压产品、抗衰老提高免疫力产品等,包括以茶多酚、茶氨酸、γ-氨基丁酸、茶黄素、茶多糖、茶色素等为主要成分的各种产品制剂。

(4)食品添加剂:包括食品、食用油抗氧化剂,及调味、调色剂等。

(5)饲料添加剂:包括鸡、猪等畜禽配合饲料(能有效降低畜禽产品胆固醇含量、消除臭味等)。

(6)日用化工品:包括空调杀菌剂、除臭剂、香波、茶香皂、茶沐浴露、茶枕头、茶纺织品,以及抗衰老、减肥美容类产品等。

5 创新创业——促进茶叶精深加工产业发展的建议

发展茶叶精深加工产业,可谓生逢其时。浙江不仅有丰富的茶叶资源,有实力雄厚的茶叶科研机构和研发力量,有一批有自主知识产权、国内外领先的技术储备,更有活跃的工商资本。娃哈哈、康师傅等大企业已成功进入茶饮料产业,浙江塔塔、振通宏等一批龙头企业已建有规模茶叶精深加工生产线。可以说,浙江发展茶叶精深加工产业已具备"天时、地利、人和"的要素条件,唯独不足的是缺少除茶饮料领域之外的茶叶精深加工终端产品产业化。只要在产品研发、产业化方面给予引导和扶持,完全可以通过精深加工将浙江绿茶的资源与技术优势转化为产品优势、竞争优势,实现由茶叶大省向茶叶强省的跨越。

一是加强规划,突出重点。把发展茶叶精深加工纳入浙江"十二五"农业、食品工业、科技等相关产业发展规划,并组织实施。借助落户杭州的茶叶科研机构和研发力量,重点围绕茶叶功能成分终端产品产业化这一目标,在茶饮料、茶食品、茶保健品、茶食品添加剂、茶叶饲料添加剂、茶日化用品等六大方面,开发一批基于茶叶功能成分的终端产品,形成新的产业。要避免产品同质恶性竞争,对部分产能已达一定规模的产品,在研发、成果转化和市场未取得大的突破之前,应适当控制同类新项目上马,避免资源浪费。

二是树立循环经济理念,合理实施专业化分工,努力打造上下游产业链。按精深加工对原料的要求,组织夏秋茶生产,保证原料供应。促进专业萃取企业发展,实行专业化生产,提高经济效益。对深加工产生的茶叶残渣,要研究循环利用的途径,变废为宝。

三是加强对茶叶精深加工新产品的研究开发,进一步提高科技含量,创新产品形式,丰富产品种类。要重视茶叶精深加工产品标准的制订,规范行业管理。质量标准是产品和产业的生命,茶叶精深加工产品没有统一标准,将会制约产业的进一步发展,制订茶叶精深加工产品行业(地方)标准应摆上议事日程。

四是重视茶叶精深加工产品的宣传、推广工作,使其深入生活、深入社会,为产业发展和人类健康事业作出贡献。

五是出台优惠政策,扶持产业发展。各级政府应出台专门的扶持措施,落实土地、税收等优惠政策,支持科技成果转化,推动茶叶精深加工产业发展。

参考文献

[1] 毛祖法. 中国绿茶产业发展现状与展望[C]. 浙江省-静冈县 2012 绿茶博览会绿茶国际论坛论文集,2012.

[2] 中村顺行. 日本的绿茶产业及其综合利用研究[C]. 浙江省-静冈县 2012 绿茶博览会绿茶国际论坛论文集,2012.

[3] 浙江省茶叶产业协会. 发展茶叶深加工,促进浙江省茶产业转型升级——浙江省茶叶深加工现状及发展对策[J]. 茶世界,2010(11):28-33.

[4] 浙江省农业厅经济作物管理局. 浙江省桑茶果生产工作年报[G]. 2010.

中国食用植物油产销现状与茶叶籽油的优势

程启坤[1] 郭庆元[2]

1. 中国农业科学院茶叶研究所,中国 浙江；
2. 中国农业科学院油料作物研究所,中国 湖北

摘 要：中国食用植物油消费增长很快,而生产发展缓慢,因此我国食用植物油的供给主要依赖进口。为了扭转这种被动局面、扩大我国食用植物油的生产,又要避免与粮争地,建议利用我国山地多的优势加快发展油茶油与茶叶籽油。

关键词：食用植物油；茶叶籽油

Current Situation of Production and Marketing of Edible Oil in China and Superiority to Develop Chinese Tea-seed Oil

CHEN Qi-kun[1] GUO Qing-yuan[2]

1. Tea Research Institute, Chinese Academy of Agricultural Sciences, Zhejiang, China; 2. Oil Crops Research Institute, Chinese Academy of Agricultural Sciences, Hubei, China

Abstract: In China, the consumption of edible oil is glowed quickly but its output rises slowly. As a result, China mainly relies on import of edible oil to meet its domestic demand. In order to change the current situation, it is necessary to expend production of edible oil of China but that will occupies the land for grain production. China has a vast land of hills and mountains, the authors suggest that development of camellia and tea-seed oil is the good way to solve the problem.

Keywords: edible oil; tea-seed oil

中国人开门七件事"柴、米、油、盐、酱、醋、茶",油与茶都是生活必需品。茶树所生茶籽可以榨油,所长芽叶可以制茶,更新台刈下来的老茶树可以当柴烧,因此茶树与人们日常生活关系密切。

人们日常生活中的食用油,都是以植物油为主。当今中国自己生产的食用植物油,远远不能满足国内人民的消费,因此中国的食用植物油主要依赖进口,使国家粮油安全面临严峻形势。本文就中国食用植物油产销现状与开发利用茶叶籽油的必要性进行讨论。

1 世界食用植物油生产消费概况

近半个世纪以来,全球食用植物油消费量增长很快,如:1996—2002 年的 6 年时间内,世界食用植物油的消费量由 7481 万 t 增至 9480 万 t,增长 26.7%,年增长 4.5%;2002—2011 年,消费量由 9480 万 t 增至 15076 万 t,9 年增加 5596 万 t,年均增长 6.6%,年增 621 万 t。1996—2011 年,食用植物油消费量由 7481 万 t 增至 15076 万 t,15 年增长了 1 倍多。

由于消费需求的强劲拉动,食用植物油生产随之快速发展。近 5 年,几种主要食用植物油的产量由 2007 年的 1.28 亿 t 增至 2011 年的 1.53 亿 t,增长了 19.1%。与此同时,国际市场上食用植物油贸易进出口量也在不断增加。2007—2011 年,进口量由 5026 万 t 增至 6000 万 t,增长 19.4%;出口量由 5397 万 t 增至 6270 万 t,增长 16.2%。

近 10 多年,几种主要植物油的产量、贸易量增长速度,以棕榈油最快。从 1998—2011 年,棕榈油产量由 1920 万 t 增至 5057 万 t,增长 1.63 倍;菜籽油产量增长 98%;大豆油、葵籽油产量分别增长 57% 与 74%。

2 中国食用植物油的产销与供给概况

中国随着人们生活水平提高,植物油的消费量逐年增加。1995—2002 年增长 61.7%,2007 年达到 2334.3 万 t,比 2002 年增加 7870 万 t,增长 50.8%。最近 5 年(2007—2011 年)消费量增加 577.8 万 t,增长 24.8%。2011 年的消费量达到 2912.2 万 t。1995—2011 的 16 年间,中国植物油消费量由 956.9 万 t 增至 2912.2 万 t,增长了 3 倍多,然而我国食用植物油消费量的强势增长,并未能拉动植物油生产的相应发展。据农业部的统计,1999—2009 年的 10 年内,几种主要油料播种面积没有增加反而减少,产量只增 3.3%,基本处于徘徊状态。如果只靠国内生产,远远满足不了国内人民食用植物油消费的增长速度。

在这种供需矛盾尖锐的形势下,只有从国外大量进口食用植物油。据统计,2009年减去同期出口量,净进口植物油籽4540.68万t,净进口植物油93.75万t。

2011年全世界植物油籽进口量10749万t,中国进口5716万t,占全球油籽进口总量的53.8%;2011年全球食用植物油进口6000万t,中国进口植物油(未包括进口植物油籽油)937.5万t,占全球植物油进口贸易的15.6%;2011年全球大豆进口量96762万t,中国进口55500万t,中国进口占全球进口量的61.1%。

由于我国进口植物食用油数量激增,中国这个人口大国食用植物油60%以上依赖进口,从而刺激了国际贸易中食用油价格的上涨。美国出口大豆油价格由2000年的每吨311美元涨到1147美元,增长2.69倍;马来西亚出口的棕榈油价格,由2000年每吨235美元涨到1058美元,增长3.5倍;鹿特丹转口的葵籽油由2000年每吨428美元涨至1639美元,增长2.82倍。这就是我国粮油安全面临的严峻局面,因此,摆在中国人民面前的现实说明,加紧我国食用植物油的生产已是刻不容缓的任务。

通过近十几年的实践证明,要想扩大油菜、大豆、花生的种植面积,就要与粮争地。而粮食生产是第一位的任务,不能再缩小种植面积了。因此大幅度增加草本植物油料种植面积已不可能。为此,国家从战略上考虑,必须从扩大山地木本食用植物油种植面积、提高产量着手。

世界上利用山地种植的木本食用植物油料,主要是棕榈油、橄榄油和油茶油。根据我国所处地理位置和环境考虑,大面积扩大棕榈油和橄榄油的可能性较小。而油茶在中国江南、华南、华中和西南的大部分山区都能种植,为此国家决定要大力发展油茶生产。从油脂的脂肪酸分析,目前人们认为,最有益于健康的植物食用油是橄榄油,油茶油最接近橄榄油,有些有益成分如不饱和脂肪酸、维生素E、茶多酚、角鲨烯、黄酮等都比橄榄油还略高。因此,大力发展油茶油大有希望。

2010年世界四大木本油料植物油,有3种年产已超过300万t,其中棕榈油5627万t,椰子油368万t,橄榄油302万t,唯有油茶油还不到30万t。

3 我国油茶油产业现状

油茶,俗称山茶,属山茶科常绿林木,是世界四大木本油料植物(油茶、油棕、油橄榄、椰子)之一,是我国长江流域及其以南地区特有的木本油料树种。我国栽培油茶已有两千多年的历史。历代皇室多以油茶油作为御膳用油,独具特色的中国饮食文化是华夏文明的重要组成部分,油茶与油茶油则是我国饮食文化的一颗璀璨明珠,她维护世世代代人民的身体健康,孕育灿烂的华夏文明。

油茶油是优质食用油,具有良好的营养保健功能。油茶油的不饱和脂肪酸(油酸、亚油酸)达到90%,是诸油品中最高的,不含胆固醇,不含芥酸、山嵛酸等难以消化吸收的成分。常食用油茶油能减轻血管硬化,降低心脑血管疾病的发病率,提高身体免疫系统功能。

我国油茶产业分布广泛,主要产地为湖南、江西、广西、浙江、福建、湖北、广东、贵州、安徽、云南、重庆、四川、河南、陕西等14个省(市、区)、642个县(种植10万亩以上的142个县)。2008年全国油茶面积4500多万亩,油茶籽产量97.55万t,产茶油26.25万t,平均亩产茶油5.79kg。2009年全国油茶籽产量116.93万t,比2008年增产19.87%,油茶籽产量超过10万t的省有湖南(41.9万t)、江西(27.0万t)、广西(13.4万t)。全国现油茶加工企业659家,设计加工能力424.8万t油茶籽,其中加工能力500t以上企业178家,具精炼能力的200多家,重点企业43家,年产油茶油近30万t,茶粕近70万t,茶皂素近2万t,发展呈良好势头。

目前,油茶产业发展中也还存在阻碍油茶产业快速发展的诸多问题:①山茶林产权数次变动,产量较低,一般亩产5kg左右,多数山茶林管理粗放,亩产值200元左右,与其他作物相比,经济效益低;②油茶加工企业多数规模较小,加工设备与加工技术落后,不少企业不能进行精深加工,企业经济效益不高,资源浪费较大;③油茶高产技术科技成果推广应用进展缓慢,油茶籽精深加工与综合利用,制油设备与制油技术尚待深入创新研究;④社会消费群体对油茶油的认知度不足,每当谈到要吃好油、吃优质营养油时,只知道橄榄油,不知油茶油。我国现有油茶林面积仅占全国森林面积的1.73%,约为5000万亩。若把种植面积扩大1倍,达到1亿~1.2亿亩,也只占全国森林面积的3.5%,远低于地中海沿岸橄榄树的比例和印尼、马来西亚棕榈林的比例,因此我国油茶林发展潜力很大。

4 茶叶籽油的优势与发展可能性

茶树与油茶树同属山茶科植物,自然界存在的山茶科植物很多,有山茶花、茶树、油茶树等。山茶花只供观赏,不采种榨油。茶树在长江流域多为灌木树型,没有明显的主干,生长出的嫩芽叶可以制茶,生长的果实中的种子称"茶子",富含油脂和一定量的茶多酚,可以榨油,因此这种茶叶籽油中约含1%的茶多酚。油茶树为乔木树型,有明显的主干,生长出的嫩芽叶不能用来制茶,生长的果实中种子称"油茶子",也富含油脂可以榨油,且油脂中不饱和脂肪酸含量是所有食用植物油中最高的,但这种油中茶多酚含量比茶叶籽油含量低。

茶树除了长叶子之外,也会开花结实,茶树每年4~6月生出花蕾,然后陆续开花,要到第二年8~9月才能长成饱满成熟的果实,每个茶果中有1~3粒茶籽。历经1年多时间才长成的茶籽,内含物丰富,含有淀粉、蛋白质、脂肪、氨基酸、胡萝卜素、维生素E、茶多酚、皂素和矿物质等。茶籽含油丰富,茶叶籽仁中含有25%~35%的油脂,一般通过压榨可以获得10%~15%的液态油,通过精炼的茶叶籽油色泽淡黄、液态、有清香。

2009年12月《中华人民共和国卫生部2009年第18号公告》中正式批准茶叶籽油为新资源食品。

茶叶籽油属植物油脂,植物油脂与动物油不同之处是:植物油脂以含不饱和脂肪酸为主,为液态。茶叶籽油的化学特征是含有丰富的不饱和脂肪酸,其不饱和脂肪酸含量要占脂肪酸总量的83%左右。这些不饱和脂肪酸包括单不饱和脂肪酸和多不饱和脂肪酸两大类。单不饱和脂肪酸主要是油酸,茶叶籽油中单不饱和脂肪酸含量约为59%;多不饱和脂肪酸主要是亚油酸和亚麻酸,茶叶籽油中亚油酸约为23%,亚麻酸约为1.2%;茶叶籽油中这种不饱和脂肪酸的含量及组成比例与橄榄油相类似。产于欧洲地中海沿岸国家的橄榄油被人们称之为高档食用油,因此,茶叶籽油也可称为东方橄榄油,是目前我国植物油中的高档食用油。

茶叶籽油除了富含不饱和脂肪酸之外,茶叶籽油优于橄榄油之处有3点。

(1)多不饱和脂肪酸中的欧米伽3(ω3)脂肪酸(即α-亚麻酸)比较丰富。ω3是深海鱼油的重要成分,具有降低甘油三酯、预防脑痴呆、脑退化、预防骨质疏松等多种功效。因此经常食用,对预防血管硬化,降低罹患心血管疾病、骨质疏松和老年痴呆症的风险以及润肤等极有益处。

(2)亚油酸(ω6)与亚麻酸(ω3)的比例为3.5~4∶1,符合世界卫生组织建议数,对健康更有利。

(3)茶叶籽油中茶多酚含量比橄榄油高得多,茶叶籽油中含有约1%的茶多酚。茶多酚是一种天然抗氧化剂,对人体保健具有降血脂、防止血管硬化、增强免疫、防辐射损伤、清除自由基、延缓衰老、预防癌症等多种功效。这1%的茶多酚还可以对茶叶籽油起到很好的保护作用,防止脂肪酸的氧化,从而大大延长了茶叶籽油的保质期。

(4)茶叶籽油含有 159~200mg/L 的维生素 E。维生素 E 是人体维持正常生长发育必不可少的物质,也是一种天然抗氧化剂,具有润肤和抗衰老的功效。

茶叶籽油具有的这些化学特征,为进一步发展茶和茶叶籽油产业提供了依据。在此基础上再来分析一下种植茶树、发展茶叶籽油的一些优势。

(1)能增加产值、比较效益高,茶树芽叶可制茶,种子可榨油。一亩茶园,茶叶产值通常可达 2000 元以上;茶树结子,根据茶树品种不同有多有少,一般情况下,一亩茶园可采收茶籽 100kg 左右。茶叶籽富含油脂,一般含油可达 25%~30%。茶叶籽的压榨出油率一般为 12% 左右,因此 1 亩茶地可产茶叶籽油 12kg 左右,按每千克茶叶籽油 50 元计算,1 亩茶园茶叶籽油产值可达 600 元左右。加上茶叶产值,1 亩茶园的总产值可达 2600 元左右。与种只能采子的油茶相比,比较效益高得多。

(2)全国茶园面积大,茶叶、茶子产量相对稳定。2010 年全国茶园面积 2955.3 万亩,可采面积(也能采收茶子)2139.15 万亩。按保守估计每亩可采茶子 30~50kg 计算,可采茶子 64 万~106 万 t;按出油率 12% 计算,每年可产茶叶籽油 7 万~12 万 t。这将是对我国食用植物油一个很可观的补充。如果全国茶区都重视采收茶叶籽,亩产茶叶籽有可能达到 100kg 左右,每年可产茶叶籽油 30 万 t 左右,可与油茶油产量相当。

另外,还可利用茶树品种资源的多样性,专门选育出多结茶籽并具有高出油率的茶树品种,建立一批主要用于采茶籽的专用茶园,或许也是我国茶业生产结构调整和转型升级的措施之一。

因此建议国家有关部门尽快出台相关规划与政策措施,鼓励茶农采收茶子投售,鼓励油脂加工企业生产茶叶籽油;建议科研部门加强茶园既采叶又采子保证茶叶、茶子双丰收的系列栽培技术研究,以及专门采子茶园品种及相应栽培技术研究,加强茶叶籽油提炼与饼粕深加工利用的研究,不断提高质量与生产效益;建议加强茶叶籽油保健作用的研究与宣传,共同促进茶叶籽油产业发展。

参考文献

[1] 农业部. 中国农业统计资料[G]. 2010.

[2] 中国茶业年鉴编辑部. 中国茶业统计资料[G]. 2011.

[3] 郭庆元. 我国食用植物油供给与山茶油产业发展探讨(内部资料). 2012.

[4] 中国粮油学会油脂分会. 中国茶叶籽油产业发展高峰论坛成果汇编[G]. 2011.

茶染纺织品的制备及其保健功能研究[①]

郭 丽[1]　林 智[1]　马亚平[2]　吕海鹏[1]　谭俊峰[1]　祁尚雄[2]　陈金择[2]

1. 中国农业科学院茶叶研究所，中国 浙江；
2. 浙江省绿剑茶业有限公司，中国 浙江

摘　要：以红茶、绿茶和茶多酚为染料，制备茶染丝巾、茶染袜子与茶染毛巾，测试其染色性能、抗菌能力、除臭功效和抗紫外线性能。结果表明：红茶纺织品、绿茶纺织品、茶多酚织品的色泽分别为黄红色、黄绿色和浅黄红色，耐摩擦色牢度均在3级以上；茶染棉袜对金黄色葡萄球菌、大肠杆菌和白色念珠菌的抑制率在80%以上，去除氨气能力是普通纺织品的10倍；茶染丝巾的紫外线防护能力可达50$^+$级，且茶多酚丝巾防护能力最强。

关键词：茶染纺织品；红茶；绿茶；茶多酚；抗菌；除臭

Study on the Tea Dyeing Textiles and Their Health Functions

GUO Li[1]　LIN Zhi[1]　MA Ya-ping[2]　LV Hai-peng[1]
TAN Jun-feng[1]　QI Shang-xiong[2]　CHEN Jin-ze[2]

1. Tea Sciences Institute, China Academy of Agricultural, Zhejiang, China;
2. Zhejiang Reen Sword Tea Co., Zhejiang, China

Abstract: Make tea dyeing textiles (silk scarves, socks and towels) with black tea, green tea and tea polyphenols, and test their dyeing performances, antibacterial ability, deodorant ability and anti-ultroviolet ability. Results showed that the color of silk scarves, socks, towel dyeing with black tea, green tea and tea polyphenols was respectively yellow red, yellow green and light yellow red, and their rubbing fastness were class 3 or above. And the inhibitory rate (staphylococcus aureus, escherichia coli and candida albicans)

[①] 基金项目：中央级公益性科研院所基本科研业务费专项（0032012027）和国家茶叶产业技术体系茶叶加工新技术新产品研发子课题（CARS-23-09B）。

of tea dyeing socks were more than 80%, and their removing ammonia abilities were ten times as many as ordinary textile. The anti-ultroviolet ability of tea dyeing silk scarves were grade 50^+, and the strongest was made with tea polyphenols.

Keywords：tea dyeing textiles；black tea；green tea；tea polyphenols；anti-microbial；deodorization

自古以来,茶叶就被用作天然植物染料。它是一种酸性染料,通常是用其浸提液来染色。有关研究表明,茶叶染料中具有染色功能的物质主要是茶色素(即儿茶素及其衍生物),它不仅能染色,还具有抗氧化、抗病毒、抗菌等多种保健功能[1]。因而,用茶叶染料染色的纺织品,不但色彩古朴、淡雅,带有茶香味,而且具有一定的保健功效。研究证实,采用茶叶染料染整的棉织物,具有良好的抑菌效果和服用性能[2-6];采用茶多酚整理的丝织物,具有优良的抗紫外线性能(UPF值可达50)和防臭功能(氨气和甲醇的消除率分别达80%和90%以上)[7-8]。笔者进行茶染研究时,也发现棉纤维进行茶多酚抗菌整理后,可有效抑制金黄色葡萄球菌等有害微生物的侵害[6,9]。但是,茶叶种类不同,制备的染料性质也不同,染色效果和保健功效也会有差异。为此,本文开展了红茶、绿茶和茶多酚等染料在棉织物和丝织物上的染色性能研究,并进行了茶染纺织品的保健功效测验。

1 材料与方法

1.1 材料与设备

染料用红茶和绿茶:分别由三门云峰绿毫茶厂(浙江三门)和浙江省绿剑茶业有限公司(浙江诸暨)制作;

茶多酚:纯度98%(儿茶素总量78.97%,EGCG 52.47%,咖啡因0.52%);

棉袜:市售白色成人棉袜(棉80%,氨纶20%);

毛巾:市售白色毛巾(34cm×76cm,棉100%);

丝巾:白色丝巾(26cm×85cm,桑蚕丝100%);

水浴锅(DK-S26)、电子天平(灵敏度0.0001g)、色差仪CM-3500d及印染助剂等。

1.2 方 法

1.2.1 茶染纺织品的制备

根据染料用红茶、绿茶与茶多酚的理化性质及织物的着色特性,配制适量浓度的红茶染液(A)、绿茶染液(B)和茶多酚染液(C)。每个纺织品组为1条毛巾、2双袜子和1条丝巾,分别用A、B、C染液染色,以制备茶染纺织品。其中,染色温度为60℃,染色时间为30min,浴比为1∶30。染色结束,纺织品用纯水清洗残液,自然晾干。

1.2.2 茶染纺织品的检验

茶染纺织品的色度采用色差仪测定(D65光源,10°观察角);色牢度、抗菌能力和抗紫外线能力委托给方圆检测,分别按GB/T 3921—2008、GB/T 3920—2008、GB/T 20944.3—2008和GB/T 18830—2009等标准测试;除臭能力采用纳氏试剂比色法测定。

2 结果与分析

2.1 茶叶染料在纺织品上的着色效果

红茶染料、绿茶染料和茶多酚上染的纺织品,分别呈黄红色、黄绿色和浅黄红色,如图1所示。

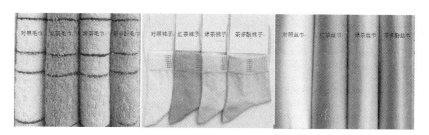

图1 茶染纺织品

由不同茶叶染料制备的茶染纺织品间,L值、a值的差异不显著,b值、c^*值的差异达到显著水平(表1);红茶毛巾的明亮度最弱,比明亮度最强的绿茶毛巾低了22.87%;当色度a值为负数时,织品显绿色,并且a绝对值越大,呈色越深。因此绿茶毛巾呈色最深;而红茶染料和茶多酚着色的纺织品 a 值为正数,因而这类织品呈红色。红茶丝巾 a 值最大,上色较深;各茶染纺织品 b 值大于0,因而均带黄色,且以绿茶毛巾 b 值最大,着色最深;同时,绿茶毛巾 c 值最大,得色量最多,因而绿茶毛巾着色效果较佳。

表 1 茶叶染料的染色性能

样 品	色 度			
	L	a	b	C^*
红茶丝巾	67.34±0.58	8.35±0.30	24.83±0.21	26.20±0.29
红茶袜子	68.08±0.77	9.97±0.21	18.36±0.35	20.89±0.38
红茶毛巾	62.31±1.35	10.25±0.33	20.42±0.47	22.84±0.53
绿茶丝巾	77.18±0.29	−5.18±0.08	18.72±0.34	19.42±0.33
绿茶袜子	78.89±3.53	−3.87±0.34	24.86±0.92	25.16±0.95
绿茶毛巾	80.79±0.90	−6.59±0.18	29.38±0.95	30.10±0.94
茶多酚丝巾	73.36±0.45	2.53±0.07	9.07±0.20	9.42±0.21
茶多酚袜子	72.21±3.47	1.99±0.36	9.53±0.63	9.74±0.65
茶多酚毛巾	70.08±1.28	2.21±0.22	15.60±0.39	15.75±0.41
显著水平	—	—	*	*

注:表中"—"表示差异不显著,"*"表示差异显著。

由表 2 可知,茶染纺织品干摩和湿摩时的耐摩擦色牢度均在 3 级以上,可达到纺织品的服用要求。并且绿茶和茶多酚染色的纺织品,色牢度比红茶染色的效果更好。

表 2 茶染纺织品的色牢度

样 品	耐摩擦色牢度(级)	
	干 摩	湿 摩
红茶丝巾	4	4
红茶袜子	4	4
红茶毛巾	4～5	4～5
绿茶丝巾	4～5	4～5

续表

样　品	耐摩擦色牢度（级）	
	干　摩	湿　摩
绿茶袜子	4～5	4～5
绿茶毛巾	4～5	4～5
茶多酚丝巾	4～5	4～5
茶多酚袜子	4～5	4～5
茶多酚毛巾	4～5	4～5

2.2　茶染袜子的抗菌与除臭性能

由表3可知，茶染棉袜能有效抑制有害菌如金黄色葡萄球菌、大肠杆菌和白色念珠菌的滋生，且对金黄色葡萄球菌和大肠杆菌的抑制力强于白色念珠菌；茶染棉袜水洗10次后，抗菌能力仍可达A级指标（图2）。同时，茶染棉袜还具有除臭功效（图3），棉袜G1和G2的除臭能力是普通棉袜（CK）的10倍。

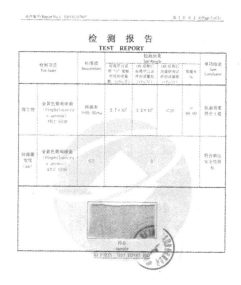

图2　茶染棉袜的抗菌效果

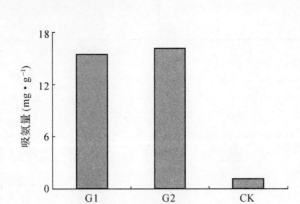

图 3 茶染棉袜的除臭能力

表 3 茶染袜子的抗菌能力

样 品	抑菌率(%)		
	金黄色葡萄球菌	大肠杆菌	白色念珠菌
G1	99.99	99.08	84.32
G2	99.99	99.16	88.92

注：样品 G1、G2 为茶多酚染色棉袜，以下同。

2.3 茶染丝巾的抗紫外线性能

茶染丝巾的紫外线防护系数(UPF)均大于50，防护等级达到50+；茶染丝巾中茶多酚丝巾的抗紫外线能力最强，是绿茶丝巾的1.73倍；茶染丝巾对紫外线的透射率均低5%，且UVA透射率高于UVB；绿茶丝巾和茶多酚丝巾对UVA的透射率高于红茶丝巾，但茶多酚丝巾对UVB的透射率低于红茶丝巾和绿茶丝巾。

表 4 茶染丝巾的抗紫外线性能

样 品	抗紫外线性能		
	UPF 值	T(UVA)AV(%)	T(UVB)AV(%)
红茶丝巾	167.00	0.93	0.51
绿茶丝巾	117.67	2.85	0.65
茶多酚丝巾	203.77	2.45	0.37

3 讨 论

(1)茶染袜子、茶染毛巾同属于棉织物,但二者棉纤维含量不同,茶染得色量也不同。本研究中相同茶叶染料上染的毛巾与袜子,均为毛巾得色较深,说明了氨纶吸附茶叶染料的能力不及棉纤维。因而,考察茶染棉织物的抗菌、除臭功效时,选择茶染袜子作代表。

(2)茶染丝巾的UPF值均大于50,因而具有很强的紫外线防护能力,这与杨诚等[7]的研究结果是一致的。但是,杨诚等染色用的茶多酚浓度(5%)是本研究的5倍,可能是茶多酚纯度不同造成的。茶染丝巾中茶多酚丝巾的防护能力最佳,说明抗紫外线能力与茶多酚含量有关。按此推算,相同分量的红茶、绿茶染色丝巾,理论上应为绿茶丝巾的防护能力强。然而,事实却是红茶丝巾强于绿茶丝巾,其原因有待进一步查明。

参考文献

[1] 宛晓春,李大祥,夏涛. 茶色素及其药理学功能[J]. 天然产物研究与开发,2001,13(4):65-70.
[2] Nobuyuki H, Yoshio M, Katsuhiko F, et al. Reports of the Hamamatsu Industrial Research Institute of Shizuoka Prefecture[C]. 2001(11):60-64.
[3] 陈艳妹,赵建平. 茶多酚在棉织物上的抗菌整理研究[J]. 印染助剂,2012,27(12):16-17.
[4] 张彩玲,贺娟,于湖生. 棉机织物绿茶染色及染后织物抗菌性的探讨[J]. 山东纺织科技,2008(4):54-56.
[5] 王海英,昌子煊. 棉织物的茶多酚抗菌整理[J]. 印染,2007(20):26-27.
[6] 林智,郭丽,马亚平等. 一种茶叶植物染料染色棉纤维的方法:中国zl201010281680.0[P]. 2012-04-25.
[7] 杨诚,唐人成. 蚕丝的茶多酚媒染性能[J]. 印染,2010(8):11-15.
[8] 钱红飞. 茶多酚在蚕丝染色中的应用与抗紫外线性能[J]. 纺织学报,2012,33(2):68-72.
[9] 郭丽,林智,马亚平等. 茶多酚用于棉纤维染色工艺研究[J]. 中国茶叶,2012(2):15-16.

韩国爱茉莉太平洋公司含茶化妆品和保健品的研究与开发

郑镇吾

爱茉莉太平洋公司研发中心,韩国

摘 要:爱茉莉太平洋公司成立于上世纪 30 年代。公司在济州岛建立茶叶种植场。那里气候温暖,火山土肥沃,被认为是种植茶树的最佳地区。近年来,公司重视天然产品开发,而绿茶一直是公认的安全有效的产品。公司开发出一系列以绿茶为原料的化妆品、护肤品、食品和饮料。公司不仅仅生产茶叶和袋泡茶,还是韩国最早开展茶保健功能性成分产品开发的公司,以茶提取物作为添加物,开发出减肥产品(含高纯度儿茶素),减缓疲劳的产品(含茶氨酸)等。这些茶提取物保健品的年销售额增长率达到 30%。公司还开发了化妆品和个人护理产品,如含绿茶提取物茶多酚的爱茉莉护肤霜、绿色泡沫洁面乳等。

Introduction of Products Used Green Tea in Amore Pacific, Research and Development of Various Products Containing Tea

Jinoh Chung

Amore Pacific R&D Center, Tea Research Team, Korea

Abstract: AP was established in 1930s and started green tea business in year 1976 with our first president's vision that we have to survive the vanishing tea culture. We have set up green tea farm in Jeju Island, where it is considered the best location for cultivating tea with warm temperature and rich volcanic soil.

In recent years, we have strengthened the program on natural materials historically known to be safe and highly effective, especially green tea. We have developed wide range of products with green tea, including cosmetics, personal care products, food and beverage through various researches.

We launched supplements which include concentrated health functional ingredient instead of general tea products such as tea leaves and tea bags, for the first time in Korea. Supplements contain tea extract (debittered, highly concentrated catechin) for weight-loss and L-theanine for giving relaxation. Then annual sales of these supplements are increasing over 30% each year.

We also have developed cosmetic and personal care products including green tea extracts such as Amore Pacific-Time response cream containing tea polyphenol, and O'sulloc-Green cleansing foam, containing brewed tea water.

Through constantly scientific research on tea ranging materials for anti-obesity to chemotherapy, we have developed products for improving human health.

孟加拉小农茶叶生产与茶叶扶贫，保障食品供给政策

Mosharraf Hossain[1]　Syed A Hasib[2]
1. Tetulia 茶叶公司主席；2. Tetulia 茶叶公司主管

摘　要：茶是世界上最价廉物美的饮料，对亚洲、非洲、拉丁美洲等发展中国家的国民经济发挥重要作用。此外，茶叶还具有药用作用，日本、中国、印度、澳大利亚等国家都开展了该方面的研究。孟加拉在世界茶叶生产中只占很小的份额，其大型茶叶种植场多数为国有或跨国公司所拥有，茶叶现已成为孟加拉出口创汇重要商品之一。由于人口增长和人均收入提高，以及城市化步伐加快，孟加拉国内茶叶需求近年来增长很快，但茶叶生产无法满足消费增长。因为从1992年以来，孟加拉未开展过规模化茶叶生产项目。茶叶生产在孟加拉具有很重要的地位，全国有1500万人直接或间接从事茶业行业工作。但孟加拉又是土地资源匮乏的国家，全国人均耕地只有 $0.105hm^2$。孟加拉与印度接壤的地区，是孟加拉最贫穷的地区，那里的土地不适合种植粮食等作物，但是非常适合种植茶叶。经过实地考察和调查，TTCL管理委员会开始帮助当地农民发展茶叶生产，并提出"茶叶扶贫，保障食物供给"政策。经过2~3年推动，取得较大成绩。TTCL管理委员会建立大型茶叶加工厂，直接从农民那里收购茶叶鲜叶。农民从出售的茶叶鲜叶获得收入，食物供给得到保障，住房、教育等得到改善。得到孟加拉政府和一些非政府机构支持该政策，并积极和TTCL开展合作，推动该项目得到持续发展。

Food Security and Poverty Alleviation through Smallholding Tea Plantation in Bangladesh

Mosharraf Hossain[1]　Syed A Hasib[2]
1. Chairman, Tetulia Tea Company Ltd. ; 2. Director, Tetulia Tea Company Ltd.

1 Background

Tea from its original home of china come to my country – Bangladesh, as British Colonial business enterprise, beginning with the tea estates named Malnicherra in the year 1854.

The spread was uninterrupted ever since, and popularity of the drink is always on the rise and now touching almost the with of china that it is one of the basic flavours of daily life.

Bangladesh today has more than 156 large tea plantations in Sylhet, Moulvibazar, Hobigonj, Chittagong hilltract and Panchagarh district.

As employment opportunity, it is very large and all in poor masses of remote areas. It offers permanent job to 1,17,728 workers with additional casual employment of another 30,000 workers and giving full range of living conditions for their including their non working dependents of another 2,59,000 human heads permanently.

Tea in Bangladesh is an equal opportunity employment sector having 50% of work force being female. The total employment span of tea from plantation to retail marketing is estimated to be somewhere near 800000 persons in Bangladesh. The livelihood of a vast mass of poor people is increasingly getting linked to tea and it is in good progress prosperity of its dependents.

There is a sudden boom in the consumption of tea in Bangladesh registering annual growth of more than 3.5%. It has outpaced the rise in production and has already consumed all the exportable surplus and has come to a stage of certainly requiring to import tea to meet the domestic demand.

The scenario of rapid rise in consumption vis-à-vis the production situation as Fig. 1.

The above graph shows our tea production, domestic consumption and export from 1975 to 2010. The graph clearly shows that in 2010 domestic consumption was very close to our production though only about 35% of the total population is under the coverage of tea consumption drinking only 1.50 cups of tea a day on average. These figures are also changing fast and even if this trend continues yet. Bangladesh will have to enter into the world tea market as

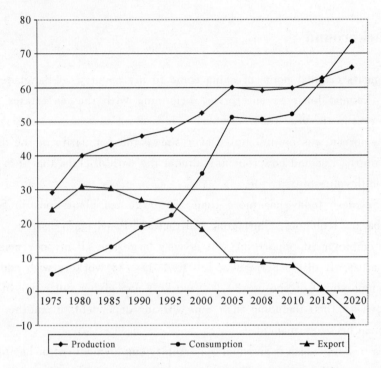

Fig 1. Production, Domestic Consumption and Export of
Bangladesh Tea 1975—2008 (Actual) & upto 2020 Based on Current Trend

a tea importer by 2016. This can be seen in the same graph based on the current trend of production and local consumption which is most likely to change faster with the increase in population, urbanization and change in quality of life with increased income level.

Bangladesh is a poor country and as such it would be extremely painful if the country enters into the world tea market as a tea importer though having high potentials/scopes to increase tea production. Now and in future there is no scope to open new tea estates, therefore, GCAL sponsors took the initiative to open new area for tea cultivation under the concept of "Tea for Poverty Alleviation and Food Security" through smallholding tea cultivation in line with the flow chart furnished later in order to save the country from the burden of importing tea by spending hard earned foreign exchange. But the primary objective has been identified "Food Security" for marginal farmers by using their fallow land under commercial agriculture by smallholding tea plantation.

2 Emergence of Tea for Food Security

We know tea as a business owned by the elite class for centuries and currently in Bangladesh all 156 tea estates in organized sector owned by national, multinational companies and rich entrepreneurs of the country. GCAL and TTCL have given a new dimension in the ownership and management structure of the tea industry by inducing marginal and poor farmers into the ownership of small tea plantations ranging from 0.35 to 5.00 acre of land and thereby tea plantation business is now owned by the marginal farmers also like the elite class of the country. Traditionally tea plantation has been known as an instrument of exploitation of labour forces as the labourers have been working like a captive labour force in the plantation industry in this country since 1854. Through Bangladesh tea industry is more than 160 years old yet the industry could not ensure food security for it's workers since calorie intake by a tea worker is almost half of similar agriculture worker.

The productions have wither to been restricted to the control of large plantations holding, pre-dominantly owned by well structured multinational interests. But in the year 2000 A.D and onward, at the back drop of a surpassing demand, production of tea took a new dimension. Social welfare organization in Northern part of the country known to be the poorest of the poor, by organizing them to grow tea in small holdings in their homestead. This scheme, initially seemed to conflict the interest of large national and multinational growers, has now overcome nearly all the hurdles and are a recognized sector now increasing their numbers and bulk of production in multiple production.

It has come out to be one of the finest ways to address poverty alleviation, generating employment for everyone in the whole family and for the season spreading almost over the year and the crop being very perennial. Great relief for the poor, big pull in the employment pool and an increasingly bigger contribution in the supply of tea to the national demand.

A highly enterprising band of entrepreneur named Tetulia Tea Company struggled their way to alleviation poverty and has established small holding tea in Northern Bengle with all its infrastructural support service. Today it is a success story and a reality that tea is grown in small holdings by the poor and marginal farmers. Small growers have to-day a significant market share selling their tea under the source hammer auction and fetching a price often in close completion with the big names.

It is now a phase of intensive activity in the area of small holding tea, growing to alleviate poverty of marginal farmers and the scenario is one of great success. Couple of NGO's have extended their helping hands through TTCL and establishment are cropping up for a geared up coverage. Poverty has receded to a good extent and the barren landscape of Northern Bengal is gradually changing to a lush green cover.

3 Objectives of Tea for Food Security and Poverty Alleviation

Optimum commercial utilization of fallow land in a land hungry country like Bangladesh and produce a cash crop by which the farmers can buy food/rice for his/her family's yearly requirement.

Create a sound financial base for marginal and poor farmers in order to address basic problems relating to their food, housing, health, education and clothing by themselves.

Creating micro-entrepreneurs in the rural areas by utilizing fallow land and other local resources in tea plantation business instead of creating daily labourers like the captive labourers in traditional tea estates.

Creating sustainable employment opportunity in the rural areas for marginal land owners and for landless mass population in the poorest district of the country.

Improving micro-environment in the project area by tea and shade tree plantation (Table 1).

Table 1　Comparison between traditional large tea estate and smallholding tea plantation for food security

Traditional large tea estate	Smallholding tea plantation
Land owned by the State / Govt. and leased to big companies or rich entrepreneurs almost free of cost.	Farmers own the land either by purchase or inherit from parents but inadequate to secure food for the whole year.
Size of the estates varies from 500 acres to 6500 acres in Bangladesh.	Size varies from 0.35 acre to 5 acres.
Workers and staff are employed by the owners / company.	Land owner is self employed on his plantation both as worker and entrepreneur.
Employees do not own the plantation as such they have no sense of belonging-resulting poor productivity per unit area and per worker.	The farmer owns the plantation as such he has the sense of belonging-resulting higher productivity per unit area and per worker/entrepreneur.
Land utilization around 46%.	Land utilization 100%.
The entrepreneurs usually own factory along with the plantation.	No factory required by small/marginal farmers.
High overhead cost by the companies.	No overhead cost by farmers.
Less profit margin per unit product (finished tea) and per unit land.	High profit margin per unit product (green leaf) and per unit land.
Gross return from one acre by selling green leaf is US$ 34570 (2325 kg GL) (Average: 1276 kg MT/hm^2).	Gross return from one acre by selling green leaf is US$ 1300 @ US$ 0.24/kg of green leaf. (Yield 3000 kg MT/hm^2).
Worker earns US$ 393-430 per year including all benefits. (Needs extensive study).	Worker (the entrepreneur himself) earns US$ 982 per year after all expenses from one acre plantation.
National GDP growth rate is 5%~6% while tea production growth rate is only 1~1.5.	Average production growth rate is 55%~60% since 2005.
With the current growth rate of 1.00%, the traditional tea industry is unable to cope with the increasing domestic consumption growth rate of 5%.	This sector if patronized by the Government or any donor then it is possible to cope with the domestic growth rate and leave sufficient tea for export.

3.1　Case study-1

Abdul Haq Tetulia, Age 46 years, Family members-5, owned 2.5 acres of land with sandy loam and acidic soil. Half of the area remained fallow and used to grow paddy in half of the area during rainy season with yield of 354 kg of rice which could provide food for the family for three months only. Used to work as day labour for supplementary cash income to buy food.

Worked as nursery worker in TTCL project in year 2000. He was motivated to become a tea grower under this project. Rooted tea plants of clone variety supplied to him from this project on credit. Self and wife working planted tea on half of his fallow land in early 2002. During third year of plantation, from April to December, 2005 supplied 3775 kg green leaf to the project's tea factory for which he earned US $ 395 and with this amount he could buy 5281 kg of rice (3.57 times more than previous production) making his family food secure. His income has increased to US $ 1245 with larger volume of green leaf during 2011 and his income will continue increasing till 2016 with increase of production.

3.2 Case study-2

Jamal Uddin-aged 32 years – a day labourer owning only 0.35 acre of land. Gained experience in tea plantation by working in another smallholder farmer's plantation close to his plot. Planted tea on his 0.35 acre land in April, 2008 and harvested 600 kg of green leaf by December '08 earned US $ 85 apart from his usual wage earning as day labourer. The three member family of Jamal is happy with income from his 0.35 acre of land which was not an asset for him as the land was not bringing any income for him. In 2011, he produced 1750 kg green leaf and earned US $ 451 @ US $ 0.25/kg which made him food secured family.

绿茶和茶多酚对笼养蛋鸡脂类代谢的影响

宛晓春　周裔彬　胡经纬　邵　磊　尚岩岩

安徽农业大学农业部茶叶生物化学重点实验室,中国 安徽

摘　要：在蛋鸡饲料中添加不同浓度的绿茶粉(2g/kg,6g/kg)和茶多酚(0.5g/kg、1g/kg、1.5g/kg、2g/kg),分析其对鸡血中总胆固醇、总甘油三酯、高密度脂蛋白和低密度脂蛋白及鸡胸肉、鸡腿肉和鸡蛋中总胆固醇、总甘油三酯和脂肪酸的影响。将1620只蛋鸡随机分成9组,每组180只,3个平行,采用笼养60d。添加6g/kg绿茶粉和1g/kg茶多酚,饲养56d时,经统计分析表明,鸡血和肌肉中的总胆固醇、总甘油三酸酯和血中低密度脂蛋白降低,鸡血中高密度脂蛋白提高;鸡蛋中总甘油三酯降低,但胆固醇没有变化。分析结果显示,绿茶和茶多酚对蛋黄中的豆蔻酸、硬脂酸、油酸、亚油酸、二十碳四烯酸、总的脂肪酸没有影响,但增加总的不饱和脂肪酸的含量。通过对鸡粪中总脂肪酸含量的分析,与对照组相比,绿茶和茶多酚均不同程度增加总脂肪酸的排泄。综合结果显示,绿茶和茶多酚的添加,能够改善脂类代谢及脂肪酸的成分。

关键词：绿茶;茶多酚;总胆固醇;总甘油三酸酯;高密度脂蛋白;低密度脂蛋白;蛋鸡

Effect of Green Tea and Tea Catechins on the Lipid Metabolism of Caged Laying Hens

WAN Xiao-chun　ZHOU Yi-bin　HU Jing-wei
SHAO Lei　SHANG Yan-yan

Key Laboratory of Tea Biochemistry & Biotechnology, Ministry of Agriculture, PRC, Anhui Agricultural University, Anhui, China

Abstract: The present study was conducted to evaluate the effects of supplementing different levels of green tea or tea catechins on total cholesterol (TC), total triglycerides (TG), high-density lipoprotein (HDL) and low-density lipoprotein (LDL) levels in the plasma of laying hens, as well as the TC,

TG and total fatty acid (TFA) levels in breast and thigh meat and egg yolk. A total of 1,620 laying hens (16 weeks old) were randomly assigned to nine dietary groups. Each treatment had 180 birds with three replicates. The basal diet was respectively supplemented with green tea powder (0, 2, 4, 6, 8 g/kg) and tea catechins (0.5, 1, 1.5, 2 g/kg). Hens were then reared in battery cages for 60 days. Supplementation with 6 g/kg green tea and 1 g/kg tea catechins at day 56 ($P<0.05$) lowered TC, TG and LDL levels in both plasma and meat ($P<0.05$), increased plasma HDL levels ($P<0.05$) and decreased yolk TG levels ($P<0.05$), where statistical significance was considered quadratically. TC levels in yolk ($P>0.05$) were not affected. The addition of green tea and tea catechins had no effect on the levels of myristic acid (C14∶0), stearic acid (C18∶0), oleic acid (C18∶1), linoleic acid (C18∶2n-6), arachidonic acid (C20∶4n-6) and TFA content in yolk, but increased ($P<0.05$) the unsaturated fatty acid content. Dietary green tea and tea catechins increased the excretion of TFA in feces compared to the controls. These results suggest that the addition of green tea and/or tea catechins to the diet of laying hens can improve lipid metabolism and the fatty acid composition of egg yolk.

Keywords: green tea; tea catechins; total cholesterol; total triglycerides; high-density lipoprotein; low-density lipoprotein; laying hens

Chicken meat and eggs are palatable, digestible, and nutritious for people of all ages. The low price of chicken products, compared to beef and mutton, has contributed to their worldwide popularity. However, the cholesterol and triglyceride present in chicken meat and eggs may constitute a health hazard, if consumed regularly over long periods of time (Bragagnolo et al., 2003). According to the American Heart Association, high cholesterol is a major risk factor for coronary heart disease, heart attack and stroke (Angelin et al., 2010). Studies have consistently indicated that high triglyceride levels are associated with an increased risk of ischemic stroke (Shahar et al., 2003; Psaty et al., 2004; Freiberg et al., 2008), coronary health problems (Sarwar et al., 2007) and symptoms of depression (Herva et al., 2006; Mccaffery et al., 2003). Although the cholesterol and triglyceride levels in processed products can be reduced through various formulations, this is not a cost-effective process and may also reduce product quality (Hand et al., 1987; Hensley et al., 1995). Several studies have been carried out with laying hens and broilers in an effort to reduce the cholesterol and fat contents in eggs (Elkin et al., 1999; Raes et al., 2002) or in broiler meat (Ahn et al., 1995).

During the past decade, green tea has received considerable attention over claims that it has specific health benefits and antioxidant properties. The benefits of green tea have been ascribed to functional polyphenols known as tea catechins; these include epicatechin (EC), epigallocatechin (EGC), epicatechin-3-gallate (ECG), and the main flavonoid, epigallocatechin-3-gallate (EGCG) (Graham, 1992). Rat studies have shown that ECG extracted from green tea lowers cholesterol and triglyceride concentrations in plasma (Lee et al., 2008). Some studies have suggested that dietary EGCG (Bursill et al., 2006) and green tea (Zhong et al., 2006) reduce cholesterol and triacylglycerol in human plasma. Tang et al. (2001) showed that adding 200−300 mg/kg tea catechins to chicken feed delayed lipid oxidation in chicken breast and thigh meat. However, little information is available about the effect of green tea and tea catechins on cholesterol and triglycerides levels.

The objectives of our research were (1) to evaluate the effect of high doses of dietary green tea and tea catechins on cholesterol and triglycerides levels in chicken meat and eggs, and (2) to investigate the possibility of using green tea or tea catechins as an additive in chicken feed in the future.

1 Materials and Methods

Dietary compounds: Tea catechins were purchased from Anhui Redstar Pharmaceutical Co., Ltd. (Hefei, China). According to the company's certificate (Lot No: 200807011, QB2154−95), the composition was as follows: 415g/kg EGCG, 113g/kg EGC, 94g/kg ECG, 54g/kg EC, 70g/kg gallocatechin gallate (GCG) and 3g/kg caffeine. The composition of green tea, also obtained from the same source, was determined (Wu et al., 2004). As a percentage of wet weight, the composition was found to be 45.8g/kg moisture, 20.8g/kg ECG, 17.6g/kg EC, 58.3g/kg EGC, 200.1g/kg EGCG, 10.3g/kg

GCG and 105.1g/kg caffeine. The basal diet, the composition of which is shown in Table 1, was provided by Huainan Xinyang Breeding Co., Ltd. (Huainan, China). Commercial test kits for cholesterol, triglycerides and C19 : 0 were obtained from Sigma-Aldrich Shanghai Trading Co., Ltd. (Shanghai, China). Standard samples were used at concentrations of 0.0125, 0.025, 0.05 and 0.1mg/mL to construct standard curves to assess levels of cholesterol and triglyceride. Other reagents were purchased from Zhihong Biotech Co., Ltd. (Hefei, China).

Table 1 Composition and nutrients of diets fed to laying hens[A]

Ingredients	Compositions(g/kg)	Nutrition	Contents
Corn	625	ME (kCal/kg)	10.83
Soybean meal	250	Calcium (g/kg)	32
Wheat bran	26.5	Total phosphorous(g/kg)	6
Limestone	60	Crude protein (g/kg)	161.2
(Phosphorous)P/(Calcium) Ca	1.5	Methionine (g/kg)	15.8
Salt	3.5	Cystine (g/kg)	28.9
Additives	20	Lysine (g/kg)	90.5
		Threonine (g/kg)	65.2
		Tryptophan (g/kg)	17.0

[A] Ingredient and nutrient composition reported on an as-fed basis.

Additives supplied (per kg of diet): vitamin B_{12}, 16g; menadione, 4mg; riboflavin, 7.8mg; pantothenic acid, 12.8mg; niacin, 75mg; folic acid, 1.62mg, Vitamin A, 15,000 IU; Vitamin D3, 800 IU; Vitamin E, 30 mg; Mn, 80mg; Fe, 60mg; Zn, 90mg; Cu, 12mg; Se, 0.147mg.

Animal Management: The cohort comprised 1,620 Lohmann (LSL) laying hens, whose beaks had been trimmed at one day old and who had received relevant vaccinations according to the manufacturer's protocols (Shanghai Hile Bio-pharmaceutical Co., Ltd, Shanghai, China). At 16 weeks of age, hens of similar body weight were randomly divided into nine groups with 180 hens in each. The hens were housed in battery cages (60cm * 60cm * 44cm) with five hens per cage, and the experiment began at 18 weeks of age. A control group (C) was fed a basal diet. The other eight groups were fed basal diets supplemented with either 2, 4, 6 or 8g green tea powder/kg feed, or 0.5, 1, 1.5 or 2g tea catechins/kg feed. Feeds were prepared weekly and the daily intake was 110~115g per hen. The birds had ad libitum access to water. The building was an open-sided house with ceiling fans and 60W incandescent light bulbs that produced a light intensity of 20Lux, on an illumination cycle of 16h light and 8h darkness. The experiment was carried out at Huainan Xinyang Breeding Co., Ltd. (Huainan, China) and lasted for 60 days.

Sampling procedure: The hens and eggs were weighed every 2 weeks during the experimental period. Nine laying hens in each group were slaughtered every 2 weeks during the rearing period, and their blood was collected into tubes containing EDTA (final concentration, 1g/L). Plasma was isolated by centrifuging at 1,500g for 10min at 4°C, and 1-mL aliquots were frozen at -20°C for later analysis (Lee et al., 2003). After feathers were removed, breast and thigh meat were immediately cut from each carcass, vacuum packaged and stored at -20°C until analysis (Tang et al., 2001). Every 2 weeks, 50 eggs were collected from each group and maintained at refrigerator temperature for analysis. Before analysis, samples were heated at 105°C for 2h in an oven to calculate cholesterol and triglyceride levels on a dry weight basis. Feces that fell onto a smooth plastic plate placed under the cages were collected prior to slaughter and cooled. Fecal samples from three cages in each group were lyophilized and then vacuum packaged for analysis.

Determination of total cholesterol and total triglycerides in plasma samples: The concentrations of TC, TG, HDL and LDL were measured enzymatically using an auto-analyzer (Cobas Bio, Roche, Basel, Switzerland) at the Biochemical Research Clinic Laboratory of Anhui Medical University (Hefei, China). Very low-density and low-density lipoproteins were precipitated from plasma with phosphotungstic acid/$MgCl_2$ (Sigma-Aldrich; Catalogue number 352-4) according to the method of Lee et al. (2003), and the supernatant (HDL) was assayed for cholesterol.

Determination of total cholesterol and total triglycerides in chicken breast and thigh: Lipids were extracted from thigh and breast muscle according to Folch et al. (1956). Briefly, meat samples (5±0.01g) were homogenized with 100mL solution of chloroform-methanol (2:1, V/V) for 2min, filtered, placed in separator funnels and mixed with 20mL saline solution (0.88% KCl). After samples were separated into two phases, the top aqueous methanol fraction was discarded and the bottom lipid chloroform fraction was washed with distilled water/methanol (1:1, V/V) and evaporated under nitrogen gas flow. The cholesterol and triglyceride content were determined using Sigma commercial kits (Sigma-Aldrich) according to the manufacturer's instructions.

Determination of total cholesterol, total triglycerides and fatty acid composition in egg yolk: Fifteen eggs (5 large, 5 medium and 5 small eggs) were obtained for each group. Eggs were boiled for 10min before the yolk was removed. All yolks were then lyophilized and ground into powder, and yolks from the same group were mixed and vacuum packaged for analysis. The constituents of the yolk samples (5g±0.01g) were extracted with chloroform-methanol (2:1, V/V) according to the methods of Folch et al. (1956) and Bragagnolo et al. (2003) with some modifications. A 3mL aliquot of the lipid extract was dried under nitrogen and saponified with 4mL of absolute ethanol and 10mL 50% KOH (w/w) for 90min at 65℃. The sample was then cooled under running water, followed by the addition of 5mL of distilled water. The unsaponifiable sample was extracted twice with 10mL of hexane, from which 3mL was dried under nitrogen. Cholesterol and triglyceride content was determined using Sigma commercial kits according to the manufacturer's instructions. Fatty acids were extracted from the yolk according to AOCS (1990). First, 2mL of 0.01mol/L NaOH in dry methanol were added to 10mg of yolk. The samples were shaken and heated for 10min on a heating block at 60℃. Next, 3mL of BF3 reagent were added and the samples were reheated for 10min. The samples were cooled under running water and 2mL 20% NaCl and 2mL hexane were added. The test tubes were shaken vigorously and centrifuged for 5min at 440g at 18℃. The fatty acid methyl esters were transferred to new test tubes and evaporated under nitrogen gas. The dry samples were redissolved in hexane for gas chromatographic (GC) analysis. The fatty acid methyl esters obtained were analyzed with a Shimadzu (Tohoku, Japan) GC, using a Hewlett Packard (Palo Alto, America) equipped with a flame ionization detector (FID) and fitted with a fused capillary column (50m × 0.25mm

i. d. , coated with Carbowax 20M [0.20μm film thickness]). The apparatus was programmed with an initial temperature of 150°C for 4min, with incremental increases of 1.3°C/min, until a final temperature of 210°C was reached. Injector and detector temperatures were 220°C and 240°C, respectively. The carrier gas was hydrogen (1.5mL/min) and the make-up gas was nitrogen (25mL/min). C19:0 (Sigma) was used as an internal standard for fatty acid quantification.

Determination of total fatty acids in feces: Concentrations of neutral and acidic sterols, the only TFAs present in the feces, were determined according to Chan et al. (1999). Stigmasterol and hyodeoxycholic acid (0.5 mg each) were used as internal standards for quantification of fecal neutral and acidic sterols, respectively. Briefly, 300mg fecal samples were dried and ground before the experiment, and were then mildly hydrolyzed and extracted with cyclohexane. The neutral sterols in the cyclohexane phase were converted into their trimethylsilyl derivatives. The acid sterols in the aqueous phase were saponified, extracted, and also converted into their trimethylsilyl derivatives. Both neutral and acidic sterol trimethylsilyl derivatives were subjected to GC analysis.

Statistical analysis: In each group, three replicates were performed for all measurements. Results are expressed as the mean ± SEM. A statistical package SPSS12.0 (SPSS Inc., Chicago, IL, USA) was used for all analysis. One-way analysis of variance (ANOVA) was used to determine the significance of differences among the groups. A value of $P<0.05$ was considered statistically significant.

2 Results and Discussion

Blood: No significance differences were observed between groups at day 0. Plasma TC, TG, HDL and LDL concentrations gradually increased from 0 to 42 days (Table 2). Plasma TC and TG levels were lower in groups that were administered green tea and tea catechins, and the best results were quadratically obtained with 6g/kg green tea at 42d ($P=0.04$) and 56d ($P=0.03$), and with 1g/kg tea catechins at 42d ($P=0.04$) and 56d ($P=0.04$). Plasma

TC and TG levels of the group given 6 g/kg green tea were 0.94 mmol/L and 3.13 mmol/L lower ($P<0.05$), respectively, compared to the control group at day 42. For the group given 1g/kg tea catechins, TC and TG levels were 0.85 mmol/L and 2.26 mmol/L lower ($P<0.05$), respectively. Plasma HDL levels were quadratically higher in the groups given green tea ($P=0.03$) and tea catechins ($P=0.04$) at 56 d. LDL levels were 0.22 mmol/L lower ($P=0.03$) in the group given 6 g/kg green tea and 0.18 mmol/L lower ($P=0.04$) in the group given 1g/kg tea catechins at 56d. These groups showed a quadratic effect. The decrease in plasma TG levels was greater than the decrease observed in plasma LDL levels. The effect of green tea on TC, TG, HDL and LDL levels was similar to that of tea catechins, but the changing trend did not show a linear relationship among the supplemented levels during breeding periods.

Chicken breast and thigh: The levels of TG and TC in breast and thigh gradually increased from 0 to 42d (Table 3). In both thigh and breast meat at 42 and 56d, TG levels and TC levels were significantly less than control values ($P<0.05$), but TC levels were only slightly reduced. The decrease in TG levels was significantly greater than that of TC. Our results indicated that 6g/kg green tea had a greater effect than 2 or 8g/kg, and that 1g/kg tea catechins had a greater effect than 0.5 or 2g/kg. However, these treatment groups did not display a quadratic dose response during breeding period.

Egg yolk: There was no significant difference between groups for yolk TC levels ($P>0.05$) (Table 3), but TG levels decreased by 2.42 mg/g at day 42 and 2.48mg/g at day 56 for 4g/kg green tea ($P<0.05$), and by 2.69 mg/g at day 42 and 56 for 6g/kg green tea ($P<0.05$). Tea catechins at a dose of 1g/kg were the most effective in reducing TG levels, since they led to a decrease of 2.74mg/g and 2.76mg/g at days 42 and 56, respectively. The effect of green tea was similar to that of dietary tea catechins, but these groups did not show a quadratic dose response during breeding period.

Table 2 Plasma TC, TG, HDL, and LDL content of laying hens[A]

Indices		0[B]	Green tea groups (g/kg)[C]				SEM	P value		Tea catechins groups (g/kg)[D]				SEM	P value	
			2	4	6	8		Linear	Quadratic	0.5	1	1.5	2		Linear	Quadratic
TC (mmol/L)	0 d	3.30	3.22	3.27	3.35	3.27	0.25	0.12	0.14	3.16	3.33	3.22	3.24	0.13	0.14	0.19
	14 d	3.49	3.52	3.42	3.49	3.80	0.10	0.13	0.10	3.41	3.28	3.39	3.56	0.11	0.10	0.21
	42 d	4.63[c]	4.36[b]	3.78[b]	3.69[b]	4.21[c]	0.09	0.10	0.04	4.15[c]	3.78[b]	3.91[b]	4.16[c]	0.08	0.12	0.04
	56 d	4.65[c]	4.22[b]	3.98[b]	3.84[b]	4.21[c]	0.07	0.15	0.03	4.37[c]	4.09[c]	4.18[c]	4.27[c]	0.10	0.11	0.04
TG (mmol/L)	0 d	10.58	10.22	10.66	10.24	10.35	2.76	0.57	0.29	10.96	10.05	10.22	10.89	2.15	0.46	0.27
	14 d	12.87	12.40	11.28	11.56	12.25	0.34	0.19	0.15	12.14	12.77	11.55	11.94	0.31	0.21	0.17
	42 d	16.39[d]	16.27[d]	14.13[b]	13.26[b]	14.29[b]	0.09	0.10	0.02	15.52[b]	14.13[b]	14.69[b]	14.69[b]	0.10	0.11	0.05
	56 d	16.72[d]	16.62[d]	14.64[b]	14.23[b]	14.63[b]	0.21	0.13	0.03	15.96[b]	14.29[b]	14.72[b]	14.81[b]	0.19	0.14	0.04
HDL (mmol/L)	0 d	2.58	2.45	2.61	2.46	2.50	0.37	0.22	0.26	2.73	2.39	2.45	2.70	0.36	0.22	0.29
	14 d	3.02	3.07	3.21	3.18	3.12	0.32	0.17	0.13	3.08	3.26	3.38	3.37	0.34	0.26	0.18
	42 d	3.46	3.53	3.81	4.01	3.45	0.23	0.13	0.05	3.63	3.94	3.85	3.69[a]	0.20	0.21	0.05
	56 d	3.42[a]	3.69[b]	4.02[b]	4.14[b]	3.43[a]	0.38	0.12	0.03	3.64[a]	4.17[b]	4.05[b]	3.90[a]	0.37	0.17	0.04
LDL (mmol/L)	0 d	0.81	0.82	0.82	0.83	0.84	0.35	0.28	0.27	0.84	0.88	0.87	0.86	0.36	0.29	0.23
	14 d	0.83	0.80	0.79	0.81	0.81	0.24	0.17	0.19	0.87	0.82	0.86	0.89	0.27	0.26	0.21
	42 d	0.91[d]	0.81[c]	0.73[b]	0.71[b]	0.82	0.19	0.13	0.05	0.81[c]	0.73[b]	0.76[b]	0.83[c]	0.15	0.14	0.06
	56 d	0.91[d]	0.82[d]	0.74[b]	0.69[a]	0.83	0.23	0.10	0.03	0.82[c]	0.73[b]	0.77[b]	0.85[c]	0.21	0.11	0.04

[A]Values are the mean ± SEM. Each value is an average value from three replicate groups, and each replicate sample is determined in triplicate. TC: total cholesterol; TG: total triglycerides; HDL: high-density lipoprotein; LDL: low-density lipoprotein. [a]–[d]The values in the same row with different letters are significantly different ($P<0.05$). [B]indicates control laying hen fed with a basal diet; [C]indicates a basal diet supplemented with 2, 4, 6, 8 g green tea powder/kg feed, respectively; [D]indicates a basal diet supplemented with 0.5, 1, 1.5, 2 g tea catechins/kg feed, respectively.

Table 3 Change in TC and TG levels of chicken breast, thigh (g/kg meat), and egg yolk (mg·g^{-1} yolk)[A]

Samples	Indices		Green tea groups (g/kg)[C]					SEM	P value		Tea catechins groups (g/kg)[D]				SEM	P value	
			0[B]	2	4	6	8		Linear	Quadratic	0.5	1	1.5	2		Linear	Quadratic
Breast	TC	0 d	3.56	3.53	3.58	3.54	3.49	0.02	0.19	0.21	3.57	3.51	3.56	3.53	0.02	0.23	0.18
		42 d	4.18c	3.87b	3.73b	3.69b	3.94b	0.03	0.13	0.06	3.81b	3.76b	3.83b	3.87b	0.03	0.16	0.08
		56 d	4.22c	4.08c	3.91c	3.78b	4.12b	0.04	0.11	0.05	3.84b	3.79b	3.91b	3.94b	0.04	0.11	0.06
	TG	0 d	17.78	17.82	17.80	17.79	17.81	0.02	0.20	0.11	17.78	17.81	17.76	17.79	0.01	0.23	0.15
		42 d	21.48b	19.39b	18.32a	18.11b	19.29b	0.06	0.12	0.01	18.67a	18.71b	18.78b	19.16b	0.07	0.12	0.02
		56 d	21.65b	19.75b	18.43b	18.25b	19.31b	0.10	0.10	<0.01	18.76a	18.72b	18.93b	19.28b	0.11	0.13	0.01
Thigh	TC	0 d	2.34	2.38	2.37	2.32	2.35	0.02	0.22	0.14	2.36	2.37	2.34	2.36	0.03	0.21	0.19
		42 d	3.18b	3.05b	2.84a	2.78b	2.89b	0.03	0.13	0.10	2.80b	2.77b	2.87a	2.96a	0.04	0.19	0.12
		56 d	3.22b	3.12b	2.96a	2.83b	3.08b	0.05	0.11	0.09	2.98a	2.85b	3.01b	3.02b	0.06	0.12	0.10
	TG	0 d	12.84	12.81	12.79	12.85	12.83	0.05	0.24	0.16	12.81	12.85	12.82	12.83	0.03	0.25	0.18
		42 d	17.39d	16.20c	14.63a	14.27a	14.72a	0.12	0.11	0.02	14.68a	14.65a	14.83a	14.81a	0.11	0.10	0.03
		56 d	17.85d	16.34c	14.76a	14.34b	14.84a	0.13	0.10	<0.01	14.74a	14.63a	15.04b	15.12b	0.12	0.11	0.01
Egg yolk	TC	0 d	3.69	3.65	3.69	3.63	3.61	0.01	0.09	0.16	3.63	3.62	3.65	3.62	0.01	0.17	0.22
		42 d	3.78b	3.69b	3.66b	3.68b	3.64b	0.02	0.16	0.10	3.70b	3.65b	3.60b	3.62b	0.02	0.14	0.13
		56 d	3.74b	3.72b	3.61b	3.66b	3.69b	0.01	0.12	0.08	3.64b	3.64b	3.66b	3.68b	0.01	0.14	0.12
	TG	0 d	19.35	19.56	19.34	19.46	19.16	0.03	0.23	0.12	19.26	19.15	19.15	19.21	0.16	0.14	0.19
		42 d	22.62d	21.48d	20.22b	20.04b	21.18c	0.12	0.13	0.03	20.16b	19.68b	20.63b	21.31c	0.11	0.10	0.04
		56 d	22.67d	21.52d	20.18b	20.09b	21.24b	0.11	0.10	0.01	20.72c	19.81a	21.07b	21.83c	0.11	0.10	0.02

[A] Values are the mean ± SEM. Each value is an average value from three replicate groups, and each replicate sample is determined in triplicate. TC: total cholesterol; TG: total triglycerides. a–d The values in the same row with different letters are significantly different ($P<0.05$). [B] Indicates control laying hen fed with a basal diet; [C] indicates a basal diet supplemented with 2, 4, 6, 8 g green tea powder/kg feed, respectively; [D] indicates a basal diet supplemented with 0.5, 1, 1.5, 2 g tea catechins/kg feed, respectively.

Table 4　Fatty acid content (mg·kg^{-1} yolk) of eggs laid at 26 weeks[A]

Fatty acids	0[B]	Green tea groups (g/kg)[C]				SEM	P value		Tea catechins groups (g/kg)[D]				SEM	P value	
		2	4	6	8		Linear	Quadratic	0.5	1	1.5	2		Linear	Quadratic
C14:0	16.9	17.1	16.2	16.1	16.7	4.2	0.12	0.16	16.6	16.2	16.7	16.8	4.3	0.11	0.21
C16:0	1104[d]	1095[c]	1084[b]	1021[b]	1087[c]	63.2	0.16	0.04	1065[c]	1019[c]	1071[c]	1086[c]	62.5	0.18	0.04
C16:1	182[a]	180[a]	191[c]	201[c]	187[a]	31.5	0.24	0.03	190[b]	195[b]	186[a]	186[a]	30.2	0.21	0.03
C18:0	356	354	357	351	357	41.3	0.31	0.16	356	352	351	349	42.5	0.27	0.19
C18:1	1684	1679	1685	1687	1682	78.1	0.46	0.05	1679	1678	1683	1683	67.9	0.37	0.05
C18:2n-6	497	501	506	517	513	20.5	0.27	0.06	501	510	510	513	16.9	0.26	0.07
C18:3n-3	1.7[a]	2.1[b]	2.6[c]	4.3[d]	2.7[b]	0.1	0.11	0.04	3.8[c]	4.1[d]	3.7[c]	3.5[c]	0.2	0.12	0.03
C20:4n-6	96.0	94.3	96.5	92.8	94.6	2.7	0.13	0.29	95.0	94.7	94.0	94.5	3.4	0.15	0.21
C20:5n-3	nd	nd	nd	nd	nd				nd	nd	nd	nd			
C22:4n-6	16.5[b]	15.3[a]	18[d]	18.5[a]	17.2[c]	1.4	0.11	0.04	16[b]	18.2[c]	17.4[c]	16.8[b]	1.6	0.13	0.04
C22:6n-3	9.5	9	10	11.6	10	0.2	0.12	0.05	11	12.5	10.6	10.3	0.5	0.12	0.05
ΣSAFA	1478	1467	1458	1390	1462	69.5	0.48	0.07	1436	1387	1439	1452	71.3	0.30	0.08
ΣMUFA	1868	1859	1876	1885	1869	78.5	0.61	0.05	1869	1873	1870	1870	66.7	0.47	0.07
Σn-6A	610[a]	611[a]	622[b]	630[c]	635[c]	40.6	0.21	0.03	623[b]	620[b]	622[b]	625[b]	34.1	0.26	0.04
Σn-3A	12[ab]	11[a]	13[b]	16[c]	13[b]	0.1	0.14	0.04	15[bc]	17[d]	15[bc]	14[c]	0.1	0.14	0.02
TFA	3965	3950	3964	3920	3978	70.3	0.48	0.24	3935	3897	3945	3960	70.6	0.57	0.23

[A]Values are the means of 15 eggs, each replicate with 5 eggs. [a-d]Values with different letters within the same rows are significantly different ($P<0.05$). nd, not detectable. C14:0, myristic acid; C16:0, palmitic acid; C16:1, palmitoleic acid; C18:0, stearic acid; C18:1, oleic acid; C18:2n-6, linoleic acid; C18:3n-3, alfa-linolenic acid; C20:4n-6, arachidonic acid; C20:5n-3, eicosapentanoic acid; C22:4n-6, adrenic acid; C22:6n-3, docosahexaenoic acid. SAFA, saturated fatty acids; MUFA, monounsaturated fatty acids; PUFA, polyunsaturated fatty acids; TFA, total fatty acids. [B]Indicates control laying hen fed with a basal diet; [C]indicates a basal diet supplemented with 2, 4, 6, 8 g green tea powder/kg feed, respectively; [D]indicates a basal diet supplemented with 0.5, 1, 1.5, 2 g tea catechins/kg feed, respectively.

The fatty acid content of yolks (Table 4) was determined at day 56. Consumption of dietary supplements had no effect on myristic acid (C14:0), stearic acid (C18:0), oleic acid (C18:1), linoleic acid (C18:2$n-6$) or arachidonic acid (C20:4$n-6$) levels ($P>0.05$), but led to a decrease ($P<0.05$) in the level of palmitic acid (C16:0) and increased ($P<0.05$) levels of palmitoleic acid (C16:1), alfa-linolenic acid (C18:3$n-3$) and $n-6$, $n-3$ when compared with the control. No significant differences were found amongst treatment groups in terms of saturated fatty acid (SAFA) or total fatty acids (TFA) except in the group that received 1g/kg tea catechins, where levels were slightly lower. C16:0 levels in yolk were lower than the control group by 83mg/kg yolk in the 6g/kg green tea group ($P<0.05$) and by 85mg/kg yolk in the 1g/kg tea catechins group ($P<0.05$). Yolk levels of C16:1, C18:3$n-3$, adrenic acid (C22:4$n-6$), $n-6$ and $n-3$ polyunsaturated fatty acids (PUFA) were greater than the control group in both the green tea and tea catechins groups.

Total fatty acids in fecal excretion: The levels of TFA excreted in the feces averaged across all groups at day 56 are shown in Fig. 1. Compared with the controls, the excretion of total fat in feces was greatest in the groups that received supplements, with the exception of 2g/kg green tea. The fat in feces excreted increased by 1% for 6g/kg green tea and by 0.8% for 1g/kg tea catechins when compared with the controls.

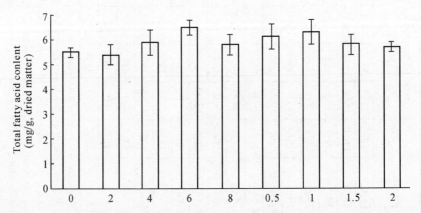

Fig. 1 Total fatty acid content excreted in faces

Although there have been numerous studies on the effects of green tea and tea catechins on lipid metabolism in humans (Lee et al., 1995; Wan, 2006) and rats (Lee et al., 2008; Kuo et al., 2005; Muramatsu et al., 1986), there have been relatively few studies into their effect in chickens. Furthermore, most of the published reports have focused on the antioxidative effect of these supplements on lipid oxidation in chicken meat systems. Tang et al. (2001) reported that a tea catechin supplement at levels of 50 to 300mg/kg feed could reduce lipid oxidation in chicken breast and thigh meat during frozen storage for 9 months, and that the most effective level was 200mg/kg. The addition of 200 or 400mg/kg tea catechins to chicken feed was found to inhibit lipid oxidation in chicken meat and improve its surface color (Mitsumoto et al., 2005). However, direct evidence for the regulatory action of green tea and tea catechins on lipid metabolism in chickens has not previously been reported.

The addition of $2\sim 8$g/kg green tea and 0.5 to 2g/kg tea catechins partially inhibited the increase in plasma levels of TC, TG and LDL, meat and yolk levels of TC and TG, and also partially enhanced plasma HDL levels. Bursill et al. (2006) reported that administration of 0.5 to 2% (w/w) crude catechin extract from green tea significantly lowered plasma cholesterol and enhanced plasma HDL in rats. Based on this experiment, our results indicated that the effects of dietary green tea and tea catechins on TC, TG, HDL and LDL levels were not proportional to levels of supplement given, suggesting that there may be an optimum concentration of supplement.

Compared with the control group, there was no significant difference in yolk TC or TFA levels, although there was a small effect on yolk unsaturated fatty acid content. The results of fecal analysis indicated that the excretion of TFA in hens fed the dietary supplements was higher than in the control group, with the greatest effect from 6g/kg green tea and 1g/kg tea catechins. However, we still do not have a clear understanding of the metabolic pathway underlying the observed effects of these dietary supplements, nor do we have a physiological theory to explain them. Therefore, further studies are needed to elucidate how dietary green tea and tea catechins influence the lipid metabolism of laying hens.

3 Conclusion

In summary, dietary green tea and tea catechins can affect the lipids of laying hens, which significantly reduce the TG contents in plasma, breast and thigh meat and egg yolk, and increase the excretion of TFA in feces. In addition, it is interesting to note that the addition of green tea and tea catechins can slightly enhance the levels of unsaturated fatty acids in egg yolk. However, the effect is not proportional to the concentrations of the added green tea and tea catechins.

4 Acknowledgements

We thank Huainan Xinyang Breeding Co., Ltd. for providing the breeding rooms and feed. We also gratefully acknowledge the experimental assistance of the Biochemical Research Clinic Laboratory from Anhui Medical University. This work was supported by the Earmarked Fund for Modern Agro-industry Technology Research System in Tea Industry (nycytx—26), by the Chinese Ministry of Agriculture.

References

[1] Ahn D U, Wolfe F H, Sim J S. Dietary-linolenic acid and mixed tocopherols, and pack-aging influences on lipid stability in broiler chicken breast and leg muscle [J]. Journal of Food Science, 1995, 60: 1013—1018.

[2] Angelin R S, Antonio T, Barbara D, et al. Cholesterol, triglycerides, and the Five-Factor Model of personality [J]. Biological Psychology, 2010, 84: 186—191.

[3] AOCS. Fatty acid composition by gas chromatography. Official methods and recommended practices[M]. American Oil Chemists Society, 1990.

[4] Bragagnolo N, Rodriguez-Amaya D B. Comparison of the cholesterol content of Brazilian chicken and quail eggs[J]. Journal of Food Composition and Analysis, 2003, 16: 147—153.

[5] Bursill C A, Roach P D. Modulation of cholesterol metabolism by the

green tea polyphenol (一)-epigallocatechin gallate in cultured human liver (HepG2) cells[J]. Journal of Agricultural and Food Chemistry, 2006, 54: 1621—1626.

[6] Chan P T, Fong W P, Huang Y. Jasmine green tea epicatehins are hypolipidemic in hamster (Mesocricetus auratus)[J]. Journal of Nutrition, 1999, 129: 1094—1101.

[7] De Winne A, Dirinck P. Studies on vitamin E and meat quality 2. Effect of feeding high vitamin E levels on chicken meat quality[J]. Journal of Agricultural and Food Chemistry, 1996, 44: 1691—1696.

[8] Elkin R G, Yan Z, Zhong Y, et al. Select 3-hydroxy-3-methylglutaryl-coenzyme a reductase inhibitors vary in their ability to reduceegg yolk cholesterol levels in laying hens through alteration of hepatic cholesterol biosynthesis and plasma VLDL composition [J]. Journal of Nutrition, 1999, 29: 1010—1019.

[9] Folch J, Lees M, Sloane G H. A simple method for the isolation and purification of total lipids from animal tissues [J]. Journal of Biological Chemistry, 1956, 226: 497—509.

[10] Freiberg J J, Tybjaerg-Hansen A, Jensen J S, et al. Nonfasting triglycerides and risk of ischemic stroke in the general population [J]. Journal of the American Medical Informatics Association, 2008, 300: 2142—2152.

[11] Graham H N. Green tea composition, consumption, and polyphenol chemistry [J]. Preventive Medicine, 1992, 21: 334—350.

[12] Hand L W, Hollingsworth C A, Calkins CR, et al. Effect of preblending reduced fat and salt levels on frankfurter characteristics [J]. Journal of Food Science, 1987, 52: 1148—1151.

[13] Hensley J L, Hand L W. Formulation and chopping temperature effects on beef frankfurtersp [J]. Journal of Food Science, 1995, 60: 55—58.

[14] Herva A, Räsänen P, Miettunen J, et al. Co-occurrence of metabolic syndrome with depression and anxiety in young adults: the Northern Finland 1966 birth cohort study [J]. Psychosomatic Medicine, 2006, 68: 213—216.

[15] Kuo K L, Weng M S, Chiang C T, et al. Comparative studies on the hypolipidemic and growth suppressive effects of Oolong, Black, Pu-erh,

and green tea leaves in rats [J]. Journal of Agricultural and Food Chemistry, 2005, 53: 480—489.

[16] Lee K W, Everts H, Lankhorst E, et al. Addition of β-ionone to the diet fails to affect growth performance in female broiler chickens [J]. Animal Feed Science and Technology, 2003, 106: 219—223.

[17] Lee M J, Wang Z Y, Li H, et al. Analysis of plasma and urinary tea polyphenols in human subjects [J]. Cancer Epidemiology, Biomarkers & Prevention, 1995, 4: 393—399.

[18] Lee S M, Kim C W, Kim J K, et al. GCG-RichTea Catechins are effective in lowering cholesterol and triglyceride concentrations in hyperlipidemic rats [J]. Lipids, 2008, 43: 419—429.

[19] McCaffery J M, Niaura R, Todaro J F, et al. Depressive symptoms and metabolic risk in adult male twins enrolled in the National Heart, Lung, and Blood Institute Twin Study [J]. Psychosomatic Medicine, 2003, 65: 490—497.

[20] Mitsumoto M, O'Grady M N, Kerry J P, et al. Addition of Tea, 2005.

[21] Catechins and vitamin C on sensory evaluation, colour and lipid stability during chilled storage in cooked or raw beef and chicken patties [J]. Meat Science, 69: 773—779.

[22] Mojca S, Vekoslava S, Antonija H. The cholesterol content of eggs produced by the Slovenian autochthonous Styrian hen [J]. Food Chemistry, 2009, 114: 1—4.

[23] Muramatsu K, Fukuyo M, Hara Y. Effect of green tea catechins on plasma cholesterol level in cholesterol-fed rats [J]. Journal of Nutrition Science and Vitaminol, 1986, 32: 613—622.

[24] Psaty B M, Anderson M, Kronmal R A. The association between lipid levels and the risks of incident myocardial infarction, stroke, and total mortality: the Cardiovascular Health Study [J]. Journal of the American Geriatrics Society. 2004, 52: 1639—1647.

[25] Raes K, Huyghebaert G, Desmet S, et al. The deposition of conjugated linoleic acids in eggs of laying hens fed diets varying in fat level and fatty acid profile [J]. Journal of Nutrition, 2002, 132: 182—189.

[26] Sarwar N, Danesh J, Eiriksdotti G. Triglycerides and the risk of coro-

nary heart disease: 10,158 incident cases among 262,525 participants in 29 Western prospective studies [J]. Circulation, 2007, 115: 450—458.

[27] Shahar E, Chambless L E, Rosamond W D. Plasma lipid profile and incident ischemic stroke: the Atherosclerosis Riskin Communities (ARIC) study [J]. Stroke, 2003, 34: 623—631.

[28] Tang S Z, Kerry J P, Sheehan D, et al. Antioxidative effect of dietary tea catechins on long-term frozen stored chicken meat [J]. Meat Science, 2001, 57: 331—336.

[29] Wan Y F. Metabolism of green tea catechins: an overview [J]. Current Drug Metabolism, 2006, 7: 755—809.

[30] Wu L Y, Juan C C, Ho L T, et al. Effect of green tea supplementation on insulin sensitivity in sprague-dawley rats [J]. Journal of Agricultural and Food Chemistry, 2004, 52: 643—648.

[31] Zhong L, Furne J K, Levitt M D. An extract of black, green, and mulberry teas causes malabsorption of carbohydrate but not of triacylglycerol in healthy volunteers [J]. American Journal of Clinical Nutrition, 2006, 84: 551—555.

遮荫栽培条件下超微茶粉制造新工艺及其应用

刘勤晋[1]　颜泽文[2]　席阳红[2]

1. 西南大学茶研所；2. 四川叙府茶业集团

摘　要：茶叶功能医学研究进展揭示其具有抗氧化、增强人体免疫功能的保健作用。采用精细现代材料科学技术成果，使用集空气动力学、材料力学、精密机械制造技术于一体的超微气流磨粉碎经特殊栽培的茶叶，可使其粒度达 $10\mu m$ 左右，从而大大提升茶叶表面能和生化活性。本研究将微米级茶粉用于食品加工和日用化工，使茶叶附加值提升，为茶叶深加工及其综合利用开拓新的空间。

关键词：遮阴茶园；气流粉碎；空气磨；微米；茶食品

New Producing Technology and Application of Ultra-micro Tea Powder Under the Condition of Shading Cultivation

LIU Qin-jin[1]　YAN Ze-wen[2]　XI Yang-hong[2]

1. Tea Research Institute of Southwestern University;

2. Xufu Tea Group of Sichuan Province

Abstract：Study on the medical function of tea has discovered that anti-oxidation function of tea is definite and it will help to strengthen immune system of human being. It is also proved that, specially-cultivated tea can be smashed into particles of 10 microns by ultra-micro airflow. Thus the surface energy and biochemical activity of tea has been promoted incredibly. Micron-level tea powder which is produced thru this method has been applied into food and chemical industry successfully. Add-value of teas has been increased and a new area to process and utilize tea has been explored.

Keywords：shading tea garden; air-flow smash; air mill; micron; tea food

在我国国民经济快速发展，农业产业化取得多项成果的大好形势下，茶产业亦得到迅速发展。2011 年据国家统计局数据我国茶园面积达到 190 万 hm^2，总产量 160 万 t，出口 30.4 万 t，分别保持世界第一和第二地位，但茶叶平均单产和产值却相对低于日、印等国。其主要原因在于产品结构仍以传统市场散茶为主，而适合国内外市场需求的高品质、清洁化、深度加工产品却十分缺乏，如利用精细化工技术生产的微米—纳米级超微茶粉和采用零排放技术生产的茶叶提取物（茶多酚、儿茶素、茶氨酸）等。在国际上，发达国家由于以上技术成熟，茶叶已在食品、饮料、保健品、日用化工产品等行业大量使用。如近邻日本年产茶 8.5 万 t，利用以上技术开发的深加工和多元化利用茶产品近 500 种，仅利用茶叶 4000t，占茶叶产量 5%，产值达 500 亿日元（相当 35 亿元人民币），相当于每千克深加工品价值达 875.0 元人民币，真正达到提高茶叶附加价值的目标，使该国茶产业由于能源和用工成本增加所面临危机得以缓解，茶产业在日本这样经济高度发展国家同样得到可持续发展。

对于茶叶深加工，我国政府高度重视。近年来茶产业已实现跨越式发展，已有良种茶园近百万公顷，高、中档名优茶在国内茶叶市场拥有较强的竞争优势，并以仅占全年 20% 产量实现 85% 产值；其余则生产供应大众消费的低档茶，如四川宜宾地区估计每年尚有 500t 茶叶未采摘下树或成为很不值钱的边茶原料。随着农村土地流转的实施和大批劳动力进入城市，茶区采茶劳力呈现十分紧缺的态势，手工采茶成本大幅上升，机械采茶势在必行，配合机采而进行的茶叶加工工艺改革也必须紧紧跟上。

纳米—微米粉碎技术是现代高新材料科学的重大突破，即应用物理方法使物料瞬间破碎至纳米—微米的粒径，使物料纯度提高、粒径变小，其分散性、吸附性、表面活性均大幅提高，有利于物料功能化学性质的充分发挥，使食品色、香、味、口感及保健作用都得到有效提高。

本项目的实施，使经遮荫栽培措施之茶鲜叶持嫩性增强，经粉碎末后各种茶叶粉体变为 $8\sim20\mu m$ 的超细粉末。由于颗粒比表面积增大，表面能提高，表面活性增强，表面与界面晶体结构发生变化，使超细茶粉产生与原茶完全不同的物理化学特性，例如极佳的分散性、吸附性、溶解性和保健功能活性等，从而拓宽茶的多元化用途，特别有利于食品和日用化学用品开发。

采用本项目开发的茶粉用于加工传统糕点和新型食品，使茶食品色、香、味口感及所含功能成分均有显著提高，其感官及理化指标均超过日本同类产品，现已申请实用新技术专利 2 个，其技术与产品获 2011 中国国际茶品牌创新食品开发金奖。

1 研究材料、方法及主要内容

1.1 材料与方法

1.1.1 基地选择

以四川叙府茶业黄山天宫庙基地有机茶园 3.3hm^2（海拔 1200m）和金秋湖基地无公害茶园 2hm^2（海拔 600m）为试验园，严格按照良好农业操作规范要求，建设黑色尼龙网大棚 125 座。

1.1.2 茶园准备及采摘管理

茶园选择：选择本公司海拔 1000m 以上良种无性系茶园 3.3hm^2，金秋湖茶园 2hm^2，其地点相对集中、交通方便，地势平坦、土层深厚。经检测土壤农药及重金属均未超标。茶园树势良好，人工管理基础较好。

冬春基肥：在开春前，雨季未到来时，以大豆、茶籽饼粕或人畜粪尿作为基肥，每亩保证纯氮 60～80kg。

有机农药：在开春前喷洒石硫合剂、川楝素等有机农药，以防除可能产生之病虫害。

搭好棚架：采用"就地取材"方法，用竹木材料搭成高 1.65m、宽度适当的遮阴篷，铺设黑色塑丝网，使茶园遮光度 70％～95％。搭设时间为茶树发芽时最好，即茶芽苞片脱落、叶片露出一芽一叶初展时。前 10d 遮光 70％～80％，后 10d 遮光 80％～95％，共遮光 20d 左右。至茶树新梢长至 20～25cm，上有鲜叶叶片 5～6 片时最佳，以鲜叶叶片色泽鲜绿光亮、叶质柔软、稍带白毫、嫩梗黄绿色、无黄片老叶者为上。

鲜叶采摘：鲜叶达到采摘标准后，及时采用采茶机或采茶剪采下枝叶送往茶厂，中途保持鲜叶不落地，不沾泥土、杂物或其他容易污染的工具，置于篮中不紧压、不损伤，4h 内必须送至茶厂进行加工。

1.2 茶粉原料处理及初加工技术

微米茶粉乃用于食用及日用化工（化妆品）之精细化学产品，对产品理化指标及安全卫生有极高要求。因而像传统制茶一样原料仅经过高温处理远远不够，必须采用强力水冲方式洗脱茶鲜叶表面留存的灰尘、泥土及其表面附着物，及时通过蒸汽杀酶,使其保持鲜绿色泽、特有清香和甘醇滋味。其原料处理及茶粉初加工工艺如图1所示。

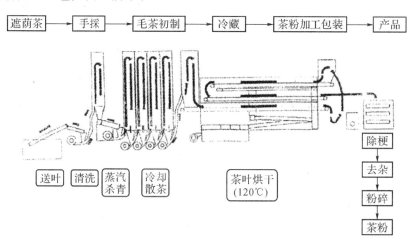

图1　原料茶工艺流程

1.2.1　鲜叶清洗

进厂鲜叶立即采用清水喷雾法进行淋洗,使其去掉表面灰尘,并立即进行沥干、待制。

1.2.2　蒸汽杀酶

对于进厂原料,应摊放在干净竹箎垫或晒席上,及时进行蒸汽杀酶,杀酶温度应在100℃以上,压力0.2Pa,时间15～20s。

1.2.3　快速冷却

经高温杀青后,鲜叶应置于高3～4m、宽2m、长10m的透气网中快速冷却、迅速干燥。冷却可置2～3个鼓风机吹冷。

1.2.4　热风干燥

经冷却连梗茶叶,迅速送入120℃热风烘干机中进行干燥,使含水量保持3%～4%,即进行精制(除梗—去杂—粉碎—包装),如毛茶初制与精制不同步进行,则须将脱梗毛茶迅速密封、冷藏待制。

1.2.5 茶粉的保存与包装

高级食用茶粉制成后,其含水量不得≥5%,由于其吸附性、分散性特高,故应小心置于牛皮纸袋中(每袋 5kg)抽气充氮密封保存,分零拆装小袋应在灭菌清洁环境中进行,以保证食用级茶粉不会造成二次污染。

通过应用本公司自行研发的 CG-110 型贮青机、CS-400 型蒸汽杀酶机等配套设备反复试验,制得原料茶。经与日本同类产品比较,其氨基酸总量比日本高级超微茶粉高 11.98%,比日本高级玉露茶高 40%;而儿茶素总量低 33.72%(表1)。

表1 遮荫绿茶粉与同类日本产品的比较

成 分	遮荫茶(叙府1号)	露地茶(叙府2号)	日本高级茶粉	日本高级玉露
儿茶素总量(mg/g)	74.40	127.33	121.77	111.76
氨基酸总量(%)	2.98	2.53	2.66	2.14
茶氨酸含量(%)	2.01	1.61	1.97	1.59
咖啡因含量(mg/g)	19.78	27.81	32.99	32.69

2 纳米—微米级茶粉加工技术研发

微米级茶粉加工技术既不同于日本传统"抹茶"加工技术,也有别于国际流行超微茶粉加工技术。它消化吸收日本"抹茶"工艺之精华,采用先进的茶叶栽培、加工工艺和现代化设备,应用微米级超微气流粉碎技术加工 8~18μm 分级茶粉获得成功,应用于四川特色食品的现代加工,并使之实现产业化。

通过科技查新,发现本项目研究开发的经过气流超微粉碎系统将片状茶原料粉碎细度达到 1200~1500 目的超微茶粉,在所检索的文献中未见相关文献报道。

2010 年 9 月,高细度超微茶粉生产线投入生产,完成了较定型的超微茶粉精细化加工技术。其关键技术是:①最佳气流速度参数的确定、控制;②最佳气流压力参数的确定、控制;③最佳气流流量参数的确定、控制;④最佳气流温度参数的确定、控制;⑤气流参数与产品粒度关系的确定、控制。

本项目的技术难度在于茶粉品质的提升、茶食品品质的提升。

经过对微米茶粉与市售普通茶粉的比较,结果见表2。

3 微米茶粉在茶食品应用研发

本项目制得微米茶粉以后,进一步研究超微茶粉在糕点类食品中的应用。经采用小型糖果糕点机械如砂糖粉碎机、凤凰卷机、桃酥饼干成形机、全自动包馅机、月饼糕点成形机、月饼糕点自动排盘机等设备,建立了1条茶食品糕点生产线,同时还建立了1条糖果生产线;配套购置和安装了超微茶粉及茶食品检测检验仪器设备等。

此期间内生产绿茶酥饼、绿茶蛋卷、绿茶含片、绿茶月饼、绿茶米花糖、绿茶花生糖、绿茶挂面等系列茶食品获得成功,产品批量生产并投入市场,获得一致好评。2011年获中国国际品牌"金芽奖",国内各大主流媒体热评如潮。目前,本项目已正式申请实用新型专利两项。

表2 微米茶粉与普通茶粉性能比较表

项 目	市售普通茶粉	普通超微茶粉	项目微米超微茶粉
制作原料	普通绿茶鲜叶	有机栽培,遮荫覆盖茶园采摘	有机栽培,遮荫覆盖茶园采摘,经清洗
性 能	细微,但有粗糙感	良好的分散性、吸附性、溶解性	各指标更优
感官性能	黄绿、香型杂、苦涩味重	浅绿色、清香	天然翠绿色至深绿色,清香、鲜爽淡雅
内含物质	正常	蛋白质、氨基酸、叶绿素含量稍高,茶多酚含量低	蛋白质、氨基酸、叶绿素是普通茶粉的2倍左右,茶多酚含量低
营养价值	低	好	高
用 途	作为绿茶粉冲泡,但沉淀多	冲泡少沉淀,食品添加剂、日化美肤产品	冲泡少沉淀,使食品色、香、味、口感及保健作用都得以更充分发挥
市场反应	不足	良好	国内未知,在日本市场反应热烈
产品安全性	原料无遮荫覆盖、无清洗	有遮荫覆盖	有遮荫覆盖,经清洗,粉碎过程管道封闭式加工,《GB/T 18526.1—2001 速溶茶辐照杀菌工艺》

4 小 结

本研究经3年来努力,获得以下成果:

(1)形成了茶粉原料的设施栽培成套技术;

(2)完成了3种茶粉生产工艺流程定型配套;

(3)完成了茶粉研磨技术的装备选型配套;

(4)年产超微茶粉60t,年产值600万元;

(5)本项目生产的超微茶粉20%用于茶食品开发,成套工艺技术成熟稳定,年开发生产茶食品(包括绿茶酥饼、蛋卷、米花糖、花生糖、月饼、含片等)200t,年产值1200万元。3年来,项目现总产值2150万元,实现利润846万元(税后),获得较好经济和社会效益。

茶叶在日本食品中的应用概况

Masashi Omori[1]　Yumiko Uchiyama[1]　Kasumi Tsukidate[1]
Mayumi Yamashita[2]　Yuki Okamoto[2]　Miyuki Katoh[3]
1. 日本大妻女子大学；2. 日本和洋女子大学；3. 日本香川县大学

摘　要：和中国一样，很久以前日本就有饮茶和食用茶叶的习俗。茶的利用方式经千百年演变，现在与人们生活密切相关。然而茶渣一直被认为是废弃物，很少关注其利用。茶叶内含物在第1次冲泡时会溶解60%，第2次冲泡达到80%。对饮茶爱好者来说，如果仅冲泡1次，茶叶中其余40%的水溶性物质就会被废弃。如果茶叶用于食用，那么茶叶中所有的营养成分都可得到利用，包括废弃的40%。此外，茶叶在冲泡过程中，粗纤维、脂类、非水溶性蛋白质、淀粉、微量元素等无法析出，茶叶用于食用，以上所有营养成分均可得到利用。家庭利用茶渣可采用冷藏方法，干燥后研磨适合做烹调配料。微波炉可用来干燥茶渣且能很好的保留茶叶色泽。日本茶叶食用方式包括黄豆煮茶叶、茶叶汉堡包、茶叶蛋糕、茶叶水饺、油炸食品配料等。

关键词：日本；茶叶；应用

All Purpose Utilization and Application of Japanese Tea

Masashi Omori[1]　Yumiko Uchiyama[1]　Kasumi Tsukidate[1]
Mayumi Yamashita[2]　Yuki Okamoto[2]　Miyuki Katoh[3]
1. OtsumaWoman's University Chiyoda, Tokyo, Japan;
2. Wayo Woman's University Ichikawa, Chiba, Japan;
3. Kagawa University, Kagawa, Japan

Abstract：Tea is eaten and drunkfor long time ago in Japan. Utilizaton methods of tea are directions changed in thousands of years, and it has stuck them to the life. However, about used tearecognisedas "garbage" and concern was seldom paid for the use. If it infusedepending on how to make tea for usual, about 60% of tea components will infusein the first, and it will infuseto

80% in the second infusion. In the person of a tea lover, since it drinks only in the first, about 40% of soluble ingredientswill be thrown away very much. If this is made edible, since the whole circle can be taken in including these 40% thrown away, it will become nutritionally very effective. In addition, in infusing, a fiber, lipid soluble vitamin, insoluble protein, starch, bindingminerals, etc. do not begin to soluble, but if the whole will be called edible used tea leaves, these can also all be taken in now and it will be called nutritionally very desirable food "tea". When utilizationof used tea at home, freezing and keeping used tea after infusingin the first place, and the second are good to dry and grind to use for cooking suitably. It is desirable, without also spoiling the color of camellia shell, if a microwave oven is used as a dry technique. I would like to utilize as a tea dish taking advantage of various ideas, such as food boiled down in soy, a hamburger, a cupcake, dumplings and tempura.

Keywords：Japan；tea；application

　　茶には，緑茶や紅茶，ウーロン茶など多くの种类がある．わが国においては，緑茶の生産が大部分で紅茶，ウーロン茶の製造は商業用としては行われていない．昭和の初期までは輸出用として紅茶も生産されていたが，戦後は主な生産国の生産量の低下により，輸出量も一次増加したが生産国の復興と共に，高度経済成長以後減少した．また，国内消費量も昭和40年代後半をピークに，以後漸減の傾向を示している．その反面，紅茶の輸入量は，食の洋風化と近年の紅茶ブームで，一定の増加の傾向にある．また，一方では国産紅茶の製造を試みる農家も現れ，それなりの評価を受けてはいるが，既存の紅茶に比べて風味は今一つ物足りなさがあるが，珍しさから一部流通している．

　　近年，ソフトドリンクとしての无糖の茶缶飲料，ペット飲料が普及し，その利便性，嗜好性から，その消費は急激に増加した．現在は，所謂カメリア族の茶だけでなく，トウモロコシ，ハーブなども含めた茶系飲料が，次々と各メーカーから発売されている状況である．

　　茶の嗜好性には非常に独特のものがあり，茶道と共に嗜好の文化を作っていると思われ

　　る．春夏秋冬を問わず，一人で茶をたしなむなどということは，他の飲料にはあまりみら

　　れないところである．茶の有する生理，薬理効果が次々に明らかにされ

る中,茶は飲んで

よし,食べてよし,とその利用法も种々试みられている.茶を食べる.简便なものとして

は茶を食卓に置き,軽くつぶしてじゃこ,ゴマ,のりなどとふりかけとして用いる.

これに茶を注いで茶漬けにしても结构美味しくいただける.茶がゆ,茶饭,茶そば,茶入

り蒸しパン,茶入りクッキー,抹茶羊羹,等は一般的であると思われるが,抹茶魚そうめんの吸い物などは面白い.抹茶入りの手作り魚そうめんを和風すまし汁にした料理で,かざり麸などと盛ると大変きれいで,来客時に使うこともできる.

白身魚,4人分として200g,卵白1ヶ分,塩5g,片栗粉8～20g,抹茶約0.8gを用意する.白身魚は鯛,ひらめなどがよく,たらなどは繊維が大きく,麺にする時にうまくいかない.白身魚をできるだけ薄くそぎ切りにし,包丁でつぶすようにしてからすり鉢でよくする.卵白は少し泡立て,塩と共に少しずつすりこむようにして白身魚に混ぜる.白身魚に片栗粉と抹茶を少しずつ均一になるように混ぜる.

水または酒を固さをみながら加え,里ごしする.これを2～3mmぐらいの口金の絞り袋で熱湯に細く絞り出し,サッとゆでて冷水にとる.すまし汁を別に作っておき,これに魚そうめんを入れ,かざり麸を使って盛りつけサーブする.

白身魚を使う代わりに市販のすり身を使うと,もっと手軽に作ることもできる.鶏のから揚げや豚肉のマスタードソースなどにもよく合い,また,抹茶の量を适宜増減すると,魚そうめんの色が変わるので,嗜好に合わせて調節するとよい.

一般に茶は一煎目で可溶性成分の約60％が溶出され,二煎目では約80％まで溶出される.茶の美味しさの点からは一煎目がやはり美味で,残りは茶売として破棄されてしまうことが多い.このように茶缶飲料やペット飲料の消費が急増すると,それに伴い茶売の扱いも

大きな問題となる.茶には,各种有効成分が多く含まれている.例えば,カテキンには抗酸化性,抗癌性,抗菌性等の作用があり,カフェインには中枢神経興奮での覚醒や睡眠の遅延の作用があると報告されている.

また,桑野らは茶を食べることによる有効性を報告している.著者らは,

茶资源综合利用研究新进展

茶壳の生体への影響についてラットを用いて実験し，生体への効果，特に中性脂肪の減少に効果的であることを認め，報告した．今，茶の生活への多目的利用として，各方面で試みられ，茶を飲む，食べるということから，茶料理として，家畜，ペットの飼料として，更にウレタンフォームや衣類・おむつ・寝具，建材から香粧品など，生活のあらゆる場面での利用が試みられ，実用化もされている．ここでは主に，食べるという面から紹介したい．

1 実験方法

1.1 茶殻の調製方法

試料：煎茶は，農林水産省野菜茶業試験場にて製造したものを用いた．浸出方法は，茶の官能検査法（製茶部審査研究会，1971）および茶の淹れ方研究会での討論を参考にして茶を浸出し，茶殻を得た．すなわち煎茶 5g に 90℃の湯 100mL を加えて 1 分間浸出し，茶殻を濾別した．茶殻は 70℃の通風乾燥機で乾燥し，その後 100 メッシュに揃えて以後の実験に供した．

1.2 茶殻の化学成分の分析

水分は，105℃常圧乾燥法で，灰分は，550℃で常圧で灰化させ行った．

カフェインの定量は，津志田らの方法（津志田藤二郎，1984）に準じて行った．すなわち，茶葉 100mg に熱水 70mL を加えて 60 分間抽出し，その後冷却して 100mL に定容後濾過した．この濾液にポリビニールポリピロリドン（PVPP）を 200mg 添加し，30 分撹拌することによりカテキンを除去，濾過後高速液体クロマトグラフによりカフェインを定量した．高速液体クロマトグラフィーの条件は，カラムは GL-サイエンス製の ODSφ4.0mm×250mm，カラム温度 40℃，検出器 272nm，流量 1mL/min，溶離液 5mmol/L リン酸ナトリウムを含む 60％メタノールを用い，注入量は 5μl として行った．

ポリフェノールの定量は，岩浅らの方法（岩浅潔 1962）に準じ，酒石酸鉄法を用いて行った．

カテキンの定量は，池ヶ谷らの方法（池ヶ谷賢次郎，1989）により行った．すなわち，茶葉 100mg に 20％アセトン溶液を 30mL 添加し，内部標準物質としてホモカテコール 1mg/mL を添加，30 分撹拌抽出を行った．これを濾過後，アセトン臭のなくなるまで濃縮し，酢酸エチルで抽出を行った．この酢酸エチル層を脱水后濃縮干固し,アセトニトリルで定容し,これを高速液体クロマトグラフの試料とした.高速液体クロマトグラフィーの分析条件は,カラムはメルク制の Hiber Lichrosorb RP－185μmφ4.0mm×250mm,カラム温度30℃,検出器 280nm,流量 1mL/min,溶離液アセトニトリル:酢酸:メタノール:水(113:5:20:862)で注入量は10μLで行った.各カテキン量は,内部標准物質のホモカテコールに対す.

る比から算出した.

アミノ酸は津志田らの方法(津志田藤二郎,1987)により,茶叶1gに熱水70mLを添加し,室温で60分間抽出后100mLに定容し濾別した.濾液を高速液体クロマトグラフの試料とし,OPA 法により測定した.

1.3 調理法別茶殻添加の調製方法

表1に示すように,茶殻添加料理を作るときの可能な調理方法を検討するために,調理操作毎にわけてそれぞれの茶殻添加量を検討した.調理方法としては,「焼く」は,スポンジケーキ,クッキー,パウンドケーキ,パン,ハンバーグステーキ,だし巻きたまご,「揚げる」は,ドーナツ,天ぷら,「煮る」は,佃煮,ジャム,「蒸す」は,カスタードプディング,しんじょ,餅,カップケーキ,「ゆでる」は,つみいれ,白玉だんご,うどん,「寄せる」は,ゼリー,寒天よせ,の合計 19 品目について検討した.茶殻の添加量については,予備実験の結果よりその料理の主材料に対する添加量を検討し,ハンバーグステーキ,カスタードプディング,しんじょ,寒天よせ,ゼリー,つみいれ,だし巻きたまごは茶殻粉末の添加量を1%とした.スポンジケーキ,クッキー,パウンドケーキ,パン,ドーナツ,天ぷら,餅,カップケーキ,白玉だんご,うどんは茶殻粉末の添加量を5%とした.佃煮,ジャムは茶殻を粉末にせず,乾燥したまま用いた.

1.3.1 焼く操作の料理

スポンジケーキ(竹治栄美,1988),クッキー(岡本信弘,1992),パウンドケーキ(主婦の友,1980),パン(上野厳,1976),ハンバーグステーキ(竹治栄美,1988),だし巻きたまご(山崎清子,1991)を常法で調製した.

1.3.2 揚げる操作の料理

ドーナツ(主婦の友,1980),いもの天ぷら(竹治栄美,1988)を調製した.

茶资源综合利用研究新进展

表 1　茶殻を添加した料理

調理方法	料理名	調理方法	料理名
焼く	スポンジケーキ クッキー パウンドケーキ パン ハンバーグステーキ だし巻き卵	蒸す	カスタードプディング しんじょ 餅 カップケーキ
揚げる	ドーナツ 天ぷら	ゆでる	つみいれ 白玉だんご うどん
煮る	佃煮 ジャム	寄せる	ゼリー 寒天寄せ

1.3.3　煮る操作の料理

佃煮(山崎清子,1991),ジャム(山崎清子,1991)を調製した.

1.3.4　蒸す操作の料理

カスタードプディング(山崎清子,1991),しんじょ(竹治栄美,1988),カップケーキ(渡辺茂,1985),餅を調製した.餅の調製法は,もち米 700g は洗って,一晩水に浸漬しておいた.ざるで水をきり,電気もちつき机で餅とし,直径 6cm に丸めた.茶殻は,もちつき机が動き始めたら少しずつ加えた.

1.3.5　ゆでる操作の料理

つみいれ(山崎清子,1991),白玉だんご(山崎清子,1991),うどん(主婦の友,1979)を調製した.

1.3.6　寄せる操作の料理

ゼリー(山崎清子,1991),寒天よせ(山崎清子,1991)を調製した.

1.4　官能検査

官能検査は,女子学生 15 名を検査員として,官能検査項目は,色,味,香り,食感,外観,嗜好,総合の7項目で7点嗜好試験方法(川端晶子,1986)を用いて行った.各試料温度は,ゼリーを除き常温で行った.

結果は無添加と茶殻添加を1元配置法(川端晶子,1986)により検定を行った.

2 実験結果及び考察

2.1 茶殻の化学成分の特徴

煎茶と茶殻の分析結果を表2に,カテキン含量を表3に,アミノ酸含量を表4に示した.煎茶に比較して茶殻では,カフェインは約30%,カテキンは約40%,ポリフェノールは約50%に,アミノ酸类は約80%に減少した.

表 2 茶殻中の成分 (g/100g)

	煎 茶	茶 殻
水分	4.85	5.40
灰分	3.68	2.42
カフェイン	3.01	0.90
ポリフェノール	14.50	7.20

表 3 カテキン含量 (g/100g)

	煎 茶	茶 殻
(－)－エピガロカテキン	3.46	1.22
(－)－エピカテキン	1.16	0.46
(－)－エピガロカテキンガレート	6.22	4.33
(－)－エピカテキンガレート	2.17	1.58
総量	13.01	7.59

表 4 アミノ酸含量 (g/100g)

	煎 茶	茶 殻
アスパラギン酸	304.8	22.6
スレオニン	70.3	14.1
セリン	90.4	18.1

续表

	煎 茶	茶 壳
テアニン	1374.4	274.9
グルタミン酸	294.3	23.5
アラニン	18.7	3.7
バリン	30.5	6.1
イソロイシン	19.7	3.9
ロイシン	17.1	3.4
チロシン	14.5	2.9
フェニールアラニン	29.7	5.9
γ－アミノ酪酸	6.2	1.2
ヒスチジン	15.5	3.1
リジン	36.6	7.3
アルギニン	95.8	38.3
総量	2418.8	429.0

アミノ酸,カフェイン,ポリフェノールともに水溶性の成分であり,また,一煎目でかなりの量が溶出される(池田重美 1972)ことが示されたが,二煎目でもまだ煎茶に類似している風味を示していた.成分によって異なるが,かなりの量の成分がまだ茶壳には残されており,また,外観から判断して色素成分もかなり残存していることが認められた.

2.2 茶壳添加の官能検査結果

官能検査の結果によれば.調理方法別では,干式加熱の「焼く」「揚げる」は,色,外観,総合以外では,无添加との差は認められなかった.甘味が強い「スポンジケーキ」カップケーキ」などは,茶壳の添加量がカスタードプディングやしんじょ等の1％添加量に比較して,10％まで添加することが可能であった.これはケーキに含まれる砂糖の分量が多いものの方が茶壳に存在するカテキンの苦味を軽減することから,砂糖の甘味と茶壳に存在するカテキンの苦味の味の抑制効果ではないかと考えられた.特に茶壳を添加することにより茶の色が加わり"色""外観"の項目において好まれたと考えられた.茶壳そのものには茶の香りがまだかなり残存していることから総合評価でも好まれたものと考えられる.湿式加熱の「煮る」の佃煮,ジャムに関しては,无添加との比較を行わなかったもののジャムは好まれず,佃煮は好評であった.「蒸す」のカスタードプディングは,卵液の凝固までの時間に茶殻粉末がカラメルソースの部分に固まり,見た目がよくなかったためにすべての評価項目において无添加が好まれたものと考えられる.

「ゆでる」の白玉だんご,いわしのつみいれについては,ゆでることにより茶殻の成分が,ゆで水中へ溶出するために无添加区との差が認められなかったものと考えられる.「寄せる」調理操作のゼリーは,官能検査試料として検査に不适当であったが,パネルの感想として茶殻1%添加でも食べられないこともない,という評価であった.干式加熱の「焼く」「揚げる」に関しては,調理時間が短時間であるためと,調理することによる焦げの風味が加わることから好まれる調理方法と考えられた.湿式加熱の「煮る」「蒸す」「ゆでる」は,調理時間が干式加熱に比べ長いことから,茶殻の風味成分の溶出が多いため,あまり好まれなかったものと考えられた.また,「寄せる」に関しては,寄せるまでの時間で茶殻の粉末が沈むことにより,外観上あまり好まれなかった.

餅に関しては,茶殻添加区において茶の香りを有することから有意に好まれていた.餅を常温で放置した時のカビの発生状況を検討した結果,放置后5日目頃より无添加区に緑色のカビが発生したが添加区には認められなかった.11日后には无添加区に数個のカビが認められたが,添加区においてはその数が少なかった.茶殻を添加したものの方がカビの発生が少なく,餅の保存性の向上が認められた.このことは,原ら(原征彦,1989)の各種細菌に対する抗菌作用がカテキン類に存在していることや茶成分の微生物に対する抗菌作用(大森正司,1982)などからも,茶殻中の成分によるものと考えられた.

今后は茶殻の添加が可能となった「スポンジケーキ」「カップケーキ」「クッキー」等について物性への影響について検討を行いたいと考えている.

茶资源利用与产品开发

林志城

元培科技大学生物科技研究所,中国 台湾

摘 要:茶叶或其中所含的多酚成分具有抗氧化等保健功效。许多研究注重研究茶饮的生理活性,但有关将茶应用于开发其他产品的讨论并不多。包冰处理可增强鱼肉片的贮存品质,利用茶汤作为包冰液可以有效提高冻藏鱼肉的品质。轻度发酵的包种茶或乌龙茶其活性不低于绿茶,使用膜分离的乌龙茶中间分子量区分液表现出优异的抗氧化、抗发炎、美白、抗紫外线效果,已被开发成保养品;此外膜分离茶也能有效抑制致病菌,但不会抑制对人有益的乳酸菌,因此可开发作为胃肠调整与皮肤清洁喷剂。而红茶经乳酸菌发酵后具有抗菌与抗 UVB 活性,可开发成口腔清洁与防晒产品。由这些结果可知,茶叶的制造加工方法依茶开发用途与目的决定。

关键词:茶多酚;包冰;保养品;抗氧化;抗发炎;抗紫外线

Comprehensive Utilization of Tea Resource and Product Development

Chih-Cheng LIN

Department of Biotechnology and Food Science,
Yuanpei University, Taiwan, China

Abstract:Tea and its polyphenolic components have received increasing attention due to their antioxidant activities and beneficial effects for human health. Most researches focused on the physiological activity of tea drink. However, the utilization of tea for other application has seldom been investigated. The technology of ice glazing enhanced on the storage quality of the fish fillet significantly as compared to untreated. The combination of glazing treatment and adding green or pouchong tea extract could greatly increase the storage quality of the frozen fish fillets. The effects of light fermented tea on skin-care activity, such as pouchung or oolong tea, demonstrated better results than green tea. The membrane-fractionated oolong tea extracts, especially with medium molecular weight fractions, exert antioxidant, anti-inflammatory, whitening and anti-UV effects in a applications for cosmeceuticals. Furthermore, membrane-fractionated oolong tea extracts also exhibited antibacterial potency on pathogenic bacteria but not against lactic acid bacteria and they can be used for bowl modulation or skin cleaning. Black tea fermented with lactic acid bacteria, exhibited anti-bacterial effect and anti-UVB activity, and was developed to oral cleaning and sun block products. These results suggested that manufacturing or processing methods of tea can be developed base on the purpose of tea application.

Keywords：tea polyphenols; glazing; cosmoceutical; antioxidant; anti-inflammatory; anti-UV

机能有营养、感官与保健三项，茶主要为感官与保健两种机能。其中感官机能通常与加工有关。茶叶富含多酚等活性物质，具有许多生理活性。但依茶种与制法的不同，不同茶叶有不同的特性，制茶过程中的杀青、揉捻、发酵、拼堆与焙火等加工对最后成茶的各种机能品质影响极大。茶青经发酵程度的不同会影响茶多酚的聚合度，有些利用微生物发酵的茶不但会分解茶多酚，甚而会产生特殊结构的代谢产物。茶多酚聚合度会改变分子极性，影响抗氧化和抗发炎等活性。

茶资源具有多元利用的价值，大部分研究偏重于经口摄取后感官品评与保健功效，主要原因是仅将茶局限在饮料用途，如此也限制茶的发展。例如夏茶因涩味重、嗜好性差，所以价值较低，可以开发其他应用。其实以天然物观点，茶除饮用外应有更广泛的应用，本研究以产品开发观点，举例讨论近年以茶运用在保鲜、保养品、杀菌剂等之发展，并探讨制茶与产品间关联性。

1 茶在保鲜方面的应用

儿茶素用于食品的保鲜最早被应用于油脂的抗氧化，但几乎无人考量到肉品，主要原因是单宁与蛋白质间的作用会影响品质。冷冻产品的包冰处理以往多使用水或维生素 C 液，最大差别是茶多酚可以清除不同来源自由基同时保护油脂与蛋白质。比较绿茶、包种茶与红茶应用于鱼肉包冰液中的效果。表 1 实验证明包种茶在保鲜、抑制油脂与蛋白质氧化方面的效果表现优异，证明轻度发酵的包种茶中的低聚合多酚抗氧化与保鲜活性较不发酵的绿茶与完全发酵的茶好。研究也证明不同制茶阶段的茶液表现不同抗氧化性,虽然在揉捻阶段有高于烘焙与最后干燥阶段的抗氧化儿茶素类化合物含量,但是成茶显然对于抑制油脂和蛋白质氧化效果最好。

表 1 不同包冰液处理对冻藏鱼肉品质的影响

包冰液	VBN	TBARS	TBARS	Protein carbonyl
控制	23.3a	25.44^a	25.44^a	0.98^a
水	17.5b	14.62^b	14.62^b	0.93^a
绿茶	14.5d	7.13^d	7.13^d	0.94^a
包种茶	14.9d	6.79^d	6.79^d	0.79^b
红茶	16.8c	10.56^c	10.56^c	1.07^a

2 茶在抑菌方面的用途与产品开发

使用膜分离设备将茶汤区分出分子量为 150~1000 的乌龙茶多酚,发现可以有效抑制中毒菌如金黄色葡萄球菌、沙门菌等,但较不会影响人类益生菌或乳酸菌的生长(表2),以 leakage 动力学研究发现抑菌效果与茶中所含的 EGCG 有关,但 EC 与 gallic acid 影响甚低。将此茶多酚控制浓度制成的杀菌液,可使用于生鲜水产品食具的消毒或手部的清洁喷剂。

表 2 膜分离茶多酚对不同菌的抑制率

菌 种	抑菌性(%)
Sal. enterica	100
S. aureus	57.21±7.55
L. casei	1.85±0.90
L. plantarum	17.81±2.28
L. rhamnosus	1.81±0.13

3 茶在皮肤保养方面的用途与产品开发

使用同一种茶种制得绿茶、乌龙茶与红茶,乌龙茶不但在抗氧化和抗发炎活性上均不输于绿茶,甚至在某些活性上高于绿茶。应用在皮肤保养上,寡聚合茶多酚无论在自由基清除能力、抗发炎活性(图1)、还原能力、美白能力、抑制痤疮杆菌能力、紫外线(图2)或自由基伤害角质细胞的防护能力都较高聚合茶多酚或儿茶素的结果佳。已开发成保养乳液并上市。

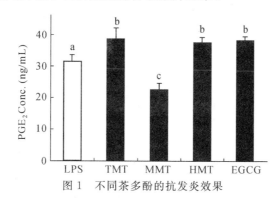

图1 不同茶多酚的抗发炎效果

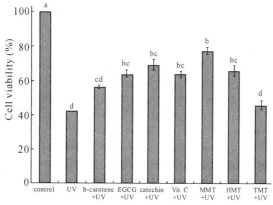

图2 不同茶多酚抑制 UVB 引发的 HaCaT 细胞损伤

4 乳酸菌发酵红茶的保健活性

虽然许多物质经微生物发酵后会使多酚释放或进行生物转换,但只有少数几种乳酸菌会发酵红茶。红茶加糖经特定乳酸菌发酵后虽然能降低抗氧化能力，但会提高抗菌能力特别是龋齿菌与金黄色葡萄球菌，可以开发成口腔清洁剂或特殊风味的保健饮料；此外也会提高抗发炎、抗 UV 活性与抑制酪氨酸酶活性能力，能运用于防晒美白的皮肤保健品和保养品中（表 3、图 3）。

表3 乳酸菌发酵红茶茶汤 48 h 后抑制 7 株病原菌之抑菌情形

菌名和菌株编号	抑菌直径（mm）					
	P2	P3	P4	C3	R3	空白对照组
E. coli BCRC 10356	5.27	4.53	5.98	6.47	3.21	5.71
S. oureus BCRC 10781	13.41	14.07	6.57	3.83	10.41	7.64
V. parahaemohyticus 1385	4.24	4.11	4.38	2.72	2.31	4.25
V. wlnificus 27562	4.57	0.00	1.16	0.00	0.88	0.00
E. aerogenes Ea05	0.00	0.00	0.00	0.42	0.70	0.00
S. sanguis 15273	8.55	7.07	9.89	7.35	4.47	6.88
S. sobrinus 14757	5.59	4.88	6.06	3.90	2.35	5.20

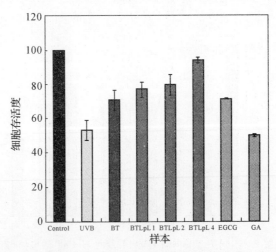

图 3 不同红茶乳酸菌发酵产物对 UVB 引发 HaCaT 细胞损伤的抑制能力

5 茶多酚的鱼油乳剂与微胶囊保健食品

鱼油具有降低心血管疾病发生等功效，但其多元不饱和脂肪酸极易氧化而劣变。将茶多酚与鱼油使用脂微粒法制成乳化液或以微胶囊化技术制成粉状鱼油,不但提高两者的功效、去除腥味,更能有效提高鱼油贮存的氧化安定性。此外,以双层包覆法制成的微胶囊速溶茶粉,物性较干燥茶粉流动性佳(表4),不但可以大幅降低茶的苦涩味,并可能维持茶多酚的贮存安定性(图4)。

表4 不同配方(固形物组成)茶多酚微胶囊物性

配 方	a	b	c	B.d.	T.d.
NO	0.42	0.08	25	0.238	0.310
SE	0.32	0.151	30	0.213	0.278
ME	0.36	0.134	25	0.223	0.301

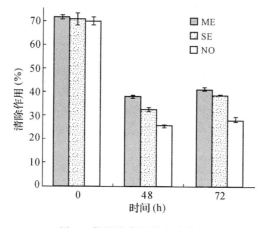

图4 茶微胶囊的贮存安定性

6 茶作为抗癌药物与放疗增敏剂

由茶青中抽出茶多酚,发现其具抑制肠癌与食道癌活性,其中紫芽茶为一般茶叶的5倍以上(表5),可将肠癌细胞阻断在细胞循环初期 $G_0 \sim G_1$ 相,并能导致细胞凋亡。分析发现主要是其含有大量花青素(是台茶12号的135倍以上)与EGCG等儿茶素所致。因此,这种滋味较差但活性高的茶种可往天然药物方向发展;此外紫芽茶其辐射敏感作用也可与Co-60合并使用,能应用于放疗辅助上。

表 5 紫芽茶的抗肠癌活性

样 本	IC50(μg/mL)	
	24h	48h
PTF	55.2±10.8[b]	23.8±6.1[b]
TTF	262.5±13.5[a]	142.6±5.2[a]

参考文献

[1] 林志城.不同制茶阶段乌龙茶液的抗氧化性与其作为包冰液对冻藏鲭鱼肉的保鲜效果[J].台湾农业化学与食品科学,2003,41(5):381-387.

[2] Lin C C, Lin C S. Enhancement of the storage quality of frozen bonito fillet by glazing with tea extracts [J]. Food Control, 2005, 16: 169-175.

[3] Seto Y, Lin C C, Endo Y, et al. Retardation of lipid oxidation in blue sprat by hot water tea extracts [J]. Journal of the Science of Food and Agriculture, 2005, 85: 1119-1124.

[4] Tsai B H, Ho S C, Kan N B, et al. The effect of drinking oolong tea on the oxidative stress of athletes at rest and post-exercise [J]. Journal of Food Science, 2005, 70(9): 581-585.

[5] Lin C C, Lu M J, Chen S J, et al. Heavy fermentation impacts no-suppressing activity of tea in LPS-activated RAW 264.7 macrophages [J]. Food Chemistry, 2006, 98: 483-489.

[6] Huang H L, Lin C C, Jeng K C, et al. Fresh green tea and gallic acid ameliorates oxidative stress in kainic acid-induced status epilepticus[J]. Journal of Agricultural and Food Chemistry, 2012, 60: 2328-2336.

[7] Hsu C P, Shih Y T, Lin B R, et al. Inhibitory effect and mechanisms of an anthocyanins-and anthocyanidins-rich extract from purple-shoot tea on colorectal carcinoma cell proliferation [J]. Journal of Agricultural and Food Chemistry, 2012, 60(14): 3686-3692.

[8] Lin C C, Hsu C P, Chen C C, et al. Anti-proliferation and radiation-sensitizing effect of an anthocyanidin-rich extract from purple-shoot tea on colon cancer cells [J]. Journal of Food and Drug Analysis: Special Issue (ICoFF2011), 2012.

台湾佳叶龙茶萃取物之生物活性成分、机能性及在食品加工上的应用

林圣敦[1]　毛正伦[2]　许清安[1]

1. 弘光科技大学食品科技系，中国 台湾；
2. 中兴大学食品暨应用生物科技系，中国 台湾

摘　要：众所周知，台湾生产之佳叶龙茶具有降血压之保健功效与渐受消费者喜爱的外观、香气与滋味。本研究室近年来针对佳叶龙茶的综合利用方面进一步深入研究。本文首先以不同浓度乙醇（0～95%）及水浴温度（25℃～95℃）针对佳叶龙茶之总酚、儿茶素、γ-氨基丁酸及茶氨酸之萃取率的影响进行探讨，然后分析萃取物之抗氧化性、抗致突变性、抗菌性及抑制血管收缩素转化酶活性，最后将所得佳叶龙茶萃取物应用在食品（速溶茶及戚风蛋糕）的开发。

关键词：佳叶龙茶；生物活性成分；机能性；速溶茶；戚风蛋糕

Bioactive Components and Functional Properties of Taiwan γ-aminobutyric Acid (GABA) Tea Extract, and Their Application in Processed Food

Sheng-Dun Lin[1]　Jeng-Leun Mau[2]　Ching-An Hsu[1]

1. Department of Food Science and Technology, Hungkuang University, Taiwan, China; 2. Department of Food Science and Biotechnology, National Chung-Hsing University, Taiwan, China

Abstract: As we all know, Taiwan GABA (γ-aminobutyric acid) tea not only has the hypertension moderation and bioactive functions but also the appearance, aroma and taste liked by the consumers. Recently, our laboratory focused on the multi-application studies of Taiwan GABA tea. This talk will concentrated firstly at the effect of extraction condition using different ethanol concentration (0～95%) and water bath temperature (25℃～95℃) on the extract components such as total polyphenols, catechins, γ-aminobutyric acid and theanine. Secondly, antioxidant, antimutagenic, antimicrobial and inhibition of angiotensin converting enzyme activities of GABA tea extracts were analyzed. Finally, the GABA tea extract obtained was applied to processed foods such as instant tea and chiffon cake.

Keywords:γ-aminobutyric acid tea; bioactive component; functional property; instant tea; chiffon cake

台湾佳叶龙茶(GABA 茶)是茶业改良场于 1994 年引进日本的技术与茶农一起加以改良,历经五六年的时光发展而成,与一般茶叶比较,其最大不同点是多了"厌氧发酵"的步骤[1]。GABA 茶是 1986 年津志田博士在研究茶氨酸的代谢过程中发现的,研究结果显示新鲜茶青以厌氧(N_2 或 CO_2)处理后,γ-氨基丁酸(γ-aminobutyric acid,GABA)及丙氨酸(alanine)含量会显著提高[2],此因茶叶中麸氨酸(glutamic acid)与天门冬酸(aspartic acid)会经由脱羧及转氨作用,分别转变成 GABA 及丙氨酸,此结果有别于一般有氧状态所制造出来的茶之成分(高量麸氨酸及天门冬酸)[3]。目前商品化 GABA 茶之 GABA 含量为每 100g 干茶需含 150mg 以上,经台湾茶叶改良场探讨不同品种制造 GABA 茶的结果,以台茶十二号、台茶十六号、台茶十七号、青心乌龙及四季春等六品种较稳定,较适宜制造 GABA 茶[1]。

在 GABA 茶的活性成分中,GABA 为非蛋白质构造的氨基酸,是中枢神经系统中非常重要的抑制性神经传送物质之一,主要存在于动物的脑中,具有降血压、改善更年期或糖尿病引起的忧郁症、帮助睡眠及解酒、促进生长激素之分泌及减轻神经退化现象等生理功效[4-5]。除 GABA 外,茶所含的多酚类化合物(主要是儿茶素)具有抗氧化、抗致突变、抗肿瘤、抗菌性及预防心血管疾病等功能[6-7]。茶氨酸亦是一种非蛋白质构造的氨基酸,其含量为茶叶之游离氨基酸中最高,可增加脑中之血清素(serotonin)、多巴胺(dopamine)及 GABA 含量,进而提升神经保护功效[8],也可降低自发高血压老鼠的血压[9]。

茶叶是台湾的重要经济作物,茶农已有能力生产无不良气味且好喝的 GABA茶,然而目前有关 GABA 茶的研究着重在制程、成分分析及保健功效[1,4,10-11],有关 GABA 茶萃取物之最适萃取条件、活性成分、抗氧化性、抗致突变性、抗菌性、抑制血管收缩素转化酶活性及在食品加工上的应用则少见。有鉴于此,为使台湾 GABA 茶加工产品能多元化、具竞争力及受消费大众的喜爱,以提高茶农的经济效益及提升食品加工业者之加工技术,本研究以台湾产制的 GABA 茶为试验材料,探讨不同浓度乙醇(0~95%)及水浴温度(25~

95℃)对 GABA 茶萃取物之萃取率、总酚、儿茶素、GABA 及茶氨酸的影响,同时依据试验所得结果建立最适萃取条件,然后分析其萃取物之抗氧化性、抗致突变性、抗菌性及抑制血管收缩素转化酶活性,最后探讨其在速溶茶及烘焙食品的应用。

1 材料与方法

(1)GABA 茶制造。以台湾南投县名间乡冬季采收的四季春所加工而成的 GABA 茶为试验材料,研磨(<0.5mm)后,以铝箔积层袋密封包装,放置在 −20℃冷冻库冻藏备用。

(2)GABA 茶萃取物制备。称取茶粉末样品 10g 于血清瓶,添加 100mL 之不同浓度(0、25%、50%、75%、95%)的乙醇溶液,于不同水浴温度(25、50、75、95℃)进行振荡萃取 30min、离心、过滤及收集上层液,沉淀物再萃取一次。然后将收集的上层液进行浓缩及真空冷冻干燥,计算样品之不同萃取条件下所得萃取物的萃取率。

(3)即溶茶制备。

(4)戚风蛋糕制备。参考文献[12]介绍的方法。

(5)品质分析。总酚、儿茶素、γ-氨基丁酸、茶氨酸及抗氧化性测定见参考文献[13]。安姆试验、抑菌试验见参考文献[14]。抑制血管收缩素转化酶活性测定见参考文献[15]。

(6)统计分析。将试验结果所得数据利用 SAS 统计套装软件进行变方分析,如果处理间有显著差异,再以邓肯式多变域分析法比较各平均值之间的差异显著性。

2 结果与讨论

2.1 萃取率

萃取率是一种用来测定萃取条件对测试样品之特定成分之萃取效率的方法。就相同萃取温度而言,本研究以 50%乙醇对 GABA 茶进行萃取所得萃取率最高(表 1),最低为 95%乙醇,此结果显示溶剂极性与测试样品之成分溶解性有关。就相同溶剂而言,茶萃取物之萃取率随萃取温度的增加而增加,此可能是由于溶剂进入细胞的扩散性增加,及形成细胞的成分发生去吸附现象所致。整体而言,以 50～95C50E 及 95C25E 之萃取率最高。

2.2 活性成分

2.2.1 总酚

就相同萃取温度而言,当乙醇浓度为 0～50％时,萃取物之总酚含量随乙醇浓度的增加而增加;然而当乙醇浓度为 50％～95％时,则相反之(表1)。就相同溶剂而言,萃取物之总酚含量亦会受到萃取温度的影响。本研究以 50～75C50E 及 75C75E 的总酚含量最高。当考虑萃取率时,GABA 茶以 50％乙醇于 50～95℃进行萃取所得总酚含量较高。

2.2.2 儿茶素

对茶青、绿茶及轻度发酵茶类而言,儿茶素是主要的酚类化合物,这些酚类化合物具有抗氧化性、抗致突变性、防癌性、抑菌性、抗过敏性及减重等[7,16-17]。另外,茶之色泽、风味及味觉亦会受儿茶素的影响[18]。本研究之 GABA 茶萃取物侦测到的儿茶素有 catechins、EGCG、EGC、EC、ECG、GC 及 GCG,其中以 EGCG 的含量最高,其次依序为 EGC、EC、ECG、GC 及 GCG,至于 catechin 及 CG 则未被侦测到。所有萃取物中,以 75C50～75C95E 及 95C25～95C75E 的总儿茶素含量(163.48～167.56mg/g)最高,其中酯型儿茶素(EGCG＋ECG＋GCG)含量(102.92～104.54mg/g)高于非酯型儿茶素(GC＋EGC＋EC)含量(61.75～63.55mg/g)。当考虑萃取率时,GABA 茶以 50％乙醇于 75℃～95℃进行萃取所得总儿茶素含量最高。具体见表1。

2.2.3 γ-氨基丁酸及茶氨酸

有报告指出,试验鼠每天摄取 4mg GABA,经 4 星期后可显著降低血压[19]。摄取含 GABA(50mg)饮料,将有助于降低精神及身体的疲劳,改善解决工作的能力[20]。茶氨酸可调节脑中去甲肾上腺素及血清素的含量,且可降低血压[8-9,21],已有研究指出抗癌剂与茶氨酸在癌症上有协同效应[22]。

表1 干萃取物和茶叶中的萃取率和总酚含量

萃取方法		萃取率 (g/100g)	总萃取酚含量	
			萃取物 (mg/g)	茶叶 (mg/g)
25℃	25C0E	19.75 0.15 JKb	205.75 0.14 O	40.64 0.32 I
	25C25E	24.45 0.45 H	304.57 3.66 M	74.47 0.46 G
	25C50E	28.43 1.10 E	438.07 10.58 E	124.54 1.79 C
	25C75E	22.41 0.53 I	398.30 18.06 F	89.26 1.94 F
	25C95E	5.41 0.30 M	316.62 3.29 L	17.13 1.11 J
50℃	50C0E	20.35 0.42 J	263.04 6.20 N	53.53 2.37 H
	50C25E	26.16 0.82 FG	357.93 2.24 JK	93.63 2.36 F
	50C50E	32.05 0.02 AB	477.63 5.44 AB	153.08 1.64 A
	50C75E	25.02 1.45 GH	463.12 5.72 C	115.87 8.15 DE
	50C95E	10.98 0.11 L	364.42 1.62 IJ	40.01 0.21 I
75℃	75C0E	24.74 0.55 H	357.25 8.35 JK	88.38 4.02 F
	75C25E	29.62 0.90 DE	377.67 1.56 GH	111.87 3.86 E
	75C50E	32.06 0.73 AB	488.69 2.33 A	156.67 4.29 A
	75C75E	29.92 1.18 CD	479.14 7.44 AB	143.36 7.88 B
	75C95E	18.79 1.48 K	370.04 0.46 HI	69.53 5.54 G
95℃	95C0E	26.76 0.23 F	346.36 6.79 K	92.69 2.63 F
	95C25E	31.29 0.08 AB	383.01 9.76 G	119.84 2.76 CD
	95C50E	32.56 0.69 A	469.86 3.32 BC	152.99 2.14 A
	95C75E	30.97 0.13 BC	449.55 3.23 D	139.23 1.56 B

a 萃取率(%)=(干萃取物重量/干茶叶重量)×100%.
b 每个值表示为平均数±标准差($n=3$). 同一列中不同字母表示显著性差异($P<0.05$).
* 数据来源:参考文献[13].

在相同萃取温度下,GABA 茶萃取物之 GABA 及茶氨酸含量会随乙醇浓度的增加而降低(表2)。本研究以 50～95C0E 的 GABA 含量较高,茶氨酸含量则以 25～75C0E 较高。就相同萃取溶剂而言,在 25℃所得萃取物之 GABA 含量较低,而在 50℃～95℃所得萃取物之 GABA 含量较高且不具显著差异。在茶氨酸方面,以水为溶剂,在 25℃～75℃所得萃取物之茶氨酸含量较高,而 75C95E 的茶氨酸含量较低。当考虑萃取率时,GABA 茶以 50%乙醇及 50℃～95℃或 25%乙醇及 95℃进行萃取所得 GABA 含量较高。茶氨酸则以 50%乙醇及 50℃～95℃ 或 25％乙醇及 75℃～95℃ 进行萃取含量较高。

依据上述活性成分试验结果及考虑萃取率,为获取最高量的儿茶素、GABA 及茶氨酸,50％乙醇及 75℃～95℃ 萃取温度是最适萃取条件,因此本研究以 75C50E 进行以下机能性试验及其在加工食品上的应用。

表2 氨基丁酸及茶氨酸在干萃取物和茶叶中的含量 （单位:mg/g）

萃取方法		γ-氨基丁酸		茶氨酸	
		萃取物	茶 叶	萃取物	茶 叶
25℃	25C0E	5.53 0.41 EFa	1.092 0.073 G	7.90 0.09 A	1.560 0.029 GH
	25C25E	5.31 0.38 F	1.298 0.115 EF	7.12 0.14 C	1.741 0.065 E
	25C50E	4.70 0.33 G	1.336 0.042 E	6.54 0.35 EF	1.859 0.028 D
	25C75E	4.57 0.22 G	1.024 0.074 G	6.77 0.11 DE	1.517 0.012 H
	25C95E	3.40 0.14 I	0.184 0.003 J	6.19 0.29 G	0.335 0.003 K
50℃	50C0E	6.42 0.26 AB	1.306 0.026 EF	8.00 0.17 A	1.628 0.001 FG
	50C25E	5.98 0.03 CD	1.564 0.056 C	7.54 0.03 B	1.972 0.068 BC
	50C50E	5.67 0.28 DEF	1.817 0.089 AB	6.45 0.15 EFG	2.067 0.047 A
	50C75E	4.81 0.19 G	1.203 0.117 F	6.15 0.34 GH	1.539 0.005 H
	50C95E	3.86 0.21 H	0.424 0.018 I	5.88 0.11 HI	0.646 0.018 J
75℃	75C0E	6.46 0.01 A	1.598 0.038 C	7.74 0.06 AB	1.915 0.027 CD
	75C25E	5.94 0.36 CD	1.759 0.052 B	7.10 0.02 C	2.103 0.059 A
	75C50E	5.67 0.12 DEF	1.818 0.078 AB	6.35 0.13 FG	2.036 0.004 AB
	75C75E	4.83 0.17 G	1.445 0.006 D	5.62 0.06 I	1.682 0.084 EF
	75C95E	3.81 0.18 H	0.716 0.090 H	4.37 0.18 K	0.821 0.031 I
95℃	95C0E	6.48 0.02 A	1.734 0.011 B	7.05 0.01 CD	1.887 0.019 D
	95C25E	6.09 0.05 BC	1.906 0.019 A	6.56 0.08 EF	2.053 0.020 A
	95C50E	5.85 0.02 CDE	1.905 0.034 A	6.24 0.32 FG	2.032 0.060 AB
	95C75E	4.85 0.14 G	1.502 0.049 CD	5.26 0.10 J	1.629 0.036 FG

a 每个值表示为平均数±标准差（$n=3$）. 同一列中不同字母表示显著性差异（$P<0.05$）.

* 数据来源:参考文献[13].

2.3 机能性

2.3.1 抗氧化性

GABA 茶以 50％乙醇在 75℃下进行萃取所得萃取物之清除 DPPH 自由基能力、还原力及螯合亚铁离子能力都随其浓度增加而增加,其 EC_{50} 值分别为 $(11.02\pm0.12)\mu g/mL$、$(66.84\pm0.68)\mu g/mL$ 及 $(1.635\pm0.023)mg/mL$。抗坏血酸、丁基羟基甲氧苯(BHA)及 α-生育醇之清除 DPPH 自由基的 EC_{50} 值分别为$(5.30\pm0.35)\mu g/mL$、$(5.64\pm0.06)\mu g/mL$ 及 $(12.29\pm3.85)\mu g/mL$。虽然 GABA 茶萃取物清除 DPPH 自由基能力低于抗坏血酸及 BHA,但高于 α-生育醇。有研究报告指出儿茶素在绿茶中赋予抗氧化效果的顺序是 EGCG ≈ EGC≫ECG＝EC＞C[23]。由此可知,EGCG 和 EGC 之清除 DPPH 能力较强,这指出了儿茶素结构中,C 环第 3 个碳位置接没食子酸盐基团在清除 DPPH 自由基的能力上扮演了重要的角色[24]。Horzic 等[25]也指出,分子结构具有最多—OH 基团就能展现最高的抗氧化活性。

抗坏血酸、BHA 及 α-生育醇之还原力的 EC_{50} 值分别为 $(12.74\pm0.37)\mu g/mL$、$(27.34\pm0.76)\mu g/mL$ 及 $(90.04\pm9.10)\mu g/mL$。GABA 茶乙醇萃取物之还原力明显高于 α-生育醇。Salah 等[26]指出,类黄酮化合物具有还原力可能是因为 C 环上 2,3-双键与 4 氧结合的功能,与 B 环上有相邻的—OH 基团。

金属离子的促氧化作用经常是造成脂质过氧化的主要因素。在多种金属离子中,Fe^{2+} 经常是最具影响力的助氧化剂[27],其会促进脂质氧化作用的进行。EDTA 之螯合亚铁离子能力的 EC_{50} 值为 $(5.01\pm0.01)\mu g/mL$,至于抗坏血酸、BHA 及 α-生育醇则测不到具有螯合亚铁离子能力。值得注意的是,本研究以水萃取物(EC_{50} 值:1.004～1.128mg/mL)螯合亚铁离子能力显著优于乙醇萃取物,而以 95％乙醇萃取物(EC_{50} 值:7.294～10.211mg/mL)最弱。

由上述结果可知,虽然抗坏血酸、α-生育醇及 BHA 是很好的抗氧化剂,且抗坏血酸及 α-生育醇是营养添加剂的来源,但它们属于食品添加物,在食品上的使用范围及用量都受到限制。然而 GABA 茶不但可作为食品成分之一,且在食品上的使用范围及用量未被限制。

2.3.2 抗致突变性

很多酚类化合物具有抑菌性,本研究 GABA 茶之 75C50E 具有高含量的总酚及儿茶素,因此在进行致突变性及抗致突变性试验之前,须先测试此萃取物之毒性,以了解其对菌株之生存及回复突变是否有影响。结果显示,无论有无添加 S9 混合物,GABA 茶乙醇萃取物在测试剂量下(0.25～5mg/plate),对沙门氏杆菌 TA98 及 TA100 之菌数均维持在对照组的 80% 以上,因此 GABA 茶乙醇萃取物对于沙门氏杆菌 TA98 及 TA100 没有毒性[28],故 GABA 茶乙醇萃取物以测试剂量 0.25～5mg/plate 范围,进行致突变性及抗致突变性分析。

本研究以 Ames 试验探讨 GABA 茶乙醇萃取物是否有致突变性,结果显示无论有无添加 S9 混合物,GABA 茶乙醇萃取物在所测试的剂量范围内(0.25～5mg/plate),对沙门氏杆菌 TA98 及 TA100 之致突变性比率分别为 0.92～1.14 及 0.96～1.02,没有超过自发性回复突变菌属两倍以上[29],所以 GABA 茶乙醇萃取物应不具致突变性。

在抗致突变性方面,本研究以 2-AA 作为间接致突变原,4-NP 及 NaN$_3$ 作为直接致突变原。针对沙门氏杆菌 TA98 而言,GABA 茶乙醇萃取物在测试剂量范围内可降低 4-NP 及 2-AA 的致突变性,且对 4-NP 及 2-AA 致突变性的抑制能力会随着添加剂量的增加而逐渐提高(图 1(a)),在 GABA 茶乙醇萃取物剂量为 0.25～5mg/plate 时,其抑制率分别为 8.31%～44.18% 及 53.03%～100%。针对沙门氏杆菌 TA100 而言,GABA 茶乙醇萃取物在测试剂量范围(0.25～5mg/plate)内可降低 NaN$_3$ 及 2-AA 的致突变性,且对 NaN$_3$ 及 2-AA 致突变性的抑制能力亦是随着 GABA 茶乙醇萃取物添加剂量的增加而逐渐提高,在剂量为 0.25～5mg/plate 时,其抑制率分别为 5.58%～35.28% 及 33.82%～100%(图 1(b))。

由上述结果可知,GABA 茶乙醇萃取物无毒性及抗致突变性,其在沙门氏杆菌 TA98 及 TA100 菌株系统中,对直接致突变剂(4-NP、NaN$_3$)的抗致突变性虽然只达到中度的抑制效果,但对需要 S9 混合物活化之致突变剂(2-AA)的抑制则有极佳的效果,此可能是因其含有酚类化合物,而有抗致突变能力,至于其作用机制仍需进一步探讨。整体而言,GABA 茶乙醇萃取物在食用上是安全的。

2.3.3　抑菌试验

本研究分析 GABA 茶乙醇萃取物对常见食品病原体——*V. parahaemolyticus*, *S. aureus*, *B. cereus*, *S. typhimurium* 及 *E. coli* 的抑制情形(见表 3)。结果显示,以 *V. parahaemolyticus* 对茶萃取物最敏感,其次依序为 *S. aureus* 及 *B. cereus*(表 3),然而 *S. typhimurium* 及 *E. coli* 的生长不受 GABA

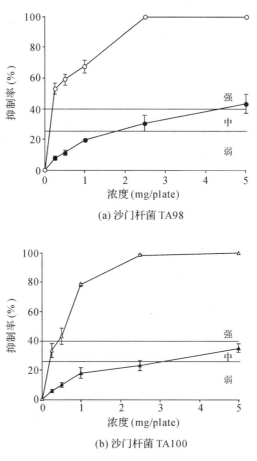

图 1　GABA 茶萃取物对沙门杆菌 TA98 和 TA100 的抑制率

数据来源：参考文献[14].

茶萃取物（1～6mg/mL）的影响。另外，S. aureus 及 B. cereus 为 G（＋）菌，V. parahaemolyticus，S. typhimurium 及 E. coli 为 G（－）菌，由抑菌试验结果，除 V. parahaemolyticus 外，G（＋）菌的生长较易被茶萃取物所抑制，此应是茶萃取物所含的酚类化合物所致，尤其是儿茶素[30-31]，至于 G（－）菌则否。

表3 GABA茶乙醇萃取物对常见食物病原体的抑制作用

菌 种	浓度(mg/mL)	细菌种数(log CFU/mL)			抑制率(%)[b]	
		0	24h	48h	24h	48h
V. parahaemolyticus	0	5.04±0.02[a]	8.74±0.05	8.69±0.02		
	1				100	100
	2				100	100
	3				100	100
	4				100	100
	5				100	100
	6				100	100
S. aureus	0	5.89±0.5	9.13±0.05	9.23±0.06		
	1	6.84±0.07	5.93±0.02	25.08	35.75	
	2	4.00±0.05	4.20±0.03	56.19	54.50	
	3				100	100
	4				100	100
	5				100	100
	6				100	100
B. cereus	0	4.90±0.05	9.96±0.08	10.02±0.11		
	1	6.45±0.04	6.62±0.04	35.24	33.93	
	2	3.32±0.04	3.40±0.03	66.67	66.07	
	3	1.24±0.03	1.29±0.05	87.55	87.13	
	4				100	100
	5				100	100
	6				100	100

[a] 每个值表示为平均数±标准差($n=3$). —：表示未检测到.
[b] 抑制率(%)=(log CFU/mL$_{control}$ − log CFU/mL$_{treated}$)/(log CFU/mL$_{control}$).
* 数据来源：参考文献[14].

2.3.4 抑制血管收缩素转化酶活性

血管收缩素转化酶（ACE）为人体调节血压之重要酵素，可将血管收缩素Ⅰ

催化形成血管收缩素Ⅱ,而引起血压的上升。因此,食品能抑制 ACE 活性将可展现抗高血压功能。本研究表明,GABA 茶萃取物抑制 ACE 活性的效果随其浓度增加而显著增加(图 2),此应与茶萃取物所含的儿茶素、GABA、茶氨酸及丙氨酸有关[32]。Persson 等[33]曾指出,茶叶对 ACE 活性的抑制作用和黄烷醇类有关,如 EC、EGC、ECG 和 EGCG,且—OH 基团数目的增加和双键氧的存在,可以增进对 ACE 活性的抑制作用。Actis-Goretta 等[34]也指出,EGC 对 ACE 的抑制程度与黄烷醇结构中—OH 基团的数量有关,其可以有效地和 ACE 蛋白质产生氢键键结。

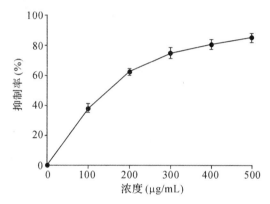

图 2　50％ GABA 茶乙醇萃取物对 ACE 的抑制率

2.4　在加工食品上的应用

2.4.1　GABA 速溶茶

在国内外食品市场上,"饮料"是与消费者最密切的食品之一,然而其具有重量重、体积大、玻璃瓶身包装者易破碎、携带不方便及运费贵等缺点。因此,已有厂商将市售饮料或传统中药开发成重量轻、体积小、包装巧及运费低的"速溶饮料",在讲求便利的今日,已成为饮料市场发展的一大趋势,同时深受速食餐饮业者所重视及消费者所喜爱,发展前景是可估的。但速溶饮在包装贮藏期间,其粉末颗粒经常出现潮解凝结成块,且出现外观、色泽及风味改变等品质变化之现象,此为产业界长期面对之关键问题,虽借助于喷雾干燥技术,亦无法完全获得解决。因此,包覆物(或载体)的选择、加工条件及包装条件等,都有探讨的必要。

在台湾,过去因加工技术的限制,导致速溶茶的风味不佳,且其在贮放期间及开封后易发生结块而丧失商品价值。加上消费者愈来愈重视身体的保健,因此,本研究团队以 GABA 茶萃取物为试验样品,具机能性(具有调理肠道、调节饭后血糖值、降低血清胆固醇及三酸甘油酯等作用)的粉末为载体,探讨"GABA速溶茶"的适当产制条件。过程如下:

佳叶龙茶→粗碎→萃取→过滤→浓缩→流动层喷雾干燥→粉末→包装→装盒→封膜。

用途:热水冲泡、冷水冲泡、直接口含、添加于乳酸饮料、以水溶解、加入米中煮饭等。

2.4.2 GABA 戚风蛋糕

蛋糕深受消费大众的喜爱,但常因含糖量高和甜度太甜而受人诟病,尤其是饮食中若摄取过多的糖常会引起肥胖,进而引发健康上的问题,例如高三酸甘油酯血症、高胰岛素血症、胰岛素抗性。故寻找对人体健康有帮助的甜味剂取代传统蛋糕常用的蔗糖遂成为一重要之研究课题。综观目前市售具机能性的甜味剂中,应以寡糖类糖浆较受重视,尤其是异麦芽寡糖糖浆(IMOS),此因使用酵素合成 IMO 的产出率可高达 50% 以上,价格低,在产品中稳定性不受产品之 pH 值的影响,保湿性高,有防止淀粉回凝功能,且生理能量值(2.7~3.2 Kcal/g)低于蔗糖[12]。其在保健功效方面,可改善排便障碍、能显著降低粪便中粪臭素含量(如吲哚及对甲酚)、可显著降低血清总胆固醇及低密度脂蛋白胆固醇含量、可促进肠道中双叉杆菌的增生、不被蛀牙菌利用[12]。

因此,本研究团队在戚风蛋糕制程中,采用不同 IMOS/蔗糖含量及烘焙时间,利用反应曲面法(RSM)探讨其对产品品质的影响,并根据喜好性品评试验,探讨 IMO 戚风蛋糕的适当产制条件,其次探讨贮藏条件对蛋糕品质的影响,最后进行人体试验[12,35-36]。结果显示,依据喜好性感官品评试验结果,IMO 戚风蛋糕的适当产制条件为:IMOS/蔗糖=22.93/77.07~56.96/43.04 及烘焙时间=30.1~31.7min(图 3),其体积及喜好性感官品质(包括蛋糕颜色、甜味、柔软性、风味及整体喜好性),受消费者的喜好程度高于全部以蔗糖所制的产品(IMOS/sucrose=0/100)。在贮藏试验方面,若以消费者型品评员对蛋糕之"有点喜欢"的喜好程度,与蛋糕之总生菌数低于 10^3 CFU/g,且不得检出大肠杆菌群及霉菌作为产品之贮藏寿命的依据,则贮藏在 5℃、25℃ 及 35℃ 的蛋糕之贮藏寿命分别为 3d、1d 及 12h。在人体试验方面,以 IMO 取代 50% 蔗糖产制 IMO 戚风蛋糕,结果显示 IMO 组受试者每天食用 71g IMO 戚风蛋糕(含 10g IMO 活性成分,热量 224kcal),经 6 星期后,其血清总胆固醇及低密度脂蛋白胆固醇、粪便中吲哚及对甲酚含量等均显著下降,且便秘情形获得改善($P<0.05$)(表4、表5),但禁食血糖、血清之高密度脂蛋白胆固醇、三酸甘油酯、钙、镁及铁含量的变化无显著差异($P>0.05$)。而食用蔗糖戚风蛋糕的受试者(每天食用68g戚风蛋糕,热量224kcal)之禁食血糖、血清脂质及矿物质、粪便之粪臭素含量的变化均无显著差异,且便秘情形未获改善($P>0.05$)。综上所述,以IMO取代蔗糖在蛋糕中所扮演的角色是可行的,且具保健功效。

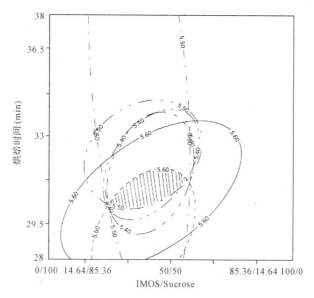

图3 最受欢迎的戚风蛋糕区域

数据来源:参考文献[12]。

本研究团队继续以 IMO 戚风蛋糕为试验样品,以 GABA 茶乙醇萃取物取代面粉 3.2%(GABA1)、6.4%(GABA2)及 9.6%(GABA3)(相当于以 GABA 茶粉取代 10%、20% 及 30% 面粉)制作蛋糕,探讨其对蛋糕之物化及感官品质的影响,并探讨蛋糕品质在贮藏期间的变化。结果显示(表6),蛋糕面糊之黏度及比重,烘焙蛋糕 GABA、caffeine、GC、ECG、EC、EGCG、GCG、EGC 含量,蛋糕之硬度及内部 Hunter a 值随着 GABA 茶乙醇萃取物取代量的增加而增加;然而,蛋糕面糊之 pH 值、烘焙蛋糕之水分、体积、外皮之 Hunter L, a, b 值及内部之 Hunter L, b 值的变化,则相反之。在喜好性感官品评方面(表7),蛋糕外皮及内部颜色受消费者的喜好程度以对照组及 GABA1 组最高,而蛋糕之甜味、风味、质地及整体喜好性受消费者喜好的程度则以 GABA1 组最高。在贮藏试验方面，GABA1、GABA2 及 GABA3 蛋糕在 25℃ 贮藏 3 天的总生菌数（1.67～4.26×10^2 CFU/g）显著低于对照组（1.06×10^5 CFU/g），所有样品于 5℃ 贮藏 7d 的样品之总生菌数均低于 9.32×10^2 CFU/g，所有样品均检测不到大肠杆菌群、大肠杆菌。所有蛋糕之硬度随贮藏时间的增加而显著增加，但样品之水分含量、水活性、Hunter L，a，b 值、GABA、caffeine 及 6 种儿茶素含量的变化不受贮藏温度及时间的影响。所有样品之整体喜好性受消费者的喜好程度随贮藏时间的增加而显著降低。若以消费者型品评员对蛋糕之"有点喜欢"的喜好程度，与蛋糕之总生菌数低于 10^3 CFU/g，且不得检出大肠杆菌作为产品之贮藏寿命的依据，则贮藏在 5℃ 及 25℃ 的蛋糕之贮藏寿命分别为 2d 及 5d。

表4　IMO 戚风蛋糕对空腹血糖、血脂和矿物质的影响

参　数	IMOS		蔗　糖	
	前	后	前	后
FBG (mg/dL)	78.0±8.6	79.4±14.3	77.7±10.2	75.3±7.1
TC (mg/dL)	220.7±24.2^a	201.7±21.1^b	215.0±12.1^a	217.8±15.6^a
HDL-C (mg/dL)	62.6±14.9	61.4±14.3	54.1±12.1	54.5±12.0
LDL-C (mg/dL)	139.9±33.0^a	121.9±27.1^b	139.4±11.9^a	142.2±15.7^a
TG (mg/dL)	88.7±38.6	92.3±46.0	107.1±40.1	105.2±41.1
Ca (mg/dL)	9.5±0.2	9.2±0.2	9.3±0.1	9.4±0.2
Mg (mg/dL)	2.0±0.3	1.9±0.4	2.1±0.4	1.9±0.3
Fe (g/dL)	163±32	154±39	158±38	151±35

所有值为平均数±标准差。同一行中标有 a、b 上标的为有显著性差异（$P<0.05$）。
数据来源：参考文献[36]。

表5　IMO 戚风蛋糕对便秘的作用

参　数	IMOS		蔗　糖	
	前	后	前	后
总便秘（Times/day）	0.58±0.16^a	0.21±0.09^b	0.62±0.19^a	0.51±0.23^a
便秘值	0.89±0.36^a	0.36±0.18^b	0.87±0.41^a	0.73±0.38^a
吲哚 (g/g dry feces)	146.70±39.62^a	69.82±20.03^b	139.64±40.31^a	130.30±41.7^a
对甲酚 (g/g dry feces)	136.01±56.32^a	42.90±19.24^b	151.27±48.30^a	137.79±42.67^a

所有值为平均数±标准差。同一行中的标有 a、b 上标的为有显著性差异（$P<0.05$）。
数据来源：参考文献[36]。

表6 GABA 茶萃取物蛋糕物化特征

物化特征		处理方法			
		对照组	GABA1	GABA2	GABA3
蛋糕面糊	黏度（cp）	12458±125[d]	15395±227[c]	16989±128[b]	19526±261[a]
	比 重	0.442±0.026[d]	0.508±0.031[c]	0.624±0.021[b]	0.702±0.017[a]
	pH	7.32±0.02[a]	7.00±0.02[b]	6.94±0.01[c]	6.81±0.01[d]
	湿度（%）	34.78±0.37[a]	34.76±0.26[a]	34.86±0.20[a]	34.52±0.31[a]
烘焙蛋糕	湿度（%）	34.03±0.82[a]	32.09±0.41[b]	30.22±0.96[c]	30.12±0.85[c]
	水 分	0.920±0.001[a]	0.929±0.008[a]	0.928±0.006[a]	0.930±0.008[a]
	体积（mL）	1570±24[a]	1385±31[b]	1237±34[c]	1201±26[c]
外皮	L	55.99±0.55[a]	47.61±0.62[b]	46.46±0.42[c]	43.24±0.42[d]
	a	11.37±0.09[a]	9.41±0.14[b]	8.90±0.20[c]	8.50±0.14[d]
	b	16.94±0.23[a]	12.56±0.23[b]	11.80±0.28[c]	10.03±0.24[d]
屑	L	72.68±2.62[a]	46.71±1.41[b]	41.52±1.15[c]	37.66±0.91[d]
	a	−0.77±0.31[c]	4.31±0.35[b]	5.43±0.35[a]	5.35±0.44[a]
	b	15.31±1.37[a]	11.83±0.74[b]	9.72±0.77[c]	7.15±0.57[d]
硬度（g）		131.6±14.3[c]	147.6±4.4[c]	169.4±9.3[b]	187.4±8.2[a]
GABA（mg/100g）		nd3	3.49±0.23[c]	7.29±0.38[b]	10.97±0.64[a]
Caffeine（mg/100g）		nd	44.57±0.98[c]	81.17±2.67[b]	119.65±2.39[a]
GC（mg/100g）		nd	1.23±0.17[c]	2.96±0.37[b]	4.82±0.59[a]
EGC（mg/100g）		nd	22.87±1.38[c]	45.86±2.12[b]	76.82±3.39[a]
EC（mg/100g）		nd	12.48±0.31[c]	22.96±0.63[b]	38.92±1.37[a]
EGCG（mg/100g）		nd	54.63±2.13[c]	101.37±3.24[b]	162.16±3.29[a]
GCG（mg/100g）		nd	0.97±0.27[c]	2.34±0.31[b]	2.94±0.53[a]
ECG（mg/100g）		nd	11.52±0.35[c]	21.07±0.58[b]	35.01±0.84[a]

L,a,b 的平均值，硬度从10开始，其他平均值从4开始，同一行标有同一字母的值无显著性差异（$P>0.05$）。nd 为未检测到。对照组、GABA1、GABA2 和 GABA3 戚风蛋糕指含 0、3.2%、6.4%和 9.6% GABA 茶萃取物

表7 戚风蛋糕喜好性感官品评

处理方式[1]	对照组	GABA1	GABA2	GABA3
外皮颜色	5.6±0.5[A2]	5.6±0.7[A]	4.8±0.8[B]	4.4±0.5[C]
屑颜色	5.6±0.5[A]	5.4±0.6[A]	4.8±0.4[C]	4.4±0.5[D]
甜 度	5.4±0.5[B]	5.6±0.5[A]	4.2±0.8[C]	4.0±0.6[D]
味 道	5.1±0.7[B]	5.6±0.6[A]	3.0±0.6[C]	2.6±0.5[D]
纹 理	5.4±0.5[B]	5.6±0.5[A]	3.8±0.4[C]	3.2±0.8[D]
总 体	5.2±0.7[B]	5.6±0.6[A]	3.4±0.5[C]	2.8±0.4[D]

对照组、GABA1、GABA2 和 GABA3 戚风蛋糕指含 0、3.2%、6.4% 和 9.6% GABA 茶萃取物。

七点评价方法中,1 为不喜欢、4 为中性、7 为喜欢,所有值表示为平均值±标准差($n=86$),大写字母代表有显著性差异($P<0.05$)。

3 结 论

本研究除建立 GABA 茶萃取物之适当产制条件外,还发现此萃取物具有良好的抗氧化性、抗致突变性、抗菌性及抑制血管收缩素转化酶活性,同时也成功地利用 GABA 茶开发出具机能性且受消费者喜爱的 GABA 速溶茶及 GABA 茶蛋糕。

致 谢

本研究承蒙国科会(NSC 89-2313-B-241-006、NSC-97-2221-E-241-001、NSC 98-2221-E-241-010)、弘光科技大学(计划编号 HKC-90-B-009)及易茗企业有限公司(HK95-010)经费补助,特此致谢。

参考文献

[1] 蔡永生. 佳叶龙茶之制造[J]. 兴大农业,2004,49,3-8.

[2] 津志田藤二郎. 血压降下作用を强化した茶の制造[J]. 茶,1986,39(12):1-6.

[3] Millin D J, Rustidge D W. Tea manufacture[J]. Process Biochemistry, 1967, 6: 9-13.

[4] 王雪芳. GABA 之理功效[J]. 兴大农业,2004,49:19-23.

[5] Abdou A M, Higashiguchi S, Horie K, et al. Relaxation and immunity enhancement effects of gamma-aminobutyric acid (GABA) administration in humans[J]. Biofactors, 2006, 26: 201-208.

[6] Almajano M P, Carbó R, López Jiménez, et al. Antioxidant and antimicrobial activities of tea infusions[J]. Food Chemistry, 2008, 108: 55-63.

[7] Gupta J, Siddique Y H, Beg T, et al. A review on the beneficial effects of tea polyphenols on human health[J]. International Journal of Pharmacology, 2008, 4: 314-338.

[8] Nathan P J, Lu K, Gray M, et al. The neuropharmacology of L-theanine (N-ethyl-L-glutamine): a possible neuroprotective and cognitive enhancing agent[J]. Journal of Herbal Pharmacotherapy, 2006, 6: 21-30.

[9] Yokogoshi H, Kato Y, Sagesaka Y M, et al. Reduction effect of theanine on blood pressure and brain 5-hydroxyindoles in spontaneously hypertensive rats[J]. Bioscience, Biotechnology and Biochemistry, 1995, 59, 615-618.

[10] 区少梅, 王雪芳. 台湾佳叶龙茶之生理功效: 不但可降血压与降血脂, 更可延年益寿[J]. 食品资讯, 2007, 217: 46-50.

[11] Wang, H F, Tsai, Y S, Lin, M L, et al. Comparison of bioactive components in GABA tea and green tea produced in Taiwan[J]. Food Chemistry, 2006, 96: 648-653.

[12] 林圣敦, 林淑妮. 异麦芽寡糖糖浆取代戚风蛋糕之蔗糖的研究[J]. 台湾农化与食品科学, 2001, 39: 76-86.

[13] Lin S D, Mau J L, Hsu C A. Bioactive components and antioxidant properties of γ-aminobutyric acid (GABA) tea leaves[J]. LWT-Food Science and Technology, 2012, 46(1): 64-70.

[14] Mau J L, Chiou S Y, Hsu C N, et al. Antimutagenic and antimicrobial activities of γ-aminobutyric acid (GABA) tea extract[C]. Yang D. International Proceedings of Chemical, Biological and Environmental Engineering: 2012 International Conference on Nutrition and Food Sciences. Singapore: IACSIT Press, 2012, 39: 178-182.

[15] 林圣敦, 许清安, 李天佑. 茶叶萃取物之活性成分及抑制血管收缩素转化酶活性的能力[C]. 中国茶叶学会. 第七届海峡两岸茶业学术研讨会论文集, 2012: 610-618.

[16] Kao Y H, Hiipakka R A, Liao S. Modulation of endocrine systems and

food intake by green tea epigallocatechin gallate[J]. Endocrinology, 2000, 141: 980—987.

[17] Lin S D, Liu E H, Mau J L. Effect of different brewing methods on antioxidant properties of steaming green tea[J]. LWT-Food Science and Technolog, 2008, 41(9): 1616—1623.

[18] Thorngate J H, Noble A C. Sensory evaluation of bitterness and astringency of 3R(−)-epicatechin and 3S(+)-catechin[J]. Journal of the Science of Food and Agriculture, 1995(67), 531—535.

[19] Abe Y, Umemura S, Sugimoto K, et al. Effect of green tea rich in gamma-aminobutyric acid on blood pressure of Dahl salt-sensitive rats[J]. American Journal of Hypertension, 1995, 8: 74—79.

[20] Kanehira T, Nakamura Y, Nakamura K, et al. Relieving occupational fatigue by consumption of a beverage containing γ-aminobutyric acid[J]. Journal of Nutritional Science and Vitaminology, 2011, 57: 9—15.

[21] Kakuda T. Neuroprotective effects of the green tea components theanine and catechins[J]. Biological & Pharmaceutical Bulletin, 2002, 25: 1513—1518.

[22] Sugiyama T, Sadzuka Y. Theanine and glutamate transporter inhibitors enchance the antitumor efficacy of chemotherapeutic agents[J]. Biochimica et Biophysica Acta, 2003, 1653: 47—59.

[23] Rice-Evans C A, Miller, N J, Paganga G. Structure-antioxidant activity relationships of flavonoids and phenolic acids[J]. Free Radical Biology and Medicine, 1996, 20: 933—956.

[24] Guo Q, Zhao B, Shen S, et al. ESR study on the structure-antioxidant activity relationship of tea catechins and their epimers[J]. Biochimica et Biophysica Acta, 1999, 1427: 13—23.

[25] Horzic D, Komes D, Belscak A, et al. The composition of polyphenols and methylxanthines in teas and herbal infusions[J]. Food Chemistry, 2009, 115: 441—448.

[26] Salah N, Miller N J, Paganga G, et al. Polyphenolic flavanols as scavengers of aqueous phase radicals and as chain-breaking antioxidants[J]. Archives of Biochemistry and Biophysics, 1995, 322: 339—346.

[27] Yamaguchi R, Tatsumi M A, Kato K, et al. Effect of metal salts and fructose on the autoxidation of methyl linoleate in emulsions[J]. Agri-

cultural and Biological Chemistry, 1988, 52: 849—850.

[28] Waleh N S, Rapport S J, Mortelmans K E. Development of a toxicity test to be coupled to the Ames *Salmonella* assay and the method of construction of the required strains[J]. Mutation Research, 1982, 97: 247—256.

[29] Ames B N, McCann J, Yamasaki E. Methods for detecting carcinogens and mutagens with the *Salmonella*/mammalian-microsome mutagenicity test[J]. Mutation Research, 1975, 31(6): 347—364.

[30] Cushnie T P T, Lamb A J. Antimicrobial activity of flavonoids[J]. International Journal of Antimicrobial Agents, 2005, 26: 343—356.

[31] Toda M, Okubo S, Ikigai H, et al. Antibacterial and antihaemolysinat activities of tea catechins and their structural relative[J]. Japanese Journal of Bacteriology, 1990, 45: 561—566.

[32] 林智, 大森正司. γ-胺基丁酸成分对大鼠血管紧张素Ⅰ转换酶素(ACE)活性的影响[J]. 茶叶科学, 2002, 22(1): 43—46.

[33] Persson Ingrid A-L, Josefsson M, Persson K, et al. Tea flavanols inhibit angiotessin-converting enzyme activity and increase nitric oxide production in human endothelial cells[J]. Journal of Pharmacy and Pharmacology, 2006, 58: 1139—1144.

[34] Actis-Goretta L, Ottaviani J I, Fraga C G. Inhibition of angiotensin converting enzyme activity by flavanol-rich foods[J]. Journal of Agricultural and Food Chemistry, 2006, 54: 229—234.

[35] 林圣敦, 蔡玉铃, 牟玉如. 异麦芽寡糖戚风蛋糕贮藏期间品质变化之探讨[J]. 台湾农化与食品科学, 2002, 40: 181—188.

[36] 林圣敦, 林柏松, 王雪芳等. 异麦芽寡糖戚风蛋糕对人体血液生化值、便秘及粪臭素之影响[J]. "中华民国"营养学会杂志, 2005, 30: 108—115.

以尼古丁受体分子抑制模式探讨儿茶素抑制抽烟诱发乳癌的分子机制

何元顺[1,2]

1. 台北医学大学医技学系,中国 台湾;
2. 台北医学大学癌症卓越研究中心,中国 台湾

摘 要:本实验室在乳癌患者($n=270$ 例)组织中发现尼古丁受体有过度表现的现象,而且发现这些尼古丁受体高表现者,其病程多集中于末期(第三至四期),愈后五年内存活率均不好。因此,有必要自天然物质或食品中寻找有效的尼古丁受体抑制物,作为缓解烟害的健康保健食品。本研究主要目的在探讨由抽烟所引的乳腺肿瘤生长可否透过茶饮料成分中多酚物质如 EGCG 摄取来获得缓解与预防。本实验室发现尼古丁($>0.1mol/L,24h$)与女性荷尔蒙(estrogen,$>1nmol/L,24h$)可以在乳癌细胞(MCF-7)诱发尼古丁受体 9-nAChR mRNA 的高度表现,我们的实验结果证实茶叶萃取物或儿茶素成分(EGCG,$1mol/L$)显著地抑制由尼古丁($>0.1mol/L,24h$)与女性荷尔蒙(estrogen,$>1nmol/L,24h$)所诱发的尼古丁受体表现。人体试验进一步证实口服茶叶粉末成分($1500mg/d$)可以抑制人类白细胞之尼古丁受体表现,我们证实 EGCG 可以抑制乳癌细胞在琼胶上形成群落的能力。进一步研究发现 EGCG 可以抑制[3H]-尼古丁与尼古丁受体结合作用。本试验结果将有助于提供乳癌患者更好的健康保健策略,透过分子生物学的学理探讨,更有助于了解茶饮料的保健原理。

关键词:抽烟;乳癌;茶多酚;尼古丁受体

Tea-polyphenol (—)-Epigallocatechin-3-gallate as a Good Candidate to Prevent Smoking-induced Human Breast Cancer Cells Proliferation through Inhibition of 9-nicotinic Acetylcholine Receptor

Ho Yuan Soon[1,2]

1. School of Medical Laboratory Science and Biotechnology, Taipei Medical University, Taiwan, China;
2. Center of Excellence for Cancer Research, Taipei Medical University, Taiwan, China

Abstract: The aim of this research was to explore whether the tea exrtact and its polyphenol (—)-epigallocatechin-3-gallate (EGCG) could be used as a potential agent for blocking smoking (nicotine, Nic)-or hormone (estradiol, E2)-induced breast cancer cell proliferation through inhibition of a common signaling pathway. We found that Nic ($>0.1mol/L$, 24h) and E2 ($>1nmol/L$, 24h) significantly increased 9-nicotinic acetylcholine (9-nAChR) mRNA and protein expression levels in human breast cancer (MCF-7) cells. We then hypothesized that tea or agents with inhibitory effects on α9-nAChR protein levels could be used to block the Nic-or E2-mediated carcinogenic signals. Our results indicated that treatment with tea extract or polyphenol (EGCG, $1mol/L$) profoundly decreased Nic-and E2-induced MCF-7 proliferation by down regulating 9-nAChR expression. We further demonstrated that oral administration of tea powder (1500mg/d) profoundly inhibited the 9-nAChR expression in human peripheral blood cells. Pretreatment with EGCG abrogated the Nic-and E2-induced 9-nAChR soft agar colony formation. We further demonstrated that combined treatment with EGCG profoundly inhibited [3H]-Nic/9-nAChR binding activity in breast cancer cells. This study provides the novel antitumor mechanisms of tea polyphenols and such results may have significant applications for chemopreventive purpose in human breast cancer.

Keywords: smoking; breast cancer; tea polyphenol; nicotinic acetylcholine receptor

尼古丁主要的摄取途径来自于吸烟,而一般吸烟者其血清内尼古丁浓度通常维持在 100~200nmol/L 左右；然而在吸烟后，其气管内黏膜细胞与肺脏内的尼古丁浓度可以达到 1mmol/L，而血清内的尼古丁浓度也可达到 $100\mu mol/L$。在之前的许多研究中，尼古丁就已经被指出与肺癌及许多其他癌症例如大肠癌、胃癌、前列腺癌等等有关，是具高度致癌可能性的危险因子。另外，一般青春期到更年期的女性（怀孕者除外），其体内主要的雌性素以雌二醇为主，生理浓度为 0.05~1nmol/L。之后研究使用之药物浓度皆使用接近于生理浓度的剂量进行实验。

我们发现，nAchR 与 ER 在与其专一性 ligand 结合后，皆会诱发 PI3K/Akt 与 MAPK 讯息传递路径。而这两者似乎也能诱发调控 nAchR α9 基因转录的转录因子，进而增加 nAchR α9 mRNA 表现量上升，并生成更多的 nAchR α9。此机制似乎具有促进细胞内许多致癌路径，并导致细胞癌化的可能。

1　茶中的多酚类（Polyphenols）

绿茶中的多酚类（Polyphenols）化合物，特别是儿茶素，占了绿茶茶叶可萃取成分之 30%~40%，其中包含了 EGCG、EGC、ECG。作用的机转包括抗氧化、自由基移除、或是酵素动力学上的诱导与修饰。另外，绿茶也可能抑制癌症起始与转移的生化指标，进而抑制细胞复制及影响细胞生长。最近的研究指出，绿茶的摄取能降低罹患癌症的风险，因为绿茶价格便宜、不具有毒性、在世界各地也是非常普遍的饮品，因此临床的试验也是值得期待的目标（Stuart et al.，2006）。

2　喝茶与抗癌

人类喝茶的历史到现在已经有五千年之久（Weisburger，2000），而从以前到现在，茶也被人们认为是一种能改善健康状况，甚至能治愈某些特殊疾病的饮品；"每日一壶茶"也早已被当作是一种健康并养生的生活方式（Schwarz et al.，1994）。

有关茶的化学预防作用主要可以分为几类（Weisburger 和 Chung，2002）：①抗氧化（Anti-oxidant）的能力；②诱发解毒酵素的能力；③在细胞生长、发育以及凋亡上的调节；④对肠道内菌群的筛选能力。

2.1 茶与抗氧化

茶中多酚类在许多生化系统中已被证实是强力的自由基清除者;而其清除自由基的能力似乎与其上的—OH 基团数目与位置有关(Lin,2002);而儿茶素在多酚类中的化学活性更是突出,已被证实可以避免 DNA 受到自由基的伤害而导致的断裂或是突变的发生(Anderson et al.,2000)。在临床方面,绿茶内多酚类化合物也被证实对于氧化性的伤害有抑制作用(Hasaniya et al.,1997;Wei et al.,1999),这对心脏病、癌症、甚至老化都有预防的效果(Weisburger,2000)。

2.2 对信息传递的抑制

MAPK 与 PI3K/Akt 路径,EGCG 已被发现具有抑制癌症细胞内多种讯息传递路径,包括 MAPK 以及 PI3K/Akt 路径。在 MAPK 路径方面,茶中的多酚类可以抑制 NIH 3T3 细胞株内 PKC、MAPK 及下游的转录因子 AP-1 的活性(Chen et al)。在 MCF-7 细胞株中,EGCG 抑制其 G1 细胞周期进行;而 EGCG 调控 G1 细胞周期的机制可能同时借由降低细胞周期进行相关蛋白如 cyclin D1、cyclin E,以及诱发 Cdk inhibitors p21 及 p27 的表现(Liang et al.,1999b)。EGCG 的在动物实验中的抗癌效果早已在许多研究中被证实,包括皮肤、胃腺、十二指肠、大肠、肝脏、胰脏与肺(Lin 和 Liang,2000;Wang et al.,1992;Xue et al.,1992;Yamane et al.,1991;Yang 和 Wang,1993)。而 EGCG 抑制癌症转移的机制可能与其对 MMP-9(matrix metalloprotease)分泌,与 FAK(focal adhesion kinase)的磷酸化有关(Lin,2002)。

3 nAchRs(Nicotinic acetylcholine receptor)的介绍

nAchRs 是位于细胞膜上对称分布且中央具有疏水性通道的 pentamers,属于 Ligand-gated ion channels 家族之一员。当它接触到神经传道物质乙烯胆碱(Acetylcholine,Ach)或是尼古丁(Nicotine)时,会促使其离子通道打开(Itier & Bertrand,2001),使得带正电的 Na^+、K^+、Ca^+ 进到细胞内,尤其是神经元会直接或间接导致 intracellular cascades 使得一些基因的活性以及神经传道物质的释放受到调控。正常生理状况下,乙烯胆碱(Acetylcholine,Ach)为 nAchRs 之受质,但一些非专一性的 nAchR agonists,如 Nicotine 及 NNK,已被证实与 nAchRs 结合产生临床生理作用(Cormier et al.,2004)。

4 研究初步结果

由实验结果,我们初步发现了 EGCG 具有抑制多数 nAchRα9 下游讯息传递路径的作用;而经由 pre-treat EGCG 的方式,可以使得 nAchR α9 的 mRNA 以及蛋白质表现量下降,此一作用似乎也与实验结果内所发现 EGCG 能抑制乳癌细胞生长有关。

5 实验方法与结果

5.1 数种天然化合物在乳癌细胞株 MDA MB-231 中能抑制 nAchR α9 mRNA 表现的最低剂量

首先要找出这几种天然化合物在本次实验中的最低有效剂量。首次实验采用较高剂量($10\mu mol/L$),之后逐次以 10 倍稀释降低其剂量。实验结果发现,在较低剂量($1\mu mol/L$)仍具有抑制 nAchR α9 mRNA 表现的能力的天然化合物有 EGCG、Curcumin、Luteolin、Quercetin、Magnolol 等数种(图 1(a))。之后考虑到 EGCG 是绿茶中含量丰富且日常生活最容易摄取到的成分(Lin,2002),而且已有许多文献报告其具有抑制各种癌症细胞株生长的能力(Lin,2002;Lin 和 Liang,2000;Wang et al.,1992;Xue et al.,1992;Yamane et al.,1991;Yang 和 Wang,1993),故选择此天然化合物来进行接下来的实验。

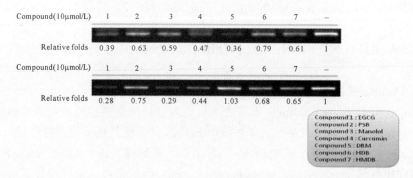

(a)MDA MB-231 乳癌细胞(2×10^6)给予各种不同天然物后 24 h 进行 RTPCR 分析

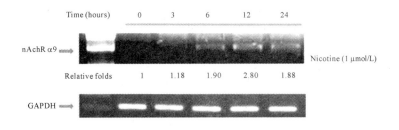

(b) 尼古丁给予乳癌细胞 3、6、12、24 h

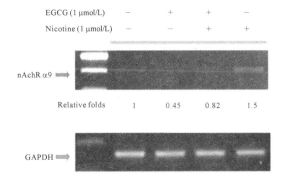

(c) 先给予乳癌细胞 EGCG(1 μmol/L)24 h,然后给予尼古丁(1 μmol/L)12 h

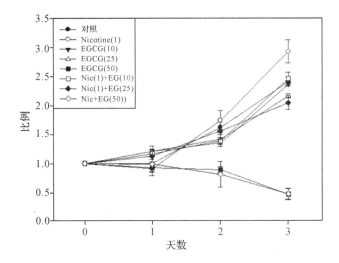

(d) 合并处理尼古丁与 EGCG 进行生长试验

图 1　天然物抑制尼古丁受体的 mRNA 表现

5.2 EGCG 具有抑制加入 Nicotine 后 nAchR α9 mRNA 表现过量的能力

1μmol/L Nicotine 能在处理 6h 开始诱发 nAchR α9 mRNA 表现增加,而在 12h 后达到最大值(图 1(b))。得知此实验条件后,我们先在 2×10^6 cells/10cm dish 中加入 1μmol/L 的 EGCG,培养 24h 后,再加入 1μmol/L 的 Nicotine,培养 6h 后收集细胞,并且以 RT-PCR 的方式观察其 nAchR α9 mRNA 表现量。

实验结果发现,对细胞前处理 EGCG,具有抑制之后经由加入 Nicotine 诱发而增加之 nAchR α9 mRNA 表现的能力(图 1(c))。而在此实验中,EGCG 的浓度(1μmol/L)也远比之前文献中所记载的实验用 EGCG 浓度为低(Ren et al.,2000),更接近正常人血液中所含的 EGCG 含量(Landis-Piwowar et al., 2007),也更具有临床研究上的可信度。

5.3 EGCG 具有依剂量增加而抑制 MDA MB-231 乳癌细胞株生长的能力

实验结果发现,虽然 Nicotine 与 NNK 能促进乳癌细胞株 MDA MB-231 生长(图 2(a),图 2(b)),本实验首先次培养 2×10^6 cells/10cm dish 实验结果发现,随着加入的 EGCG 浓度增加,nAchR α9 蛋白质表现量有逐渐减少的趋势。25μmol/L 的 EGCG 能抑制 nAchR α9 蛋白质表现量,而 50μmol/L 的 EGCG 则几乎完全抑制其蛋白表现(图 2(c))。由于 50μmol/L 的 EGCG 与其能在正常生理状况下存在的浓度比较而言似乎太高,故之后 EGCG 的实验浓度选择 25μmol/L。但在加入 EGCG 后此作用即被抑制;高剂量的 EGCG 甚至能完全抑制细胞生长(50μmol/L)(图 1(d))。

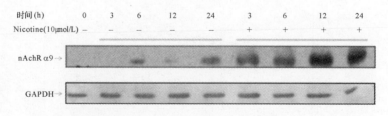

(a)乳癌细胞 MDA MB-231(2×10^6),观察 3、6、12、24h

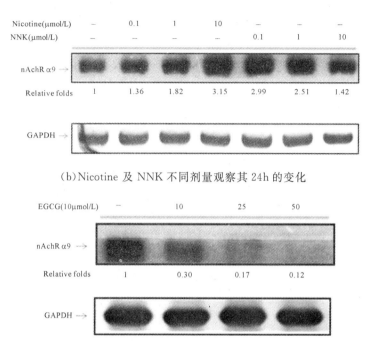

(b) Nicotine 及 NNK 不同剂量观察其 24h 的变化

(c) 不同剂量 EGCG 处理乳癌细胞 24h 观察尼古丁受体变化

图 2　尼古丁与其代谢产物 NNK 活化 nAchR α9 过度表现,EGCG 抑制其作用

5.4　在 MDA MB-231 乳癌细胞株中,EGCG 具有抑制加入 Nicotine 后,nAchR α9 蛋白质过量表现的能力

本实验首先找出 Nicotine 诱发 nAchR α9 蛋白质增加的条件。以之前的实验条件进行 Western blotting,我们发现在 MDA MB-231 细胞中加入 $10\mu mol/L$ 的 Nicotine,能随着时间增加而诱发 nAchR α9 蛋白质表现,在 24h 达到蛋白表现量极大值,之后再慢慢减少(图 2(a))。因此之后将以 $10\mu mol/L$ Nicotine 作用 24h 作为本次实验的条件。此外,我们也以同样实验条件找出了 NNK 在本次实验下的条件(图 2(b));NNK 是 Nicotine 经由肝脏代谢后所产生的化合物(Schuller 和 Cekanova,2005),可以作为本次实验的对照组。

我们先在 2×10^6 cells/10cm dish 中加入 $25\mu mol/L$ 的 EGCG,培养 24h 后,再加入 $10\mu mol/L$ 的 Nicotine 或 NNK,培养 24h 后收集细胞,并且以 Western blotting 的方式观察其 nAchR α9 蛋白质表现量。实验结果发现,在 MDA MB-231 细胞株中,EGCG 具有抑制被 Nicotine 或 NNK 所诱发的 nAchR α9 蛋

白质表现的能力(图3)。

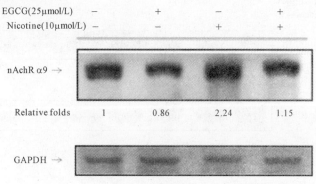

图3 EGCG抑制由尼古丁诱发之受体表现

5.5 在长期暴露在E2下的MCF-7乳癌细胞株中,EGCG具有抑制加入E2后nAchR α9蛋白质表现过量的能力

实验结果显示,随着处理E2的时间增加,nAchR α9蛋白质的表现也会跟着增加,并且在90d(约继代培养30次)后具有明显高于30d(约继代培养10次)的表现量(图4(a))。实验结果发现,长期暴露在E2下的MCF-7细胞株内nAchR α9蛋白质表现量会被E2诱发而上升,但此作用可以经由前处理EGCG而抑制(图4(b))。

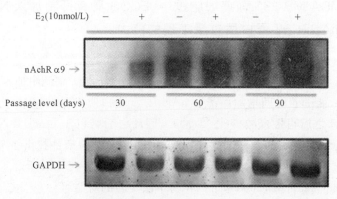

(a)不同代数的乳癌细胞细胞加入女性荷尔蒙E2 30min

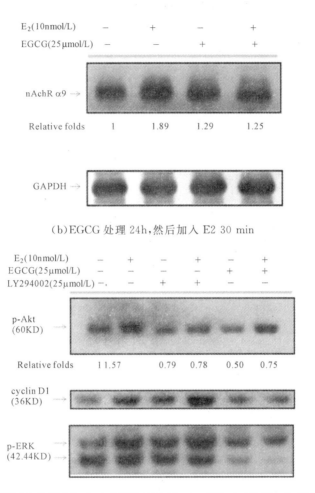

(b) EGCG 处理 24h, 然后加入 E2 30 min

(c) EGCG 处理 24h, 加入 LY294002 30 min, 再加入 E2 30 min 观察

图 4　EGCG 抑制 nAchR α9 受体之表现是透过抑制乳癌细胞 MCF-7 之 PI3K/Akt 信息传递路径

5.6　EGCG 经由 PI3K/Akt 信号通路影响长期暴露在 E2 下 MCF-7 细胞株内 nAchR α9 蛋白质的表现

实验结果发现 EGCG 与 LY294002(图 4(c)),两者都能阻止 Akt 被磷酸化(图 4(d)),进而抑制下游的讯息传递,而在加入 EGCG 后,甚至还能发现 ERK 的磷酸化也被抑制。由于之前文献也有指出 EGCG 具有抑制 MAPK 信号通路的能力(Kim et al.,2004),本次实验也间接印证了这个现象。

5.7 利用 Soft Agar Assay(Colony Formation Assay)发现 EGCG 能抑制 MDA MB-231 乳癌细胞株转化的发生

在理清了一系列 nAchR α9 细胞分子层面的机制后,最后我们将利用一个模拟活体内环境的实验-Soft Agar Assay(Colony Formation Assay),观察 EGCG 对于 MDA MB-231 细胞株转化的发生是否具有抑制作用。我们发现,EGCG 在 25μmol/L 浓度下即具有明显的抑制作用(图 5);由于 MDA MB-231 是一种具有明显恶质化特性的乳癌细胞株,若 EGCG 能抑制其形成群落的能力,则在活体试验中的可行性也是可以预期的;对于之后是否应用在临床上也具有很重要的意义。

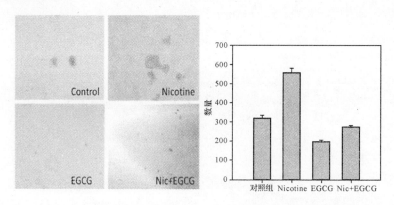

图 5　EGCG 抑制 MDA MB－231 乳癌细胞在琼胶上群落形成

6 讨 论

由于本实验室之前已经发现雌二醇(Estrogen)与尼古丁影响 nAchR α9 的共同讯息传递路径为 PI3K/Akt-ERα(Ser167),以及 MEK1/2、ERK1/2、JNK1/2、p-ERα(Ser118);而本篇研究也发现 EGCG 具有抑制此共同讯息传递路径,进而使其下游基因表现量减少的作用。然而,根据之前的文献指出,EGCG 除了能抑制数种讯息传递路径之外,也具有抑制细胞核内多种转录因子启动基因表现的作用(Stuart et al.,2006)。在之后的研究中,除了 nAchR α9 下游的讯息传递,也可以探讨其与癌症细胞转移、侵袭的关联,相信也会具有很好的结果。

本研究中探讨细胞核内转录因子的部分,发现 AP-1 是 EGCG 抑制 nAchR

α9启动子的主要作用部位；但若将各片段启动子活性相互比较后，会发现启动子主要的活性表现集中在位置偏前段处（0～44bp、0～125bp、44～250bp），而其中又以44～250bp的活性表现最为明显。另外在加入Nicotine诱发启动子活性的部分，也以44～250bp的趋势最为明显。而若将各片段中的AP-1结合部位删除后，互相比较可以发现，在125～250bp内似乎还有另一个重要的转录因子结合位；也就是说，Nicotine诱发nAchR α9启动子可能经由一个以上的转录因子调控此作用。经过软件分析后，目前推测此转录因子可能是GATA-1或是MZF-1，之后利用点突变或是删除结合序列的方式即可以证明此推论。

而之前曾研究到EGCG是否可能竞争抑制Nicotine与nAchR的结合，此实验主要是利用3H-Nicotine进行实验，之后利用β-Counter测定其放射量，观察是否加入EGCG后nAchR所结合的3H-Nicotine量会减少。然而虽然之前的研究曾经指出，EGCG能有效抑制EGF与EGFR的结合，但在我们的实验中，EGCG与nAchR之间的关系似乎并不那么单纯；推测原因可能是因为最近发表的一篇文献中指出，EGCG对于膜上许多种类受体（如EGFR、IGFR等）的影响，主要是因为其具有改变膜上脂筏构造的能力，进而造成受体无法形成二聚物构造（Adachi et al.，2007）。nAchR与细胞膜的连结是否有可能也受到EGCG作用影响，而使受体下游讯息传递直接受到抑制？甚至nAchR可能与细胞膜上其他受体连结进一步增强其作用？这些推测可能都需要进一步的实验以证明。

最近的研究发现，在细胞膜上似乎有一种67-kDa层黏连蛋白受体，对EGCG有很高的结合特异性，而这种蛋白也与EGCG的抗癌效果有所相关（Tachibana et al.，2004）。若将此蛋白过表达，是否能让EGCG有近似于肿瘤消除基因的功能；而此蛋白如何调控EGCG与其他信号通路之间的相关性？这些问题都是尚待回答的；但不论是从何论点探讨，EGCG始终是扮演一个癌症抑制者的角色，也因此这么多的研究始终以它为主轴。

茶树花新资源的活性成分
研究现状与市场前景

屠幼英　徐元骏　杨子银　李　博　金玉霞　汤　雯

浙江大学农业与生物技术学院茶学系,中国 浙江

摘　要:近10年研究表明,茶树花含有丰富的营养成分。采用高效液相色谱法测得茶树花中含有儿茶素和黄酮苷类等;茶树花中有16种游离氨基酸,茶氨酸的含量(4.93mg/g)、还原糖含量(35%～40%)和可溶性糖含量(达到21.6%)高于茶叶。茶树花安全性研究未见受试物对雌雄各剂量组动物食物利用率、血液学、血生化、脏体比和病理组织学有不良影响。在茶树花粉大鼠致畸试验中未见致畸作用。研究还发现茶树花对大鼠肥胖病和高脂血症有良好的预防作用,茶树花多酚、多糖和精油具有明显的抗氧化活性。所以,茶树花可以在食品、医药、轻工等领域应用。

关键词:茶树花;成分;抗氧化活性;减肥

Studies and Utilization in the Future Market of the Bioactive Compounds in Flowers of Tea (*Camellia sinensis*): A New Resources

TU You-ying　XU Yuan-jun　YANG Zi-ying
LI Bo　JIN Yu-xia　TANG Wen

Department of Tea Science, Zhejiang University, Zhejiang, China

Abstract:Recent ten years, many researchers reported flowers of tea (*Camellia sinensis*) plants contain many bioactive compounds. Flavones and catechins were detected by HPLC. Sixteen kinds of amino acids were found. Theanine content (4.93mg/g), sugar (21.6%) and polysaccharides (35%～40%) were higher than these of tea. The present investigation was carried out to evaluate the safety of tea flower extract by acute and genetic and subchronic toxicity studies. In the acute toxicity study, all animals gained weight and appeared active and normal, so the LD50 value must be >12.0g/kg body weight. In the Ames test, micronucleus formation test and sperm abnormalities test, tea flower extract didn't have genotoxic potential. In the subchronic toxicity study, no dose-related effects on survival, growth, hematology, blood chemistry, organ weights, or pathologic lesions were observed. Our laboratory and cooperators found tea flower have good antiobesity and lipid lowering effects, which contributed to the antioxidant abilities of polyphenols, polysaccharides and essential oils in tea flower. We suggested that tea flower and its extract will be used widely in normal food, cosmetics and functional food.

Keywords: tea flower; compound; antioxidant activity; antiobesity

我国茶园面积约 200 万 hm^2，年可产茶树鲜花约 400 多万 t。由于茶树开花结果会争夺叶片养分，致使茶叶产量和质量下降，因此少量茶园，茶农每年都人工采摘或使用除花剂除花；而大部分茶园则自动落花。总之，以往认为茶树花是茶园里的废弃物。然而，近年来，已证实茶树花具有安全、低毒、高生物活性的特点，其抗氧化能力可以与迷迭香相媲美，有"安全植物的胎盘"、"茶树上的精华"的美誉。

茶树鲜花的主要化学成分与茶叶相似，其活性成分及含量约为：总糖 30%、蛋白质 25%、儿茶素 5%、氨基酸 3%、咖啡因 2%，黄酮类物质含量较其他花卉高，微量元素锰、钴及烟酸含量也比较高。其茶多糖含量显著高于茶树鲜叶，但咖啡因和茶多酚含量明显低于鲜叶。并且不同茶树品种和发育程度的茶树花，功能性成分含量存在很大差异[1-4]。所以，增加茶树花资源的利用率就是增加茶农的收入和茶园的产出，减少天然资源的浪费，对巩固和发展茶叶产业也有重要的意义。本文对研究小组近 10 年对茶树花活性成分研究结果进行总结，并且对这些活性成分未来可应用领域提出建议。

1 LC-MS 鉴定茶树花中多酚类组成及其抗氧化活性

近年来，自由基与多种疾病的关系已愈来愈受到关注，自由基生物医学的发展使得探寻高效低毒的自由基清除剂——天然抗氧化剂成为生物化学和医药学的研究热点。迄今为止，关于茶树花的抗氧化活性鲜有报道。

对茶树花各主要成分进行不同溶剂的萃取，经液质联用分析，从茶花乙酸乙酯层中鉴定出 3 类化合物：①儿茶素类：儿茶素、表儿茶素、没食子儿茶素、表没食子儿茶素、儿茶素没食子酸酯、表儿茶素没食子酸酯、没食子儿茶素没食子酸酯、表没食子儿茶素没食子酸酯。②儿茶素衍生物:儿茶素糖苷物与儿茶素二聚体。③黄酮苷类:山奈酚、洋黄黄酮与槲皮素单糖苷及双糖苷类物质。对各萃取组分进行羟自由基与DPPH自由基的清除能力测定,结果表明,表没食子儿茶素没食子酸酯与表儿茶素没食子酸酯为茶花中主要抗氧化活性成分,在茶花抗氧化活性中起到主要的作用。

表1 茶花提取物及其萃取组分化学成分组成及其含量

组 分	化学成分及其含量(mg/g)		
	黄 酮	多 酚	咖啡因
EE	28.8±0.50	145±1.06	1.02±0.12
EEC	25.1±0.34	129±0.80	25.6±0.25
EEA	36.1±0.66	298±1.54	2.29±0.12
EEB	24.9±0.69	122±2.15	nd
EER	3.31±0.06	20.3±0.27	0.20±0.00
WE	22.2±0.31	37.4±1.06	1.31±0.15
WEC	23.9±0.55	33.8±0.99	34.4±0.64
WEA	25.7±0.19	41.3±1.25	nd
WEE	16.5±1.06	28.3±0.78	nd
WER	34.5±1.69	126±1.36	nd

注:数值表示为平均值±标准误差($n=4$)。nd,表示未检测到.

茶花醇提物组分制备:冷冻干燥茶花(浙江开花金茂茶场提供)与70%乙醇1∶10比例,50℃下回流浸提3次,每次120min,然后合并提取液,真空浓缩除去乙醇,得到乙醇层(EE)。然后再依次用等体积的氯仿、乙酸乙酯、正丁醇萃取3次,得到的各有机层真空浓缩脱去有机溶剂,依次得到氯仿层(EEC),乙酸乙酯层(EEA)与正丁醇层(EEB)。剩余水层冷冻干燥得醇提残留物(EER)。

茶花水提物组分制备:冷冻干燥茶花30g用200mL蒸馏水在100℃沸水浴中浸提30min,所得提取液浓缩干燥得深褐色的茶花水提物(WE)。然后同样依次用氯仿、乙酸乙酯与70%乙醇依次萃取,得到氯仿层(WEC),乙酸乙酯层(WEA)、乙醇层(WEE)及剩余相(WER)。

2 茶树花安全性研究

茶树花如要开发食品相关应用领域,首先要进行安全性评价。Bo Li 等对茶树花提取物进行了安全性评价,研究提取物对 SD 大鼠母体及胎仔致畸毒性,对大鼠进行灌胃,记录体重、体长、活胎数、死胎数等指标,实验结果表明茶树花提取物对孕鼠生长和胚胎发育无不良影响和致畸作用。通过 Ames 试验发现茶树花提取物对 4 个沙门氏菌株均没有诱变效应。茶树花粉对雌、雄性大鼠经口急性毒性,LD50 均大于 12.0g/kg。根据急性毒性分级,茶树花粉属实际无毒物;Ames 试验、小鼠骨髓嗜多染红细胞微核试验、小鼠精子畸形试验,结果均未见茶树花粉有致突变作用;以不同剂量的茶树花粉灌胃给大鼠 90d,结果显示,动物活动正常,雌雄性各剂量组体重与对照组比较未见不良影响,未见受试物对雌雄性各剂量组动物食物利用率、血液学、血生化、脏体比和病理组织学有不良影响;在茶树花粉大鼠致畸试验中,未见茶树花粉对孕鼠生长和胚胎发育有不良影响,未见致畸作用。结果表明,茶树花可以在食品等领域进行应用。

3 茶树花可溶性糖和氨基酸的研究

采用高效液相色谱法测定茶树花中可溶性糖的含量。结果表明:茶树花中含有 216.85mg/g 可溶性糖,其花冠、雄蕊群和雌蕊群部分(F-1)及花托、萼片和花梗部分(F-2)的含量分别为 250.94mg/g 和 112.65mg/g(表 2);茶树花中总儿茶素含量为 37.03mg/g,且 F-2 中的含量(46.90mg/g)高于 F-1(30.88mg/g)。

表 2 茶树花可溶性糖含量

样 品	果 糖	葡萄糖	蔗 糖	可溶性糖	
				HPLC 法	苯酚-硫酸法
F-1	84.14 ± 7.88	67.74 ± 5.89	99.06 ± 7.2	250.94	491.51 ± 22.12
F-2	28.27 ± 2.19	20.76 ± 2.53	63.62 ± 1.43	112.65	222.06 ± 17.02
茶树花	67.59±0.90	57.44±1.49	91.82±1.95	216.85	422.73±13.14

注:数值均为平均数±标准差($n=3$)。

采用 Zorbax Eclipse XDB-C18 色谱柱和 DAD 检测器对茶树花中的游离氨基酸进行 HPLC 分析,发现茶树花中含有 16 种游离氨基酸,含量为 10.87mg/g,其中 15 种为蛋白质氨基酸、1 种为非蛋白质氨基酸(茶氨酸),F-2

中游离氨基酸的种类较少,但其含量(16.47mg/g)显著高于 F-1(10.59mg/g)。茶树花中茶氨酸(4.93mg/g)、组氨酸(2.14mg/g)、脯氨酸(1.28mg/g)的含量较高;F-1 与 F-2 相比,F-1 中含有较多的脯氨酸(2.15mg/g),而 F-2 中茶氨酸(5.94mg/g)是茶叶和部分山茶科植物特有的酰胺类物质,也是重要的功能成分。茶氨酸具有抗肿瘤、保护神经、增强记忆、抗糖尿病、降血压、抗疲劳、缓解抑郁症、保护心脑血管、减轻酒精对肝脏的伤害、增强对流行性感冒病毒疫苗的免疫响应等生理作用。与茶叶相比,茶树花中茶氨酸的含量(4.93mg/g)高于茶叶中的含量(1.62mg/g~4.12mg/g)。

4 茶树花精油的提取和抗氧化活性

采用超临界二氧化碳萃取茶树花精油进行提取,利用气相色谱-质谱联用技术(GC-MS)对精油化学成分分析鉴定,采用磷钼络合法和 DPPH(二苯代苦味肼自由基)自由基清除法测定了茶树花精油的抗氧化活性。结果表明:萃取压力、萃取温度、静态及动态萃取时间 4 个因素对茶树花精油萃取率有显著性影响,优化得到的最佳工艺参数为压力 30Mpa、温度 40℃、静态萃取时间 10min、动态萃取时间 90min。在筛选出的最佳工艺参数下进行试验,茶树花精油萃取率为 1.2082%±0.0941%。从超临界 CO_2 提取得到的茶树花精油中香气成分主要有 59 种,其中主要的香气成分包括植酮(2.99%)、邻苯二甲酸二丁酯(4.97%)、十九烷(18.7%)、二十一烷(12.2%)、二十三烷(4.91%)、甲基丙烯酸乙二醇酯(3.07%)等。

茶树花精油具有明显的总抗氧化活性,在低浓度下,茶树花精油的抗氧化活性高于迷迭香精油而低于天竺葵、熏衣草精油和 BHA;随样品浓度的增加,各个阳性对照的总抗氧化活性均高于茶树花精油,其中以天竺葵和熏衣草精油的总抗氧化活性最高;茶树花精油对 DPPH 自由基具有明显的清除作用,且浓度与清除能力呈正相关,其 $EC_{50}=13220.09\mu g/mL$,几种对照 DPPH 自由基清除效果为 TBHQ>VC>丁香>BHA>茶树花精油>薄荷精油>天竺葵精油。茶树花精油总抗氧化能力随浓度的变化结果如图 1 所示。由图可知,随着样品浓度的增加,吸光度值相应增大,表明茶树花精油的抗氧化活性随着浓度的增大而升高。

5 茶树花多糖及其活性研究

茶多糖是继茶多酚之后发现的又一茶叶重要生理活性物质。药理研究表明,茶多糖具有降血糖、降血脂、抗凝血、抗血栓、降血压、耐缺氧、增加冠状动脉血流量、防辐射、增强机体免疫力、抗炎、抗癌等多种功效,尤其是显著的降血糖和增强免疫力,茶多糖有望成为预防治疗糖尿病及心血管疾病,增加免疫力的天然药物。茶树花内含成分丰富,和茶叶相比有高多糖、低咖啡因的特点,因此,是开发多糖类保健品和药品的天然资源。

用高效液相色谱方法分析得到的茶树花多糖组分有葡萄糖、木糖、鼠李糖、半乳糖,是水溶性、非店方、非酚类物质,是不含还原糖和核酸、含蛋白质的多糖。

茶树花多糖可明显增强免疫活性。灌胃80%纯度的茶树花多糖75、150、300mg/kg的小鼠14d后,能显著增加正常小鼠巨噬细胞的吞噬活性,对二硝基甲苯诱导的小鼠迟发型超敏感反应也有显著的增加。28d后可以抑制溴化苯诱导的肝脏脂质过氧化,显著提高小鼠SOD活性和T-AOC水平,抑制MDA含量升高。对S180肉瘤小鼠的研究发现,3种浓度茶多糖7d后,抑制肿瘤率分别为45.5%、60.9%和64.5%,增加小鼠血清的IFN-γ和IL-2的含量,可明显增强免疫活性。

茶叶中茶多糖的含量一般为20%~25%,而已有研究的实验数据表明茶花中还原糖的含量就有35%~40%,大大高于茶叶,而且茶花中也含有较多的游离蛋白质,蛋白质与多糖又极易分离,故从茶花中提取茶多糖工艺较为简单,而且易得到较高纯度,可为茶多糖的深入研究提供较好的材料,并继而为今后在医药化工等行业的广泛应用提供原料。

6 茶树花对大鼠肥胖病和高脂血症有良好的预防作用

将40只大鼠分为5组,即:基础饲料组,高脂饲料组,茶树花2.5%、5%、7.5% 3个剂量组。每两天测量体重和摄食量1次,24d后断头取血,测定总胆固醇(TC)、甘油三酯(TG)、低密度脂蛋白胆固醇(LDL-C)、高密度脂蛋白胆固醇(HDL-C)。取体脂(肾及睾丸周围脂肪)称重。结果表明,与基础饲料组相比,高脂饲料组大鼠体重、脂肪系数、TC、TG、LDL-C、动脉粥样硬化指数(AI)显著增加;3个剂量茶树花添加组大鼠体重、脂肪系数、TC、TG、LDL-C、AI均

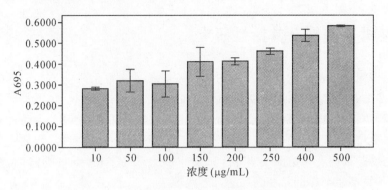

图 1 茶树花精油总抗氧化能力与浓度的关系

低于高脂饲料组,且呈显著或极显著差异;中、高剂量组大鼠体重、脂肪系数、TC、TG、LDL-C、HDL-C、AI 与基础饲料组无明显差异(表 3)。结论:茶树花对大鼠肥胖病和高脂血症有一定的预防作用,3 个剂量组中添加 5％茶树花效果最好,可开发为降脂减肥的纯天然保健食品。

表 3 茶树花对大鼠肥胖病血清脂肪指标的影响

组 别	血清(mmol/L)				
	TC	TG	HDL-C	LDL-C	AI
基础饲料组	1.75±0.11	1.03±0.18	0.82±0.14	0.93±0.15	1.48±0.39
高脂饲料组	3.75±0.15**	2.47±0.23**	0.80±0.06	2.76±0.14	3.59±0.42
低剂量组	1.77±0.16△△	1.06±0.08△△	0.96±0.12	0.82±0.12△△	0.95±0.18
中剂量组	1.90±0.08△△	1.08±0.10△△	1.09±0.10	0.82±0.12△△	0.84±0.18
高剂量组	1.89±0.08△△	1.07±0.16△△	1.05±0.10	0.84±0.13△△	0.89±0.17

注:与基础饲料组相比,** $P<0.01$;与高脂饲料组相比,△△ $P<0.01$。

综上所述,多年的研究发现茶花具有大量的生物活性物质,而且安全无毒。除本实验室研究结果,目前对于茶花皂苷的研究也取得了非常重要的结果,至今已发现有 21 种茶花皂苷单体,具有肠胃保护、抗血脂、抗血压,减肥等生物功能。并且已经开发成为可口可乐的减肥产品,在日本畅销;台湾黑松公司同样销售处于领先的位置。但是山茶花和茶树花的皂苷在分子结构上有怎样的区别还有待进一步研究,相信随着人们对茶花皂苷的认识进一步的深入,对茶花皂苷单体的药物活性进一步验证和安全试验,茶树花皂苷天然产品和茶树花其他系列产品将迅速走向工业化。

参考文献

[1] Yang Z Y, Tu Y Y, Baldermann S, et al. Isolation and identification of compounds from the ethanolic extract of flowers of the tea (Camellia sinensis) plant and their contribution to the antioxidant capacity[J]. LWT-Food Science and Technology, 2009, 166: 887－891.

[2] 李博, 屠幼英, 徐纪英. 茶树花对大鼠致畸作用的研究[J]. 茶叶, 2009, 35(2): 87－90.

[3] Li B, Jin Y, Xu Y, et al. Safety evaluation of tea (*Camellia sinensis* (L.) O. Kuntze) flower extract: Assessment of mutagenicity; and acute and subchronic toxicity in rats[J]. Journal of Ethnopharmacology, 2011.

[4] Wang L, Xu R, Hu B, et al. Analysis of free amino acids in Chinese teas and flower of tea plant by high performance liquid chromatography combined with solid-phase extraction[J]. Food Chemistry, 2010, 123: 1259－1266.

[5] Yang O, Tu Y, Dong F, et al. Liquid chromatographic-mass spectrometric analysis of phenolics and evaluation of the antioxidant compounds of extracts from flowers of the tea plant (*Camellia sinensis*), 2nd Symposium on Pharmaceutical Food Science[J]. Japan Yakugaku Zasshi-Journal of the Pharmaceutical Society of Japan. 2007, Suppl: 10－12.

[6] 陈晓敏, 屠幼英, 徐懿等. 饮用干花中微量元素和黄酮类含量的分析[J]. 食品工业科技, 2006, 27(10): 175－217.

[7] 王晓婧, 翁蔚, 杨子银等. 茶花研究利用现状及展望[J]. 中国茶叶, 2004, 26(4): 8－10.

[8] 杨子银. 茶(红茶)与茶(*Camellia sinensis*)花多酚类物质的分离鉴定及其抗氧化机理研究[D]. 杭州: 浙江大学, 2007.

[9] 李博. 茶花的安全性评价及茶黄素和茶籽黄酮苷对呼吸链酶作用机理的研究[D]. 杭州: 浙江大学, 2010.

[10] 翁蔚. 茶(*Camellia sinensis*)花主要生物活性成分研究及应用展望[D]. 杭州: 浙江大学, 2004.

高新技术在茶叶多酚提取分离中的应用及发展趋势

王洪新　娄在祥　马朝阳　尔朝娟　秦晓娟

江南大学食品学院 食品科学与工程国家重点实验室，中国 江苏

摘　要：我国茶叶深加工尤其是茶多酚、儿茶素方面经相关领域技术人员20多年的共同努力，在技术和产品上形成了"百花齐放、百家争鸣"的繁荣景象，并且已经形成了一个具有一定规模的行业，为人类健康及茶产业的增产增收作出了很大贡献。但目前仍然存在技术水平、加工规模高低不一，终端产品开发滞后，产品质量参差不齐，市场价格混乱、恶性竞争、经济效益较差等问题。本文结合研究及生产实际，简要报道高新技术在茶叶深加工中的应用、茶叶深加工现状及发展趋势，期望对该行业有所裨益。

关键词：茶叶深加工；技术；质量；价格；现状；前景

The Application and Development Tendency of Hi-tech in the Extraction and Separation of Tea Polyphenols

WANG Hong-Xin　LOU Zai-xiang　MA Zhao-yang
ER Chao-juan　QIN Xiao-juan

School of Food Sci. & Technology, Jiangnan University, The State Key-Laboratory of Food Science and Technology, Jiangsu, China

Abstract: The deep processing technology and products of tea, especially in tea polyphenols and catechins, had made great achievements in the past 20 years, and has formed a rather scale industry in China. However, there are some problems such as different technology and scale levels, different quanlity of products, unstuck price, inadequate terminal product, etc. This paper introduced the application and development tendency of Hi-tech in the extraction and separation of tea polyphenols according to the author' practice in this field.

Keywords: deep processing of tea; technology; quality; price; prospect

茶叶深加工一般是指利用低档茶、碎茶或夏秋茶老叶,采用现代提取分离纯化技术及装备,生产制造茶叶中的功能性成分,包括茶多酚(儿茶素)、茶氨酸、茶多糖、咖啡因、茶蛋白、茶黄素、茶蛋白等。目前最先进的工艺技术是综合制备上述系列产品,就像对待大豆一样"吃干榨尽"。

仅结合作者多年从事茶叶深加工的经历,就高新技术及装备的应用,技术、产品质量指标、规模和市场的现状,以及发展趋势谈谈自己的体会和经验,以期抛砖引玉。不妥之处,请各位专家指正。

1 工业化技术发展历程及现状

1.1 工业化技术发展历程

茶多酚作为天然抗氧化剂的研究最早起于上世纪60年代,但工业化生产始于上世纪90年代初。按时间先后、或者按工业化进程追溯,大致经历了以下几种工艺流程(主体技术特征)。

(1)水或水醇溶液提取→金属离子(Ca^{2+}、Pb^{2+}等)沉淀富集→硫酸溶解→有机溶剂萃取法。

(2)水或水醇溶液提取→(氯仿脱除咖啡因)→乙酸乙酯萃取。

(3)水或水醇溶液提取→乙酸乙酯萃取→水反萃取脱除咖啡因。

(4)水或水醇溶液提取→(进口、国产)树脂柱层析→有机溶剂萃取法。

(5)水或水醇溶液提取→国产大孔树脂柱层析→二次柱层析→高EGCG茶多酚。

(6)综合工艺。

1.2 植物提取物(多酚)的工业化实现——高新技术及装备的应用

植物提取物(多酚)的研究开发,最重要的环节之一就是其高效提取分离的工业化实现。从化工单元操作角度,包括:提取—过滤—浓缩—萃取—蒸发浓缩—干燥。

仅从上述技术特征及单元操作来看,与一般的传统工艺没有什么区别,似乎很容易实现。事实上,每一个单元操作如何在工业化上科学、合理、高效的实现,是要在大量的理论和实践基础上才能达到较好的效果。

1.2.1 提取

最早采用一般的"提取罐",与化工常用的"反应釜"没什么区别。但发现提取率低、操作繁琐、料液比高、能耗大,尤其是出料、出渣困难,生产效率低下。

我们在1992年建设无锡绿宝公司茶多酚项目时,率先使用了"多功能提取罐",当时是一个专利新装备。重点解决了出渣困难,提高了生产效率。但在后来的实践中发现,由于低档茶叶、碎茶原料中细末多,含有一定的泥沙,造成出料叶时间长,这是限制生产的主要问题。

所以,目前开发了"连续逆流浸提"装置:隧道式(密闭、不密闭)、环式和旋转式等。优点是不存在堵塞滤网,实现连续式生产,料液比大大降低,减少了后续处理成本,节约水电气及人工消耗。

目前工业上已经出现了超声波、微波辅助下的"连续逆流浸提"装置。可以较大程度地提高提取率,还可以考虑酶法辅助提取。

1.2.2 过滤

过滤最初主要是为了去除碎茶末、泥沙等物理杂质,有利于下道工序进行及提高产品质量。采用过三足式离心机、板框压滤机、振动筛、涡螺式压榨机等,目前这道工序以振动筛综合效果较好。

但随着对产品质量要求的提高,技术不断革新进步,现在过滤工序已经发展到分离和除去多糖、蛋白等大分子杂质,以保证后续产品的高质量。目前超滤膜技术及装备已成功用于茶叶提取物的过滤和初步分离,效果良好。尤其是针对速溶茶粉,要解决其冷后浑浊问题,这几乎是不可或缺的技术装备。但设备选型十分重要,有的厂家随便选用超滤装置,造成使用效果达不到要求,设备闲置或浪费。

1.2.3 浓缩

茶水提取物原液固形物浓度很低,为了后续溶剂萃取过程中少用溶剂以及后续浓缩干燥产品时的需要,要先进行适当倍数的浓缩。溶液浓缩技术也经历了从常压蒸发、负压蒸发、负压升膜式或降膜式、二效浓缩、三效浓缩的发展过程,每一次改进都是为了减少能量消耗、提高效率及降低生产成本。

随着装备技术的发展及节能减排的需要,最新的工业化浓缩已经发展到反渗透膜浓缩,效果良好,能源消耗大大降低,节能80%以上。浓缩回收得到的水可以循环使用。

1.2.4 萃取

经过水或者水醇提取，茶叶茶多酚(儿茶素)、茶多糖、茶氨酸、咖啡因、部分茶蛋白、色素等各种活性成分基本都被提取出来。经过简单的过滤、超滤、浓缩、干燥，即可以得到速溶茶粉(含茶多酚30%~40%)。但是，如果想得到更高纯度的茶多酚(40%~98%)，必须进一步纯化。最简便的方法之一就是溶剂萃取富集。

实验室萃取富集的手段通常使用萃取分液漏斗，因此有些工艺技术就简单模仿实验室的方式，选择许多像"反应釜"一样的萃取罐，几乎不能实现工业化，主要因素有以下几点：①简单模拟放大，需要多台萃取罐，总投资较大；②由于萃取罐体积较大，搅拌或震荡萃取效果很差；③由于萃取时大多产生乳化，很难短时间分层，即使分层后工业上要将两项分离也不易做到；④几乎做不到连续性操作，溶剂挥发带来的安全性较差。

1992—1993年建设无锡绿宝公司时，我国首次自行设计并放大，将连续逆流萃取塔用于茶多酚的萃取，可以做到密闭、连续、高效率萃取，设备可以任意放大，这是一次天然产物工业化萃取意义重大的革新。

1.2.5 蒸发浓缩回收溶剂

一般茶多酚生产中用到的溶剂主要有乙酸乙酯、氯仿、乙醇，具有较强的挥发性，易燃易爆，氯仿具有腐蚀性和毒性。

目前回收上述溶剂都是采用负压蒸发、冷凝回收。回收设备的选型及整体系要科学设计，多采用多功能刮板浓缩器。冷却水要进行热量衡算。如何在大部分溶剂回收后赶走最终残留溶剂，在具体操作上有一定技巧，最后一级冷凝往往采用冷冻水。

1.2.6 干燥

干燥方式有高速离心喷雾、真空烘箱、冷冻干燥等。真空烘箱法效率较差、溶剂残留高；冷冻干燥由于成本高，只适合附加值高的高端产品如高儿茶素产品；高速离心喷雾干燥是目前使用普遍的设备，但高速离心喷雾干燥机的选型计算、除湿功能、收料室的空气净化达到GMP等都是重要的。

1.2.7 高EGCG茶多酚的生产

由于茶叶本身儿茶素组成的限制，上述工艺得到的茶多酚含量虽然可以达到98%以上，但其中生物活性最主要的EGCG含量一般在30%~45%，一些企业靠选择茶叶原料来控制其含量，最高可以达到50%左右。再想提高只能依靠现代分离纯化技术。国外一般采用价格昂贵的葡聚糖凝胶分离材料，但对于国内企业来说，投入巨大，很难承受工业化成本。经过多年研究，我们从众多国产分离材料中筛选出适合工业化的分离材料，价格低廉，可以获得EGCG纯度60%~98%的产品。

2　产品规格及质量指标演变历程

业界对茶多酚的了解很少，所以对茶多酚的质量指标没有什么特别要求。另外，茶多酚产品也没有行业标准。一般是按照GB/T 8313—87《茶　茶多酚测定》的方法进行测定茶多酚，后来实践证明，该标准只适合茶叶原料，而不适合茶多酚，按此方法测定许多茶多酚样品的含量超过100%，有的高达140%。1993年，在经过多种方法验证、对比的基础上，无锡绿宝的企业标准后来由东亚茶多酚厂主持制定为行业标准QB 2154—1995《食品添加剂茶多酚标准》，沿用至今。

随着茶多酚工业的快速发展，尤其是应用范围的扩大及国内外用户的增加，市场对茶多酚质量要求越来越高、质量指标越来越多，产品规格、指标趋于细化和特征化。如对于咖啡因、溶剂、EGCG等含量要求多样化，尤其是随着人们食物安全意识的提高，对于溶剂残留要求变化很大。开始时没要求（当时乙酸乙酯残留高达1%~4%），后来逐渐限制，先后要求小于1000r/min、100r/min、50r/min、…、0r/min，氯仿的残留要求类似。现在有的客商明确提出生产中不得使用乙酸乙酯和氯仿。这就要求技术和装备上跟上市场需求，加大了工业化生产难度，一般企业还做不到。

表1举例说明了目前茶多酚质量指标的细化情况。

3　规模、市场现状

1995年后，随着国内外茶多酚市场的不断扩大，各主要产茶区上马茶多酚项目的要求快速增加，其间涌现了湖南金农、茶科院东方、漳州大闽、深圳华城、无锡世纪生物、海南清华源等较大的公司。尤其是2000年后茶多酚生产厂数量快速增加，单厂生产规模也呈现不断扩大态势，像遵义陆圣康源、安徽红星药业、江西绿康、无锡宏欣等。云南、四川、江苏等地也都有茶提取物为主的工厂。

表 1　茶多酚规格及质量指标

茶多酚系列	TP98	TP98	TP98	TP98	TP98	TP95	TP95	TP95	TP90	TP80	TP80	TP60	TP30速溶茶
茶多酚(%)	≥98	≥98	≥98	≥98	≥98	≥95	≥95	≥95	≥90	≥80	≥80	≥60	30
儿茶素(%)	≥90	≥90	≥90	≥85	≥80	≥75	≥70	≥70	≥70	≥60	≥60	—	—
咖啡碱(%)	<0.5	<0.5	<0.5	<0.5	<0.5	<1	<1	<3	<8	8—10	—	—	8—10
EGCG(%)	>90	>70	>60	>55	>50	>45	>40	>30	>30	>30	25	25	—
水分(%)	<3	<4	<5	<5	<5	<5	<5	<5	<5	<5	<5	<5	<5
灰分(%)	<0.2	<0.2	<0.2	<0.2	<0.2	<0.2	<0.5	<0.5	<0.5	<2	<2	<2	—
堆密度(g/mL)	0.6~0.9	0.6~0.9	0.6~0.9	0.6~0.9	0.6~0.9	0.6~0.9	0.6~0.9	0.6~0.9	0.6~0.9	0.2~0.4(烘) 0.6~0.9	0.2~0.4(烘) 0.6~0.9	0.6~0.9	0.6~0.9
粒径(目)	>80	>80	>80	>80	>80	>80	>80	>80	>80	>80目	>80目	>80目	>80目
溶解性												易溶于水	易溶于水
色泽	淡黄至无色	淡黄至无色	淡黄	棕黄	棕黄	棕黄	棕黄	淡黄	淡黄	淡黄	淡黄	淡黄	淡黄
重金属(mg/kg,以铅计)	<10	<10	<10	<10	<10	<10	<10	<10	<10	<10	<10	<10	<10
铜(mg/kg)	<2	<2	<2	<2	<2	<2	<2	<2	<2	<2	<2	<2	<2
细菌总数(个/g)	<100	<100	<100	<100	<100	<100	<100	<100	<100	<100	<100	<100	<100

有的公司经历了技改、扩产、重组等过程。据不完全统计,目前全国大约有近百家规模不一的茶叶深加工基地(工厂),一般年生产规模在10t~100t(以含量80%以上茶多酚计),茶多酚总产量在3000t左右(不包括速溶茶粉产量)。

但是,茶多酚价格从1993年算起却呈现逐年递减的态势,目前仍没有上升的明显征兆。1993年,80%茶多酚国内交货价为400~500元/kg,95%以上茶多酚价格为700~850元/kg;至2001年,80%茶多酚国内交货价为120~150元/kg,95%以上茶多酚价格为200~250元/kg,98%茶多酚仅280~300元/kg;而目前80%茶多酚国内交货价为80~120元/kg,95%以上茶多酚价格为140~150元/kg,98%茶多酚仅为170~200元/kg。上述茶多酚的价格是在质量要求愈来愈高、生产成本越来越高的前提下发生演变的。因此,如何提高茶叶深加工行业的效益是迫切需要解决的问题。近一年来茶叶深加工产品的价格在逐渐回升。

4 发展前景展望与建议

4.1 茶叶深加工行业前景广阔

我国茶叶综合深加工基础具有以下国外无法比拟的优势,同时随着茶多酚用途的不断扩大,使得我国茶叶综合深加工前景广阔。

(1)原料丰富。我国是茶叶生产大国,茶叶总产量已达160万t左右,其中低档茶、加工下脚料碎茶约占20%,尚不包括夏秋季后不被采收的老茶叶。

(2)技术水平已经达到国际先进或领先水平。经过国内众多学者、专家的共同努力,茶叶深加工的技术水平有了极大的进步,研究水平达到国际先进,由于该行业中国特色明显,工业化水平可以代表国际水平。

(3)加工综合成本具有国际竞争优势。我国的用工成本、水电气等综合成本与发达国家比较具有竞争优势。

(4)产品用途越来越广阔,市场接受度越来越高。目前,茶多酚已被广泛应用于食品(配料和添加剂)、医药、保健品、日化(牙膏、沐浴露、洗发精等)、纺织、空调净化。最新研究证明,有可能在更深入和广阔的层面上对人类健康作出贡献。如儿茶素用于制造百日咳、抗新型流感等疫苗。

4.2 茶叶深加工行业发展趋势

(1)需要考虑综合深度加工发展,进一步提高该行业的总体效益。茶叶深加工产品从20世纪60～70年代仅利用咖啡因,80～90年代仅利用茶多酚、速溶茶粉,发展到今天,综合制取速溶茶粉、茶多酚(儿茶素)、茶氨酸、茶多糖、咖啡因、茶蛋白、茶黄素等多种产品,有的还制备 γ-氨基丁酸等。

(2)需要加强应用技术及终端产品开发,提高产品的附加值,扩大应用。目前,国内用于终端产品的研究开发较少,包括投入的资金和意识。前面一些专家介绍了,国外用于保健品、食品、医药、日化,甚至环境净化、疫苗制造等。国内目前主要用于食品保鲜(抗氧化、防腐)、日化,只有个别厂家用于医药(如"心脑健")。如果不包括速溶茶粉,则自己的市场很小。保健品"绿多维"市场也没做起来。一是不愿意投入,二是投入不对路。如开发较早的产品由于受到当时原料质量限制及配方是否最优化等,不一定能达到很好的效果。

目前,我们在遵义陆圣康源公司开发的"茶多酚"食品(QS520314020008),经使用者实际验证,降血糖效果十分明显(25/14—18/9.5—14/5.8),但由于不是保健品(投入不足)而不能大力宣传。其实只要投入100万元左右就可以拿到健字号。

我们正在以下两个方面,加强应用深度研究。①酶法分子修饰、提高油溶性,应用于食用油行业。儿茶素(黄酮)在脂相或水相中的低溶解性和稳定性限制了其在许多领域中的应用(黄酮糖苷类具有低的脂溶性,而某些黄酮苷元又有低的水溶性)。对黄酮类化合物进行分子修饰,可以从根本上改变黄酮类化合物的理化特性,从而获得符合需要的生物活性效果(图1)。改善黄酮类化合物的结构、溶解性、稳定性及其生理学活性,正成为近年来的热点课题。通过酶法分子修饰,增加其油溶性。如果我国植物油人均消费量达到20kg,即达到欧美人均消费量的50%～60%,按13亿人口测算,植物油需求量将增长650万t,相当于我国目前菜籽油及花生油产量的总和。据此测算,本项目拟开发的酶催化酰化的脂溶性 EGCG,仅在精炼食用油抗氧化剂的国内市场将达到5000t/年,参照 TBHQ 的售价,将形成72800万元/年的销售。无疑,该产品还具有广阔的国际市场。②"三降"(降血糖、血脂、血压)食品开发。茶叶儿茶素类化合物有多个酚羟基,具有很好的抗氧化、降血脂、降血糖等活性,但多个酚羟基导致对肠胃刺激性增大,部分人群对其适应性差,同时其较大的水溶性导致其在血液中吸收代谢率较低,稳定性及靶向"给药"效果都不是很好,其生物活性不能得到充分发挥。对其进行分子结构修饰、改变生物学活性,有望提高其降血糖、血压、血脂的"三降"功效。已经尝试的研究方法包括酶法、化学法和化学/酶法。化学法因其反应的立体选择性低、副产物多而逐步被摒弃;酶法具有高立体选择性、高效性及环境友好等优点而受到更多的关注,成为近年来研

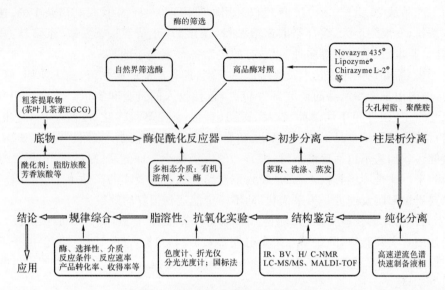

图 1　酶催化酰化黄酮工艺流程图

究热点。其中酰化和糖苷化修饰方法引起特别关注。第一种反应通过连接脂肪酸使它们具有更强的疏水性,第二种通过连接糖分子增强它们的亲水性。

国内对于儿茶素、茶多酚酰化的研究至今都是化学法为主,尚未见酶法酰化儿茶素、茶多酚及其他黄酮化合物的报道。

(3)技术装备的优化、提升及绿色、安全制造。①技术工艺的选择、优化。选择科学合理的技术工艺是工业化成功的必要前提。②设备、装备的选型及其配套,决定了是否能够完成技术工艺的要求,以及决定投入产出、产品质量。③低碳、节能、环境友好措施。热能、水资源的循环利用、节能设备的使用、废水处理技术以及清洁生产技术。④绿色、安全产品的实现。比如农残、重金属、溶剂残留指标的控制,甚至是去除技术。

以上因素决定了植物多酚工业化生产的综合成本及其竞争力。

参考文献

[1] 吴心南,王洪新等.茶叶茶黄素的酶法多相态工业化生产技术.江苏省十五农业攻关项目.

[2] 王洪新,戴军,张家骊等.茶叶儿茶素提取研究.1996年轻工部科技发展项目.

[3] 王洪新,吴心南,谢国银,张家骊等.高纯度茶多酚及儿茶素单体.江苏省九五攻关、江苏省火炬计划、无锡市科技攻关项目.

[4] 王洪新,顾峰,秦卫国等.食品天然抗氧化剂:茶多酚.江苏省科技新产品开发项目.

[5] 肖丰,王洪新,庄平等.燕麦茶奶、豆茶奶.国家星火计划项目.

[6] 王洪新等.茶叶综合深度加工关键技术、装备及产业化实施.

[7] 王洪新,聂小华.油溶性茶多酚:茶多酚脂肪酸酯的研制[J].食品科学,2004,25(12):92-96.

[8] 王洪新,胡昌云.茶叶蛋白质的改性及其功能性质的研究[J].食品科学,2005,26(6):135-140.

中国茶饮料市场未来发展趋势

罗之纲　李松原　杨千辉

康师傅控股有限公司,天津 300457

摘　要:茶饮料在全世界范围内已成为发展最为快速的饮料之一。随着中国经济成长,以及东方人特有的饮茶文化,茶饮料在中国的发展整体已处在上行通道,进而改变了全世界饮料市场的结构。在迎接茶饮料的发展同时,厂商在开发市场时应注意茶原料的开发精细化和管理,茶汤萃取新技术的开发与应用,以及加工工艺与包装技术的提升。未来中国茶饮料新品的发展将由全国性口味发展到地方性的口味,由偏向加糖型走向无糖茶与复合茶。

关键词:茶饮料;市场发展;工艺技术

The Trend of Tea Beverage Market in China

LUO Zhi-gang　LI Song-yuan　YANG Qian-hui

Ting Hsin International Group, Tianjin, China

Abstract: Tea beverage has become one of the fastest growing beverage categories in the world. Along with the growth of China's economy, and the unique habit of tea drinking in oriental culture, ready-to-drink tea beverage in China is booming recently and changes the structure of world beverage market. The manufacturers should pay more attention to the raw materials management, the application of new extraction technologies, and packaging technology. We estimate that future market will separate from national-wide taste to domestic tastes and flavors. And the sweetened type will change to non-sugar added type and composite type.

Keywords: tea beverages; market trends; processing technology

纵览全球饮料市场,欧美国家的饮料市场十分成熟,但经过多年的发展,在亚太地区的具有区域独特性的茶饮料发展快速,已经成为全球最大的饮料消费市场类别。整个亚太地区茶饮料市场蓬勃发展,虽然人均消费量 10L 以下,但在东方人群中具有极大的消费基础和偏好,茶饮料消费亚太地区占到60%,北美洲23%,欧洲16%,中东和非洲1%,南美洲和中美洲基本没有此类产品。

1 茶饮料在亚太地区的发展

1.1 日本茶饮料市场

由于经济发展较早,且同样有着饮茶的文化习惯,日本是茶饮料市场最成熟的市场,但在过去几年中整个市场出现了成长动力的趋缓。

日本茶饮料市场主要由众多饮料厂商共同推动,在饮料市场中属于主要类别,主要的营销诉求分为以下3个构面:①强调自有茶叶农场,最具有代表性的品牌是伊藤园,通过宣传自己拥有安全可靠的茶园来说明茶叶的安全来源;②强调正宗百年老店制茶技术,伊右卫门是典型代表;③以同种茶叶组合成不同的风味,开发系列商品获得更多不同消费者的支持。除了上述诉求外,日本在推广茶饮料时还会采取一些特殊的利益点,去争取消费者的认同。例如强调自己的产品使用了国产茶叶,或强调是自己茶叶研究所研究开发的复合配方产品;更进一步还有从包装特性来推广产品。此外,日本商品所惯用的季节限定或添加功能性素材等,在茶饮料中也屡见不鲜,通过强调各个季节所应喝的茶饮料特性,或是使用了罕见的茶叶原料来做茶饮料,以增强对消费者的吸引力。一些大厂商如可口可乐,更将茶饮料的健康诉求作为他们的市场策略,除了使用茶叶他们还使用茶花作为原料,甚至于强调他们使用了两倍的茶叶原料量等。从日本茶饮料的过去到现今的发展,可以发现饮用茶饮料的消费者十分注重理性的诉求点,在原汁原味的茶叶基础上生产茶饮料。

1.2 我国台湾茶饮料市场

我国台湾地区从20世纪80年代起推出了第一款茶饮料产品,但还没有发展完全成熟。台湾茶饮料的主要消费者还是年轻人和上班族,他们从长辈身上传承了一些文化认识,但同时也有喝预包装饮料的需求。与中国大陆地区相同,一般中年以上的消费者,比较习惯于传统茶叶冲泡方式饮茶,老年人是不喝茶饮料的。目前,以茶饮料为载体,添加某些具有保健功能素材的产品也受到消费者喜爱。

1.3 我国大陆茶饮料市场

目前大陆的茶饮料市场尚处在起始的阶段,国内现在对茶饮料的消费相对还很小,但成长势头十分强劲,茶饮料的品种主要集中在冰茶、绿茶和草本茶。最近的趋势是原叶茶和原味茶饮料,这是一个十分有发展潜力的领域。国内的茶叶品种十分众多,且均为消费者所认知,因此相对日本具有很强的优势。原味茶向中国茶饮料发展,如乌龙茶、铁观音等原味茶都会陆续进入市场。各种茶饮料中,目前冰红茶整体市场份额仍平稳成长,绿茶份额则略有衰退,茉莉花茶整体市场份额亦有一定成长。目前市场主要仍集中在几个主要的厂商,如康师傅、统一、娃哈哈、可口可乐、农夫山泉等。

2 中国茶饮料发展的驱动力

中国茶饮料市场的发展受到理性和感性诉求的强力支撑。在情感上,消费者认为茶饮料是与生活息息相关的必需品,另外也认为喝茶是十分风雅的事情,由茶叶做成的茶饮料在情感上受到消费者认同。在生理上,中国具有十分悠久的喝茶历史,且越来越多的实验证明茶叶中含有多种功能性成分(表1),喝茶对身体十分有益。因此以上两个动力移情到以茶为原料的包装饮品上,也推动了茶饮料在国内有更大的发展。

表1 茶叶中的功能性成分

功　能	成　分
一次功能	维生素:V_C、V_E;氨基酸、多糖类等
二次功能	口味:茶氨酸、游离氨基酸(鲜味)、茶多酚(涩味)、咖啡因(苦味)、萜类、醇类、羰基化合物、酯类、黄酮醇类、黄酮及类黄酮类、叶绿素
三次功能	抗氧化型维生素(V_C、V_E、β胡萝卜素)、茶氨酸、γ-氨基丁酸、皂素、人体必须微量元素(锌、锰、氟、硒)等

中国茶叶的品种十分丰富,各地都有自己具有特色的茶叶品种:安溪铁观音、西湖龙井、黄山毛峰、冻顶乌龙、武夷大红袍、云南普洱等。如何把茶叶的优势导入茶饮料市场,来推动中国茶饮料市场的发展,可以从以下4个方面来研究。①文化:包括饮食文化和茶叶文化;②技术:包括原料新技术、制程新技术等;③生活形态:了解消费者的生活形态的变迁,满足消费者的需求;④多样化:消费者对产品使用的多元化,包装多样化(包装容量、包装材料、包装外形设计

等变化)和贩卖形式的多样化(夏季冷藏/冬季温热)。

原料管理新技术的发展对整个茶叶的供应有十分积极的推动作用,日本将乌龙茶进入茶饮料中就是一个很成功的案例,推动了乌龙茶的发展。其中对大宗茶叶管理是未来茶饮料发展的一个重要因素。目前中国的茶饮料还处于前导期,引导消费者品尝不同茶的风味,特色茶进入茶饮料市场,加工、萃取、口味都很重要,可以从日本、甚至欧美市场发展中得到启示,要特别关注消费者对茶文化和口味的关注。

3 展 望

由于消费市场的分众化和细化需求,未来茶饮料新品的发展将由全国性口味发展到地方性的口味,如果将地方性口味的市场份额扩张到其他区域,则需要厂商的引导和对厂商的一个有力的考验。目前的口味还是偏向加糖型,但未来必将走向无糖茶,甚至是复合茶;复合茶和混合茶是未来发展趋势,包装也更加多样化,储藏时冷藏和加热将考验包装的设计。但由于茶饮料本身的形象已相当完美,未来要如日本以茶中功能性成分添加以发展健康饮品的机会则尚待观察。

总而言之,未来厂商在开发市场时,可能还会遇到各类的挑战和问题,主要是以下几个方面,一是茶原料的开发精细化和进步。包括茶叶品种/茶园管理、采摘方式、茶叶加工工艺、茶叶包装与保鲜茶叶入厂检验如何适应大宗茶叶管理和茶饮料生产。二是茶汤萃取新技术的开发与应用。三是加工工艺及包装技术的提升(过滤技术、无菌冷充填、PET阻隔层等)都会影响茶饮料的发展,未来需要由原料供应者、生产厂商及学术界共同的努力。

福鼎白茶爽保健糖开发与研究

张星海[1]　虞赔力[1]　许金伟[1]　王岳飞[2]

1. 浙江经贸职业技术学院应用工程系,中国 浙江；
2. 浙江大学茶学系,中国 浙江

摘　要：主要研究以福鼎白茶提取物为主要功能成分原料的茶爽保健糖的生产工艺技术。该项生产技术利用白茶提取物与木糖醇及其他有关辅料,通过科学配比,运用干法配料、高温化糖、控温返砂、保温浇注及冷却修饰的技术工序,生产出风味优良、口味独特的白茶爽无胶基保健糖。研究发现本工艺关键环节为温度控制（85℃为临界温度）、干法配料比（白茶提取物含量2%为临界点）和感官风味（茶味香精选型与用量）。

关键词：福鼎白茶；白茶爽；保健糖；无胶基

Cool Health Sugar of Fuding White Tea Development and Research

ZHANG Xing-hai[1]　YU Pei-li[1]　XU Jin-wei[1]　WANG Yue-fei[2]

1. The Department of Applied Engineering, Zhejiang Economic and Trade Polytechnic, Zhejiang, China；
2. Department of Tea Science, Zhejiang University, Zhejiang, China

Abstract：Fuding white tea extract raw materials for the main functional components of sugar production of tea cool health technology research. The production technology uses white tea extract and xylitol, and other related accessories, by the ratio of science, the use of the dry ingredients, high temperature of the sugar, temperature control back to the sand, holding pouring and cooling modified the technological process to produce excellent flavor and unique taste of white tea cool gum health sugar. The study found that the key link of the process control temperature (85℃ is the critical temperature), the dry ingredients (white tea extract content of 2% is the critical point) and sensory flavor (flavor of tea selection and dosage).

Keywords：Fuding white tea；white tea cool；health sugar；no gum

福鼎白茶爽保健糖中不添加作为糖载体的胶基,可含化(咀嚼)吞咽,免除食用后无处扔残胶的尴尬,在全球普遍提倡低碳、环保的今天,无胶基保健糖的推出,必将解决传统口香糖污染环境难题。福鼎白茶爽保健糖主要成分为木糖醇、白茶萃取物、柠檬酸及其天然食用色素、香料等。保健糖采用木糖醇为甜味剂,食用后对血糖无影响,符合现代人追求健康的消费观念。

木糖醇是一种天然存在的五碳糖醇,广泛存在于许多水果和蔬菜中,科学研究证实其有很多医疗保健功能,诸如预防龋齿、提神醒脑缓解压力、清胃气与口气、抗辐射、改善胃肠道环境等[1]。木糖醇口香糖防龋功能得到欧洲食品安全局(EFSA)认可;白茶萃取物具有清热润肺、平肝益血、消炎解毒、降压减脂、消除疲劳等功效,可以起到提高免疫力和保护心血管等作用[2];柠檬酸,广泛存在于很多水果和蔬菜中,即可作为风味成分,也可用作化痰药和利尿药;原料成分经过科学配比,为口香糖的爽口、保健奠定坚实的物质基础;运用木糖醇的保健功能,结合21世纪广泛应用的茶爽风味理念,制作成的福鼎白茶爽保健糖不但具有木糖醇和茶多酚的保健功能,而且口感更佳,并且体现绿色环保理念,其商业价值更具潜力。

1 材料与方法

1.1 材　料

白茶、木糖醇(纯度98％以上)、白茶萃取物、柠檬酸、枸橼酸钠、食用香精、天然色素等均为食用级原料,符合国家标准。

1.2 仪器设备

设备:超声波提取器(杭州天华食品有限公司)、喷雾干燥机(SY－6000型)、150型糖果浇注机(杭州天华食品有限公司)、熬糖锅(杭州天华食品有限公司)、保温锅(杭州天华食品有限公司)、返砂锅(杭州天华食品有限公司)、DPH-90辊板型铝塑包装机(吉首市中诚制药机械厂)、紫外可见分光光度计(752型、上海精密仪器仪表有限公司)、电子天平(AR2140型,上海梅特勒仪器有限公司)、高效液相色谱仪(安吉伦Agilent 1100)。

1.3 白茶粉提取工艺

称取适量干燥白茶置于超声波提取罐内,加入适量蒸馏水,约 m(茶叶):m(水)=1:20。设置浸提温度60℃、提取时间30min,超声波辅助提取3次。浸提完成后,合并提取液,将提取液转移至浓缩罐内进行初步浓缩,约浓缩至原体积的 1/3,将初步浓缩完的提取液用陶瓷膜过滤,滤液使用真空浓缩罐进行进一步浓缩,至浓缩液稠状为止。稠状浓缩液通过喷雾干燥机干燥,得较高纯度干燥白茶萃取物。

1.4 白茶爽保健糖生产工艺

保健糖生产工艺流程如图1所示。

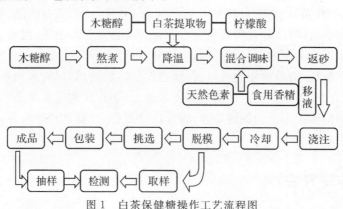

图1 白茶保健糖操作工艺流程图

1.5 保健糖生产工艺

(1)原料准备。称取 2000g 木糖醇、45g 白茶萃取物、180g 柠檬酸及枸橼酸钠置于混匀器内,充分混匀,备用。

(2)浇注机的调试与预处理。浇筑前首先应对糖果加工车间与浇注机主要部件进行消毒,车间用紫外灯消毒 2h,糖果模具用 75% 酒精喷雾消毒。消毒完毕后,打开浇注机主电源与各控制电源,打开空调机,将冷却隧道的温度控制于15℃~20℃。糖膏料斗的保温温度控制在 82℃~85℃。打开风机开关,对浇注设备进行核查,模具进行预冷,确保浇注机能正常、稳定工作。

(3)熬糖。称取 8000g 木糖醇,置于不锈钢夹层化糖锅内,打开电加热开关,将化糖锅预设温度设定于 150℃,适当打开化糖锅的搅拌机,使糖醇均匀溶解,木糖醇持续升温溶解,当糖液温度达到 135℃~140℃ 后,关闭化糖锅开关,停止加热。

（4）混合调味。将预处理过的木糖醇、茶粉混合物缓慢倒入化糖锅内与糖液混合，打开化糖锅搅拌电机，使其在醇液中充分溶解并与醇液完全混匀。打开输液泵，将化糖锅内的糖液输送到保温锅内，保温锅夹层温度控制在85℃，并打开保温锅搅拌电机，持续搅拌，使糖膏缓慢降温。当温度降至90℃～95℃时，加入预先处理好的食用香精和食用色素，持续搅拌降温，并将保温锅的夹层温度控制在85℃。

（5）返砂[3]。在搅拌降温过程中，当糖液温度降至85℃后，会有短暂的结晶升温过程，当温度再次降至85℃后，糖液表面有结晶形成，返砂基本完成，便可进行浇注。

（6）浇注。返砂完成后，打开输液泵，将保温锅内的糖膏通过保温管路转移至糖膏料斗内，开启浇注机电机开关，控制浇注速度进行定量浇注。

（7）脱模。浇注在模具上的木糖醇膏通过冷却隧道冷凝降温，模具经过脱模部分时，糖果被压板压出，掉落于输送带，经输送带传送至包装车间。

（8）筛选包装。脱模后的糖果通过输送带输送到无菌包装车间，对不规则和次品颗粒进行筛选，合格产品由自动包装机包装成成品，入库。

1.6 产品质量分析

(1)感官要求。色泽：均匀一致，符合品种应有的色泽。

形态：块形完整，表面光滑，边缘整齐，大小一致，无变形，无黏连。

组织：糖体光亮，少透明，略有弹性，不粘牙，无硬皮。

滋味、气味：符合品种应有的滋味气味，无异味。

杂质：无肉眼可见杂质。

(2)理化检测[4-6]。样品预处理：取10粒糖果，内层用滤纸，外层用塑料袋包好，然后用锤子锤碎。取10g样品于凯氏烧瓶中，按硝酸硫酸法进行有机破坏后定容至100mL。

砷：取样品消化液25mL，按GB 5009.11操作。

铅：取样品消化液20mL，按GB 5009.12操作。

铜：取样品消化液20mL，按GB 5009.13操作。

(3)微生物指标[7-10]。样品预处理：用无菌镊子夹取带包装纸的糖果，称取约25g数块，加入预温至45℃的灭菌生理盐水225mL，待溶化后检验。菌落总数按GB 4789.2操作。大肠杆菌按GB 4789.3操作。致病菌按GB 4789.4、4789.5、4789.10、4789.11操作。

2 结果与分析

2.1 操作温度分析

温度是制糖工艺中最重要的因素之一,温度控制的好坏将直接影响到糖果最终的质量与口味。在常压熬煮过程中,应将醇液中大部分的水分排出。熬煮过程中糖液的温度过高,糖膏颜色会加深,影响糖果美观;也不能将熬糖温度控制过低,否则醇膏比较软,导致木糖醇不易凝固。本工艺中木糖醇的最高熬煮温度控制在150℃左右。

其次控制混合调味时的温度,当醇液降到一定温度后才能添加香精、色素等调味辅料,过高的温度会使香精等成分分解,影响糖果口感。实际操作中,熬糖锅内醇液温度降至90℃~95℃便可添加香精与香料。

最后一道控制温度工序是浇注温度的控制,也是最关键的一步。浇注温度偏低会使糖膏黏度过大,流动性差,导致浇注效果差,糖果出现拖尾,影响外观;浇注温度过高则导致糖果流动性过大,浇注后不易凝结,模具上的糖果成黏稠状。浇注温度过高会使醇膏与模具的温差过大,冷凝后的糖膏容易产生气泡。木糖醇的浇注温度控制在85℃最佳,浇注机的糖膏料斗保温温度亦为85℃。

2.2 返砂操作分析

返砂是硬糖的主要质量变化状况,指其组成中的糖类从无定形状态重新恢复为结晶状态的现象。经吸收水汽并呈发烊的硬糖表面,在周围相对湿度降低时,表面的水分子获得重新扩散到空气中去的机会,水分扩散导致表面熔化的糖类分子重新排列形成结晶体,这是一层细小而坚实的白色晶粒,硬糖原有的透明性完全消失,这种现象称为返砂。

木糖醇糖果生产过程中,应严格控制返砂程度。本工艺须促进糖果返砂,抑制糖果发烊。当保温锅内糖膏温度第1次降至85℃时,继续搅拌,糖果出现短暂升温,此过程为糖液返砂现象。当温度第2次降至85℃时,返砂基本结束,应立即浇注,防止流动性过低,影响糖果浇注。温度随时间变化波动曲线如图2所示。

2.3 质量指标分析

(1)感官评价。块形完好,不毛糙、无气泡、边缘整齐,大小厚薄均匀且无杂质;色泽鲜艳,香味纯净适中,滋味可口,风味独特。

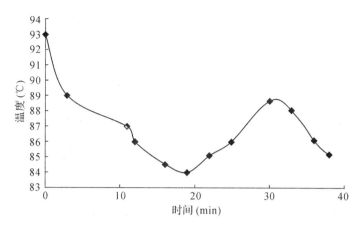

图2 返砂温度变化曲线

（2）理化分析。铅(mg/kg)≤0.5;砷(mg/kg)≤0.3;铜(mg/kg)≤1.0。

（3）微生物分析。菌落总数≤1000个/g;大肠菌群≤40MPN/100g;真菌≤25个/g;致病菌未检出。

参考文献

[1] 李香妹,马少怀.木糖醇的生理功能及含木糖醇食品的试制[J].食品工业科技,2003,24(7):51-53.

[2] 冯永强,王江星.木糖醇的特性及在食品中的应用[J].食品科学,2004,25(11):379-381.

[3] 杨贤强,王岳飞,陈留记等.茶多酚化学[M].上海:上海科技出版社,2003.

[4] GB/T 5009.11—2003 食品中总砷及无机砷测定[S].

[5] GB/T 5009.12—2010 食品中铅的测定[S].

[6] GB/T 5009.13—2003 食品中铜的测定[S].

[7] GB/T 4789.2—2008 食品微生物学检验 菌落总数测定[S].

[8] GB 4789.3—2010 食品微生物学检验 大肠菌群测定[S].

[9] GB 4789.4—2010 食品微生物学检验 沙门氏菌检验[S].

[10] GB 4789.10—2010 食品微生物学检验 金黄色葡萄球菌检验[S].

茶色素染纸技术研究

吴媛媛　王　洁　屠幼英

浙江大学茶学系，中国 浙江

摘　要：茶叶中含有大量的色素，有绿色、红色、黄色、橙色等，是理想的天然染色剂。本研究采用茶色素作为染色剂，通过单因素分析及正交试验，探讨了染色温度、时间、固色剂用量及茶色素用量对染纸效果（上染率、色差值和褪色率）的影响，获得茶色素染纸最佳工艺条件：50℃、茶色素 TF14 用量 30%（o.w.f）、海藻酸钠 12%（o.w.f）、时间 50min。

关键词：茶色素；染纸；上染率；褪色率；色差

Research on the Technology of Paper-dyeing with Tea Pigments

WU Yuan-yuan　WANG Jie　TU You-ying

Department of Tea Science, Zhejiang University, Zhejiang, China

Abstract：The tea pigments with green, red and yellow color are ideal natural dyes. In the present work, the tea pigments were used in dyed paper. The single factor analysis and orthogonal experiment were performed to evaluate the effects of the dyeing temperatures, dyeing times, amounts of fixing agent and amounts of black tea pigments on the dyeing rate, color parameters and color fastness. The optimal conditions for this technology were obtained as 50℃, 50min, Sodium alginate 12% (o.w.f), and TF14 30%(o.w.f).

Keywords：tea pigments; dyeing rate; color fastness; color parameters

我国利用植物汁液染色的历史已有 4500 多年。早在夏代，就已有使用蓝草进行染色的记录。到了明清时期，根据《天工开物》和《本草纲目》的记载，可用作染色的植物扩大到几十种[1]。大多数天然植物染料具有环保、无毒、无害、对人体无刺激的特点。有些植物染料如茜草[2]、靛蓝[3]等，其染色的织物还有防虫和杀菌作用，具有自然芳香。

工业革命的发展,合成色素以其省时省力且色彩艳丽、色谱齐全、耐洗耐晒、价格便宜等优点迅速代替了天然染料。但已有研究表明,有100多种常用染料有可能产生致癌物质,并且染色工厂对周边的环境污染也成为严重的问题,开发"环境友好型染料"和"绿色清洁染色工艺"刻不容缓。

茶叶中的茶色素(茶黄素、茶红素、茶褐素)是一种理想的天然染色剂[4]。茶色素水溶液色泽鲜艳,根据不同的浓度可配出从淡黄到深咖啡色等不同的颜色,长期保存亦不褪色或沉淀。将茶色素应用于印染工艺[5-7],其颜色自然柔和。同时,茶色素如茶黄素等具有一定的抗氧化[8-9]、杀菌[10]作用,在染纸方面有其应用价值。

本文在初步筛选了染色方法、纸张原料、固色方式的基础上,研究了染色温度、时间、固色剂用量及茶色素用量对染纸效果的影响,通过正交试验获得茶色素染纸的最佳工艺条件。

1 材料与方法

1.1 材料与设备

(1)试验材料。染纸色素:TF14(含有茶黄素14%,杭州英仕利生物科技有限公司)。

纸张:市售原浆纸板、定性滤纸(双圈牌,杭州沃华滤纸有限公司)、市售素描纸(不含酸性、无漂白剂,大千纸业有限公司)。

固色剂:海藻酸钠、明矾(华东医药有限公司)。

(2)试验设备。分析天平(上海精密科学仪器有限公司)、电热恒温水槽DK-8B(上海精宏试验设备有限公司)、紫外—可见分光光度计UV-759S型(上海精科)、台式恒温振荡器TH2-C(太仓市华美生化仪器厂)、超声波清洗器、循环水式多用真空泵SHB-IJIA(上海豫科教仪器设备有限公司)、烘箱DGX-9073B-2(上海福玛试验设备有限公司)、Hunterlab测色仪(ColorQuest XE)。

1.2 试验设计

本研究初步试验通过色差值和感官评定,确定了基本的染色方法(表面染色和浆内染色)、染色纸张原料(市售原浆纸板、定性滤纸、市售素描纸)、固色方式(海藻酸钠、明矾、无固色剂)。通过L、a、b的比较、染色均匀度以及纸张质量三方面的比较,确定浆内染色为染色方法、市售素描纸为染色原料纸、海藻酸钠作为固色剂。在此基础上，进行下面的试验。

(1) 单因素变量试验。市售素描纸 1g、TF14 (10%～60%)(o. w. f)、固色剂(3%～18%)、浴比 1∶50、染色时间(15～110min)、染色温度(30℃～80℃)、打浆 30s、抽滤 30s、烘箱干燥 1h 的条件下，分别以温度、时间、固色剂海藻酸钠用量、TF14 用量为变量进行试验，测定成品的上染率、褪色率以及色差，每组实验重复进行 3 次。

(2) 正交试验。设计染色温度、时间、染料 TF14 用量、固色剂用量的四因素正交试验 $L_9(3^4)$，研究各因素对染色效果的影响，筛选效果最佳的染色工艺参数。

1.3 试验方法

(1) 纸张茶色素上染率测定。茶色素上染率测定方法：采用 Roberts 比色法测定茶黄素、茶红素、茶褐素含量的变化[11]。TF14 中主要含有茶黄素(TF)、茶红素(TR)、茶褐素(TB)，纸张主要吸附这些色素，因此通过测定染色前后液体 3 种色素的含量比测定上染率：

$$上染率(\%) = (1 - nA_1/mA_0) \times 100\%$$

式中：m/n 为染色前染液/残液稀释倍数；A_0/A_1 为染色前染液/残液稀释 m/n 倍数后的浓度。

(2) 纸张褪色率测定。采用 100℃水浸泡的方法，检测茶黄素、茶红素、茶褐素 3 种色素的褪色量。具体方法为：将染色后成品纸投入 100℃水中，自然降温 10min 后，取出纸张，检测浸泡液中 3 种色素溶出量。染液浓度、残液浓度和浸泡液的浓度通过分光光度计测定(比色法)。

$$褪色量(\%) = m/(m_0 - m_1) \times 100$$

式中：m 为浸泡液的染料量；m_1 为残液染料量；m_0 为反应前染料量。浸泡液的染料量＝浸泡液的质量浓度×浸泡液的容积。

(3) 色差分析。纸张色差测定以空白纸张作为对照，用 Hunterlab 测色仪测定染色后纸张的色差。

以 L、a、b 3 个参数来表示物体的颜色变化性。其中 L 值表示明暗度，代表人眼对黑色、白色反应之值。完全白的物体可视为 100，完全黑的物体视为 0。a 值与 b 值分别为物体的色调及彩度，以红(+a)、绿(-a)、黄(+b)、蓝(-b)来表示。L、a、b 表示与空白对照的差值，差值越大，表示纸张颜色偏向于指标为正值时的颜色。以 L 为例，值越大，表示纸张颜色越亮；值越小，表示纸张颜色越暗。

(4)数据统计与分析。采用 Excel、SPSS V13.0 等统计软件对试验数据进行汇总与统计、方差分析、相关分析和回归分析。

2 结果与分析

2.1 染色条件对染色效果的影响

2.1.1 温度对染色效果的影响

上染率如图 1 所示,随着温度升高,上染率随之上升。30℃条件下上染率最低,50℃上染率达到最高,随后又呈缓慢下降的趋势。

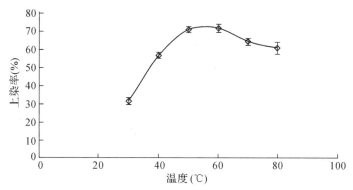

图 1 温度对上染率的影响

温度对染纸色差影响如表 1 所示,50℃下的 L、a、b、dE 的数值最高,这说明其红度、黄度比较高,颜色较好,与空白对照的色差大,染色较深。

表 1 温度对染纸色差影响

温度(℃)	L	a	b	dE
30	60.26	8.2	13.08	39.41
40	62.72	8.05	13.37	37.64
50	62.79	8.75	14.16	38.49
60	66.23	6.95	12.18	34.07
70	66.24	6.73	11.21	33.40
80	65.40	6.45	11.88	34.42

图2为染色温度对纸张褪色率影响,可以看到随着温度的升高,其褪色率上升,一方面是由于较高温度下,色素分子热运动活跃,导致褪色量较多;另一方面,在高温下,可能纸张纤维结构变得松散,更利于色素分子脱离纸张纤维。

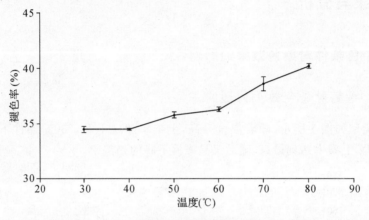

图2　温度对纸张褪色率的影响

2.1.2　时间对染色效果的影响

上染率随着时间的增加而增加,在50min时出现最高上染率(图3)。之后,随着时间的增加,上染率略有下降,可能与部分茶色素的解吸附有关。

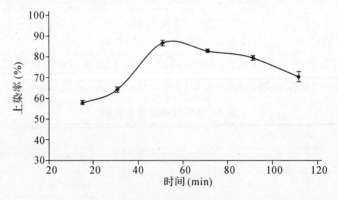

图3　染色时间对上染率的影响

表2为染色时间对色差的影响,50min条件下的纸张颜色较好,其L、a、b值均呈现出最高值。其中50min之后的L值有所下降,说明纸张的明亮度降低,可能与色素分子的氧化聚合有关。

表 2　染色时间对染纸色差的影响

时间(min)	L	a	b	dE
15	59.63	7.79	12.91	39.74
30	61.69	7.75	13.16	37.49
50	61.74	9.84	18.51	44.85
70	57.77	8.02	16.68	44.28
90	57.42	8.96	16.3	44.3
110	55.6	9.42	16.78	46.07

图 4 为染色时间对纸张褪色率影响,纸张褪色率随染色时间延长而降低;60min 以后,其褪色率缓慢下降,之后慢慢趋于平衡。

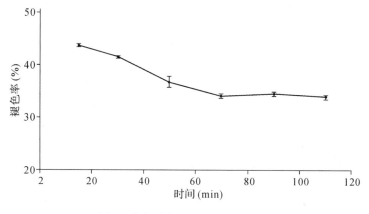

图 4　染色时间对纸张褪色率影响

2.1.3　TF14 用量对染色效果的影响

上染率如图 5 所示,最初阶段上染率随着时间的增加而增加,在 TF14 用量 30%(o.w.f)条件下,显示出最高上染率,在 30%(o.w.f)之后,随着时间的增加,上染率有明显下降。

表 3 为 TF14 用量对染纸色差的影响,可以看到随着 TF14 用量的增加,a、b 值都有所增加,其中 20%～30% 间增加幅度较大;L 值随着 TF14 用量的增加而有所降低,这说明其暗度增加,染色加深。

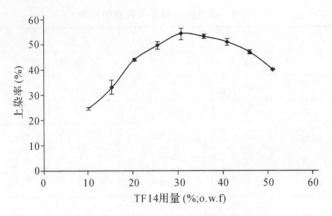

图 5 TF14 用量对上染率的影响

表 3 TF14 用量对染纸色差的影响

TF14(%)	L	a	b	dE
10	67.14	6.81	10.54	32.34
20	61.51	7.90	13.67	38.74
30	55.89	9.00	15.03	44.17
40	52.63	9.34	16.28	47.56
50	51.85	9.52	16.43	49.01
60	49.60	9.84	16.54	50.29

图 6 为染色 TF14 用量对纸张褪色率的影响，TF14 用量在 10%～30% 褪色率增加不明显，30% 用量以上褪色率显著增加，可能是因为 TF14 量越多，在纸张上吸附越多，而纸张纤维表面积是一定的，色素分子与纸张纤维表面结合数达到一个相对饱和状态后，其色素附着并不牢固，导致褪色率上升。

2.1.4 固色剂海藻酸钠用量对染色效果的影响

上染率如图 7 所示，最初阶段上染率随着时间的增加而增加，在海藻酸钠用量 9%(o.w.f)时，上染率最高；之后，随着时间的增加，上染率略有所下降。

海藻酸钠用量对染纸色差的影响见表 4。从 a 值的红－绿轴以及 b 值黄－蓝轴来看，当海藻酸钠用量为 12%(o.w.f)时，a、b 值都最大，显示出红、黄色度最高。另外，亮度 L 值随着海藻酸钠用量的增加，其值也增加，当海藻酸钠用量为 18%(o.w.f)时，其亮度最高，说明海藻酸钠的使用有利于染纸质量的提高。

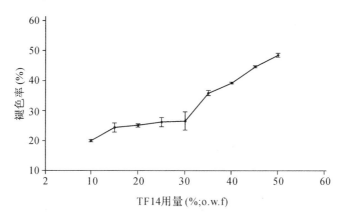

图 6 TF14 染色对纸张褪色率的影响

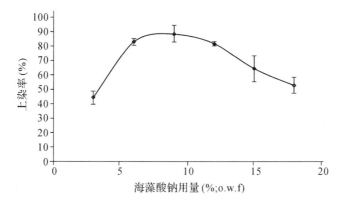

图 7 海藻酸钠用量对上染率的影响

表 4 海藻酸钠用量对染纸色差的影响

海藻酸钠(%)	L	a	b	dE
3	61.18	8.07	13.84	39.12
6	61.51	8	14	38.96
9	62.5	7.9	13.42	37.82
12	62.97	8.58	13.48	37.62
15	64.5	7.83	12.75	35.87
18	64.65	7.21	12.62	35.57

图 8 为不同用量海藻酸钠染色对纸张褪色率的影响,随着固色剂海藻酸钠用量的增加,褪色率降低。这是因为海藻酸钠有一定的固色效果,能增加色素分子与纸张纤维的结合能力,并在纸张表面形成一层薄薄的膜状物质。

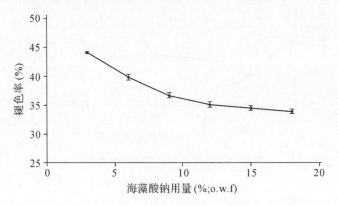

图 8　海藻酸钠染色对纸张褪色率的影响

2.2　正交试验

根据单因素试验结果,我们设计了 $L_9(3^4)$ 正交试验,正交试验因素水平及结果见表 5 和表 6。以最主要的指标——上染率为考察因子,各试验因素的影响效果依次为 B(TF14 用量)＞A(温度)＞D(时间)＞C(海藻酸钠用量)。最优组合为 A2B2C3D2,即在 50℃、TF14 用量 30%(o.w.f)、海藻酸钠 12%(o.w.f)、时间 50min 条件下,上染率最高。同时,辅助指标色差值以及褪色率在该条件下的效果最好。

表 5　正交试验因素水平表

水平	因素			
	A:温度(℃)	B:TF14 用量(%)	C:海藻酸钠用量(%)	D:时间(min)
1	40	20	6	40
2	50	30	9	50
3	60	40	12	60

表6　正交试验结果

试验号	A	B	C	D	上染率(%)
1	1	1	1	1	43.50
2	1	2	2	2	68.89
3	1	3	3	3	52.97
4	2	1	2	3	82.18
5	2	2	3	2	81.06
6	2	3	1	1	60.73
7	3	1	3	2	57.99
8	3	2	1	1	77.68
9	3	3	2	3	31.39
K1	165.36	183.67	181.91	181.91	
K2	223.97	227.63	182.46	207.94	
K3	167.06	145.09	192.02	166.54	
k1	55.12	61.22	60.64	60.64	
k2	74.66	75.88	60.82	69.31	
k3	55.69	48.36	64.01	55.51	
极差R	19.54	27.51	3.37	13.80	
主次顺序	B＞A＞D＞C				
最优组合	$A_2B_2C_3D_2$				

3 展　望

本文以天然茶色素作为染纸的染料,研究了染纸工艺参数对染纸效果的影响,通过正交试验获得最优的染纸参数,并形成茶色素染纸工艺。制得成品纸张色彩自然素雅,能很好地减弱纸张反射光,让眼睛处于一种相对放松的状态。茶叶色素染色纸张,对身体无毒无害、不致癌,对环境无污染,可广泛应用于书本纸张、个性信纸、明星片、美术用纸等。同时,由于茶色素相对合成染料的成本更高,若要完全替代合成染料还需要从降低成本,提高上染率、色牢度、颜色丰富度等方面进行改进。

我们相信随着生物工程技术的发展,天然染料会重新焕发出生命活力,使21世纪的着色产品色彩更加斑斓、更加环保、更利于人类健康。

参考文献

[1] 屠恒贤. 植物染料染色历史与现状[J]. 染整技术, 2002(2): 18—20.

[2] 李清蓉, 陈宁娟. 天然染料等同体茜素的染色热力学研究[J]. 北京纺织服装学院学报, 1999, 10(2): 23—26.

[3] 曾桥, 董文宾. 功能性使用天然色素研究进展[J]. 中国调味品, 2009, 34(5): 22—24.

[4] 夏明, 杜琪珍. 天然红色素研究进展[J]. 食品研究与发展, 2002(6): 38—41.

[5] 李凤艳, 杨巧芬, 洪彩虹等. 天然茶色素对棉纤维的铁离子媒染染色[J]. 纺织学报, 2010(31): 60—68.

[6] 李凤艳, 刘丽华. 媒染剂对茶色素染色棉纤维性能的影响[J]. 纺织学报, 2011(8): 62—66.

[7] 黄旭明, 王燕, 蔡再生. 茶叶染料对真丝绸染色性能初探[J]. 丝绸, 2005(6): 31—33.

[8] Yang Z Y, Tu Y Y, Xia H L, et al. Suppression of free-radicals and protection against H_2O_2-induced oxidative damage in HPF-1 cell by oxidized phenolic compounds present in black tea[J]. Food Chem, 2007, 105: 1349—1356.

[9] Wu Y Y, Li W, Xu Y, et al. Evaluation of the antioxidant effects of four main theaflavin derivatives through chemiluminescence and DNA damage analyses [J]. J Zhejiang Univ-Sci B (Biomed & Biotechnol), 2011, 12(9): 744—751.

[10] 金恩惠, 吴媛媛, 屠幼英. 茶黄素抑菌作用的研究[J]. 中国食品学报, 2011, 11(6): 108—112.

[11] 张正竹. 茶叶生物化学实验教程[M]. 北京: 中国农业出版社, 2009. 52—54.

茶饼干的配方优化研究

陈贞纯　金恩惠　孟　庆　屠幼英

浙江大学茶学系,中国 浙江

摘　要：本文采用正交实验研究了黄油、糖粉、膨松剂、超微绿茶粉四因素对绿茶饼干风味的影响,获得饼干最佳配方：面粉100g、黄油65g、糖粉60g、泡打粉0.5g、超微绿茶粉3g,其他配料为奶粉5g、鸡蛋20g,制得口感好,略带茶香,金黄带绿,入口较酥软且茶味醇正的绿茶曲奇饼干,并对其理化性质进行研究,发现添加超微绿茶粉可降低饼干含水量、酸度和游离脂肪含量。

关键词：绿茶粉；饼干；配方研究；理化指标

Research on Optimizing the Formula of Tea Cookies

CHEN Zhen-chun　JIN En-hui　MENG Qing　TU You-ying

Department of Tea Science, Zhejiang University, Zhejiang, China

Abstract: In this study, four factors and three levels were designed to select the optimum formula of cookies, the sensory evaluation and physical and chemical characteristics of the tea cookie were measured too. According to the results we obtained the optimum formula of cookie which is contain 100g flour, 65g butter, 60g powdered sugar, 0.5g baking powder, 3g ultrafine green tea powder, 5g milk powder and 20g eggs, and made the good taste and fresh color green tea cookie. Meanwhile, we found that adding ultrafine green tea powder would cut down the moisture, acidity and free fat contents of cookie.

Keywords: green tea powder; cookies; formulation study; physical property

近年来,我国大部分产茶省对夏秋茶不能有效利用,造成资源的极大浪费。而饼干作为消费者青睐的休闲食品,消费量逐年上升。目前中国饼干人均年消费量为1kg左右,而发达国家饼干的人均年消费量为24～35kg,中等发达国家也有12～18kg,故中国的饼干市场潜力十分巨大。同时随着人们健康观念的加强,健康食品也越来越受到消费者的青睐。

饼干含有大量的淀粉、糖粉和脂肪,多吃较腻,不适合糖尿病人和肥胖病人等多种人群食用。但是绿茶具有良好的减肥效果,并且可以降低和调节人体的血糖,尤其夏秋茶 TP 含量高,可以减轻饼干的甜度,增加抗氧化能力,延长货架期[1]。本研究采用超微绿茶粉作为添加物,参照曲奇饼干的加工工艺,试研制一种口感好、色泽漂亮的绿茶饼干,不仅可以扩大食用人群,又可以降低大量淀粉、糖粉和脂肪的摄入对健康带来的风险,同时可为我国大量夏秋茶找到可行的消费途径。因此,绿茶饼干融口感、保健于一体,作为新型健康饼干种类有良好的发展空间。

1 材料与方法

1.1 材料、试剂及设备

(1)实验材料。超微绿茶粉(四川青衣神茶叶有限公司);市售低筋面粉、黄油、精制白砂糖、奶粉、鸡蛋、糕点用膨松剂。

(2)主要试剂。甲基橙、氢氧化钠、浓盐酸、95％乙醇、无水乙醚,均为分析纯。

(3)仪器设备。DELTA-320PH 计(梅特勒－托利多仪器有限公司)、DGX-9073 烘干箱(上海福玛实验设备有限公司)、C.202 型电热恒温干燥器、电子天平 AE2009(梅特勒－托利多仪器有限公司)、Colorquest XE 光谱光度计(Hunterlab)、电热食品烘炉(HK01-302-00025)、打蛋机。

1.2 实验方法

(1)饼干制作。饼干制作流程如图 1 所示。

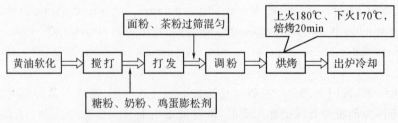

图 1　超微绿茶粉曲奇饼干制作流程

(2)饼干配方正交试验设计。加入面粉100g(以面粉为基准,其他辅料均以占面粉质量比例计算),奶粉5％、鸡蛋液20％,并采用$L_9(3^4)$正交试验对黄油、糖粉、膨松剂和超微茶粉的添加量4个因素进行筛选,其因素水平如表1所示。

表1 正交试验因素水平表

水平	因素			
	A:黄油(％)	B:糖粉(％)	C:膨松剂(％)	D:超微绿茶粉(％)
1	45	40	0	3
2	55	50	0.5	4
3	65	60	1.0	5

(3)风味审评实验。成立评定小组,根据文献制作评定标准(表2),在互不干扰的情况下打分评定产品品质。

(4)理化性质检测。参考 GB/T 5009.3—2003[2] 测定饼干水分含量;参考 QB/T 1254—2005[3] 测定 pH 值和碱度;参考 GB/T 5009.6—2003[4] 测定脂肪含量;将 ColorQuest XE 光谱光度计调至反射光模式,测定样品的色差。所有数据均采用 SAS.9.1 软件进行统计分析。

表2 风味评定标准表

项目	外形(10分)	结构(20分)	口感(40分)	色泽(15分)	香气(15分)
好	8～10(花纹清晰、整齐,不收缩,不变形,不起泡)	17～20(紧密,质地酥松,断面呈多孔状,无大孔,无杂质,无针孔)	36～40(酥化,入口很快溶化,无粗糙感,细腻不粘牙,茶味醇正)	12～15(均匀一致,呈棕金色或金黄色并带绿色)	12～15(香气浓郁,无异味,略带茶香)
一般	5～7(凹底面不超过1/3,花纹略模糊,有一定破碎)	13～16(断面无层次,呈多孔性组织,略有大孔,组织较粗糙)	31～35(松软,略粘牙,略有粗糙感,茶味较小)	9～11(色泽不太均匀,呈浅黄色,表面有少许焦煳)	9～11(香气略淡,茶香不明显)
差	<5(花纹模糊,表面起泡,破碎较严重)	<13(质地僵硬,疏松,断面有层次)	<31(口感僵涩,粗糙,粘牙,苦味略明显)	<9(呈黄褐色,不均匀,有焦糊,有明显异色)	<9(香气很淡,有异味,无茶香)

2 结果与分析

2.1 饼干配方筛选

以风味审评结果为参考,正交试验结果见表3。实验结果显示较优的水平组合为 $A_3B_2C_2D_2$。影响绿茶曲奇品质的主要因素依次为黄油＞膨松剂＞糖粉＞超微绿茶粉。

油脂早已被证明是酥性饼干中至关重要的因素[5],黄油在曲奇饼干中用量较多,且对饼干花纹的清晰度、光泽度、色泽、面团组织结构和口味等起着重要作用,故正交实验得出黄油对饼干风味影响最大[6]。而膨松剂影响着饼干的松软程度、口感等,因此其对饼干的风味具有较大的影响[7]。糖也已被证明是酥性饼干中较重要的因素[5],不仅赋予饼干香甜的滋味,也影响饼干比容和着色。而实验结果中超微绿茶粉对饼干风味的影响相对于其他三者来说较小,原因可能在于实验设计中茶粉的浓度梯度较小,而人们对茶粉的感观阈值较大。

进一步试验比较9号样品 $A_3B_3C_2D_1$ 和实验得分最高配方 $A_3B_2C_2D_2$,得出 $A_3B_3C_2D_1$ 风味更优于 $A_3B_2C_2D_2$。由此,我们选择最佳配方为:100g面粉中,黄油65％、糖粉60％、膨松剂0.5％、超微绿茶粉3％,其他配料为奶粉5％、鸡蛋20％。

表3 风味审评正交试验结果

样本号	A	B	C	D	总 分
1	1	1	1	1	74.5
2	1	2	2	2	80.5
3	1	3	3	3	77.4
4	2	1	2	3	79.6
5	2	2	3	1	81.4
6	2	3	1	2	78.6
7	3	1	3	2	80.8
8	3	2	1	3	79.6
9	3	3	2	1	83.8

续表

样本号	A	B	C	D	总 分
T1	232.40	234.90	232.70	239.70	
T2	239.60	241.50	243.90	239.90	
T3	244.20	239.80	239.60	236.60	
t1	77.47	78.30	77.57	79.90	
t2	79.87	80.50	81.30	79.97	
t3	81.40	79.93	79.87	78.87	
R	3.93	2.20	3.73	1.10	
最优选择	A_3	B_2	C_2	D_2	
影 响	A＞C＞B＞D				

2.2 不同配方对饼干理化性质的影响

对正交试验各样品进行理化指标测定,结果见表4。7号样品的pH值、水分和脂肪含量是9个处理中最高的,分别为6.58%±0.06%、4.66%±0.19%和20.57%±1.33%;9号含水量最低,为1.47%±0.26%;而6号则有最低碱度值,为(0.11±0.01)g/100g;3号脂肪含量最低,为10.37%±1.80%;在色泽方面,1号明度最好($L=65.48±0.19$),6、8、9次之,且所有样品都偏红偏黄($a>0,b>0$)。

表4 理化性质正交试验结果

样品号	测定值						
	pH	水分(%)	脂肪(%)	碱度(g/100g)	L 值	a 值	b 值
1	6.43±0.04	3.87±0.05	13.27±0.25	0.15±0.01	65.48±0.19	2.20±0.02	19.26±0.04
2	6.42±0.04	3.29±0.71	16.94±1.48	0.12±0.02	61.43±0.88	2.17±0.03	18.94±0.31
3	6.34±0.07	3.51±0.32	10.37±1.80	0.16±0.00	60.07±0.63	3.87±0.05	18.76±0.09
4	6.33±0.08	2.43±0.10	14.37±0.60	0.16±0.02	60.66±0.14	2.79±0.00	19.67±0.05
5	6.46±0.01	2.49±0.12	12.47±1.00	0.17±0.00	62.17±0.72	4.03±0.05	19.52±0.27
6	6.41±0.03	1.91±0.18	11.92±1.76	0.11±0.01	64.21±0.42	2.90±0.03	19.53±0.14

续表

样品号	测定值						
	pH	水分(%)	脂肪(%)	碱度 (g/100g)	L值	a值	b值
7	6.58±0.06	4.66±0.19	20.57±1.33	0.13±0.00	59.83±1.01	1.68±0.05	19.49±1.09
8	6.32±0.05	2.21±0.26	12.43±1.48	0.13±0.01	63.08±0.16	1.80±0.03	20.48±0.06
9	6.40±0.05	1.47±0.26	14.47±0.35	0.12±0.01	63.56±0.21	3.27±0.04	19.89±0.05

表 5 的方差分析结果表明，超微绿茶粉、膨松剂和糖对饼干 pH 值有显著影响，而所有因素均对产品的碱度有显著影响，其中超微绿茶粉和膨松剂对饼干 pH 值和碱度的影响均达极显著水平。图 2 显示超微绿茶粉添加量的增加可降低饼干 pH 值。绿茶因其含有的茶多酚、维生素 C、绿原酸等酸性物质而呈弱酸性，因此茶粉加入饼干等面制体系中会使酸性增加。选用的膨松剂为中性泡打粉，主要成分是碳酸氢钠，再配以酸性材料及淀粉，添加相对较高水平的膨松剂会使体系的 pH 值有所增大，主要原因为烘焙过程中产生碱性物质。我们考虑到偏碱性的环境对绿茶色素的稳定不利，为减少在生产中尤其是高温烘焙过程中破坏绿茶的天然色素，在加工超微茶粉烘焙食品时应尽量选择中性膨松剂或 pH 调节剂，使体系维持中性偏弱酸性条件，才有利于保持产品自然绿色的品质。

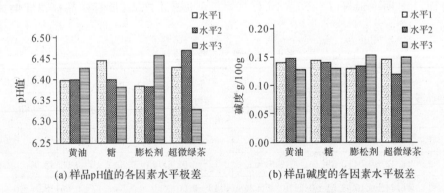

图 2 各因素水平对超微绿茶曲奇饼干 pH 值和碱度的影响

由含水量的方差分析结果可知 4 个因素对含水量都有极显著的影响。在一定范围内随着糖添加量增多，含水量逐渐降低，因为其反水化作用限制了淀粉的吸水能力，这不仅抑制面筋形成也控制饼干的含水量。同时糖和黄油与面粉形成的多孔性结构也影响水分的蒸发效率,从而决定了产品最终含水量。超微茶粉对含水率的影响呈先上升后下降的趋势。

表5还显示膨松剂对饼干的脂肪含量有显著影响,而其余3个因素对其影响达极显著水平。

方差分析表明,膨松剂和超微绿茶粉均对明度(L值)有极显著影响,而从图3可以看出,随着膨松剂和超微绿茶粉添加量增多,明度都呈下降趋势。

表5 pH值、碱度、含水率、脂肪方差分析结果

变异来源	pH值		碱度		含水量		脂肪	
	均方(自由度)	Sig(F值)	均方(自由度)	Sig(F值)	均方(自由度)	Sig(F值)	均方(自由度)	Sig(F值)
黄油(A)	0.003126 (2)	0.3061 (1.27)	0.00039104 (2)	0.0333 (5.08)	3.77173118 (2)	<.0001 (40.32)	14.06340139 (2)	0.0072 (8.98)
糖(B)	0.009337 (2)	0.0426 (3.78)	0.00039105 (2)	0.0333 (5.08)	4.40095559 (2)	<.0001 (47.04)	21.97330139 (2)	0.0017 (14.03)
膨松剂(C)	0.016381 (2)	0.007 (6.63)	0.00127572 (2)	0.001 (16.58)	3.32448787 (2)	<.0001 (35.54)	11.73190556 (2)	0.0121 (7.49)
超微绿茶粉(D)	0.045215 (2)	<.0001 (18.3)	0.04521481 (2)	0.0007 (18.08)	1.19038071 (2)	0.0004 (12.72)	27.24291806 (2)	0.0008 (17.4)
误差	0.00247 (18)		0.00007693 (18)		0.09355431 (18)		1.5660028 (18)	

结果显示4个添加成分对 a 值(红绿程度)具有极显著的影响。图3显示,随着糖的添加量增加,a 值逐渐增大,即红色更明显。烘焙食品呈现的红棕色,主要是在高温烘焙过程中发生的美拉德反应和焦糖化作用,而这一过程除与烘焙温度和时间有关外,还与糖的种类和含量有关系。蔗糖作为一种非还原糖主要以焦糖化作用对饼干着色,因此相同时间温度下随着蔗糖增加饼干红度也随之增加。

黄油对样品的黄蓝程度有极显著影响,并且随着黄油添加量增加 b 值逐渐上升(黄色更加明显)。膨松剂有显著影响,但其影响趋势恰与黄油相反,随着膨松剂增加 b 值呈下降趋势。

3 总结与展望

通过正交设计和风味审评方法,我们筛选出了添加超微绿茶粉的饼干配

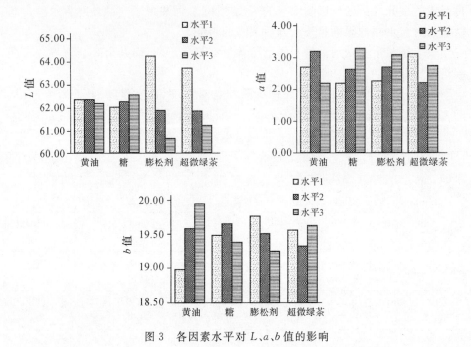

图 3 各因素水平对 L、a、b 值的影响

方：即 100g 面粉中，黄油 65％、糖粉 60％、泡打粉 0.5％、超微绿茶粉 3％，其他配料为奶粉 5％、鸡蛋 20％。制作的饼干风味独特，茶香明显，色泽金黄嫩绿。

添加茶粉可使饼干酸度增加、碱度下降。弱酸的环境有利于绿茶色素稳定，故在加工超微茶粉烘焙食品时应选用中性膨松剂或 pH 调节剂，使体系维持中性弱酸性条件，才有利于保持产品自然绿色的品质。添加超微茶粉可降低饼干游离脂肪含量，高油脂能改善饼干口感，而更易被消费者接受，但却对健康无益，超微茶粉能在不影响产品质量的前提下，减少摄入过多脂肪，同时使产品爽口不易产生油腻感。黄油和糖对饼干的色泽有正面影响，即添加量越大饼干越具有金黄且棕红的特点；相反超微绿茶和膨松剂对饼干色泽有负面相关，添加量过大显著降低了饼干明度。

本文在给出绿茶饼干配方的同时研究了各添加成分对饼干理化性质的影响，但对其稳定性及成品饼干中有效成分的检测分析有待进一步研究和探讨。

参考文献

[1] 周友亚. 茶多酚在食品工业中的应用[J]. 广州食品工业科技，2001，(3)：79—81.

[2] 中华人民共和国卫生部. GB/T 5009.3—2003 食品中水分的测定. 2003.

[3] 中华人民共和国国家发展与改革委员会. QB/T 1254—2005 饼干试验方法. 2005.

[4] 中华人民共和国卫生部. GB/T 5009.6—2003 食品中脂肪的测定. 2003.

[5] Pareyt B, Talhaoui B, Kerckhofs G, et al. The role of sugar and fat in sugar-snap cookies: Structural and textural properties [J]. Journal of Food Engineering, 2009(90): 400—408.

[6] 刘传富, 董海洲, 侯汉学. 影响饼干质量的关键因素分析[J]. 食品工业科技, 2002, 23(8): 87—89.

[7] 史宁. 食品加工中膨松剂的应用(综述)[J]. 食品与健康, 2002(2): 47—48.

附: 基金项目: 浙江省青年科学基金项目(LQ12C20007)、浙江省教育厅资助项目(Y201226032)、浙江省茶产业重点科技创新团队项目(2011R50024)。

茶多酚在护肤品中的应用

江和源[1]　龙　丹[1,2]　张建勇[1]　王伟伟[1,2]

1. 中国农业科学院茶叶研究所,中国 浙江;
2. 中国农业科学院研究生院,中国 北京

摘　要:茶叶含有茶多酚等功能成分,具有许多医药保健功能,被广泛应用于食品、医药、日化、建材、养殖等行业。对皮肤细胞和组织的作用机理和功效表明,茶多酚具有良好的防晒、防皱抗衰老、修复美白、收敛保湿、祛痘等作用,可应用于护肤品等人体外用型健康产品,具有良好的市场开发潜力。

关键词:茶多酚;护肤品;防晒;抗衰老;修复美白

Application of Tea Polyphenols on the Skin Care Products

JIANG He-yuan[1]　LONG Dan[1,2]
ZHANG Jian-yong[1]　WANG Wei-wei[1,2]

1. Tea Research Institute, Chinese Academy of Agricultural Sciences, Zhejiang, China;
2. Graduate School, Chinese Academy of Agricultural Sciences, Beijing, China

Abstract:Tea have many kinds of potential health benefits attributed to tea polyphenols which are mainly composed of catechins. Tea compounds, such as tea polyphenols, were universally used as food, medicine, cosmetics, feed, etc. Results from in vitro and in vivo studies showed that tea polyphenols had beneficial functions for human skin, like sun protection, anti-aging, skin whitening, water keeping, antimicrobials, and it could be used external healthy products for human skin.

Keywords:tea polyphenols; sunblock; anti-aging; skin whitening

茶消费有几千年的历史,现已成为风靡全球的三大饮料之一。茶叶悠久的生产和饮用历史,是与它的保健功能分不开的,其根本原因在于所含有的多种功能成分,尤其是茶叶中的茶多酚(儿茶素)等特征性成分。

1 茶叶中的茶多酚

迄今为止,从茶叶中分离并得到鉴定的化合物种类达到 700 种以上,其中有机化合物有 500 种以上。茶叶中具有特殊功能的成分,一般源自茶叶中的次生代谢产物,或称天然产物。

茶多酚是茶叶中最具特征性的次生代谢产物,是一类富含多羟基的酚性物质,在茶叶中含量可达到干物质总量的 18%~46%,其中儿茶素类物质的含量约占多酚类总量的 70%。已发现茶叶中的儿茶素主要有 12 种,其中含量较多的儿茶素有 L-EGCG、L-ECG、L-EGC、L-EC,D,L-C,D,L-GC 等。除儿茶素外,茶多酚还包括酚酸类、羟基-4-黄烷醇类、花色苷类、黄酮醇类、黄酮等化合物。

近年来的生物学活性研究表明,茶多酚具有抗氧化[1]、抗辐射[2]、抗突变[3]、抗肿瘤[4]、延缓衰老[5]、减肥降脂[6]、保护神经[7]、护心健脑[8]等功能。儿茶素类物质作为茶叶中主要功能性成分的地位已得到医学界学者的广泛认同,其高纯度的各类单体已在国内外逐渐实现商品化。

随着现代科学技术的发展与进步,茶多酚及其儿茶素单体成分也得到了广泛利用,被开发成多种类、强功效的保健产品,不仅可以以饮料或膳食补充剂等形式通过内服的方式发挥作用,在外用时也能取得良好的效果。例如,儿茶素所具有的抗氧化、吸收紫外线、抑菌、螯合金属离子等作用,可用于化妆品、日化用品等途径[9]。

2 茶多酚的防晒作用

随着近些年来太阳黑子活动日益频繁,大量的有害射线将危及肌肤健康。因此,适当的防晒措施已不仅仅是爱美女士的必修课,也是全民护肤不可或缺的步骤之一。国内外科学研究表明,茶多酚具有吸收紫外线、修复紫外线操作细胞等防晒作用。

紫外线(UV)按波长可分为 UVA(400~320nm)、UVB(320~280nm)、UVC(200~280nm)等 3 种类型,其中 UVB 可以到达人体皮肤的表皮层,会引起皮肤红斑甚至灼伤;UVA 能够穿透表皮层到达真皮层,直接表现为晒黑皮肤,其长期慢性的积累效应会诱发多种皮肤疾病。当人体皮肤接受了较大累积剂量的紫外线辐射后,其中的 UVB 可能对 DNA 产生较大的影响,使其吸收能量而发生突变;由 UVB 或 UVA 诱导产生的活性氧簇(ROS),一方面会间接损伤细胞核及线粒体的 DNA,导致细胞功能的异常或凋亡,另一方面可能氧化损伤蛋白质和脂质,引起相应的细胞功能及结构的异常[10]。此外,紫外线的照射会引起自由基的骤增,继而诱发脂质过氧化产物——丙二醛(MDA)的大量产生,MDA 则会加速脂褐素的生成,并且会交联蛋白质、DNA、脂质等大分子物质,从而引发一系列皮肤光老化症状[11]。

茶多酚中的儿茶素类化合物,具有良好的防晒作用。首先,从抵制紫外线照射的防护角度来看,儿茶素类作为一类含有羰基、并具有共轭结构的芳香族化合物,可以直接吸收紫外线,特别是对 200~330nm 波段的紫外线有较强的吸收[12],因此具有广谱防晒的作用。其次,从修复紫外损伤细胞生物活性的角度来看,儿茶素类可以通过抑制自由基的产生、直接清除自由基等方式来预防或阻断自由基反应,以及激活生物体的自由基清除体系等途径[13]来调节皮肤的自由基平衡,从而保护肌体免于受到自由基的损伤。我国现已有以 EGCG、ECG 等儿茶素,以及儿茶素衍生物为主要功效成分的防晒类产品专利。

3 茶多酚的防皱抗衰老作用

抗衰老已经成为当代美容营养品市场的热门话题,与我国人口逐步进入老龄化有着一定的关系,这种发展趋势也出现在世界其他国家。据统计,2009 年我国 60 岁以上老年人口已经占到总人口的 12.5%,预计 2050 年时,每 5 个人中就有 1 个老年人。目前全球抗衰老产品的市场份额大概有 1000 亿美元,估计到 2015 年会增长到 1760 亿美元。茶多酚由于良好的抗氧化、清除自由基等能力,在抗衰老美容品方面也具有广阔的应用前景。

皮肤光老化,是指由于长期的日光照射导致皮肤衰老或加速衰老的现象,临床表现为皱纹出现和色斑沉积,其机理有别于遗传因素及不可抗因素导致的皮肤自然老化[14]。紫外线诱发脂质过氧化反应中的过氧化物降解产物 MDA,能够交联胶原蛋白,使得胶原坚化、长度变短、失去膨胀能力,外观上表现为皮肤硬化、弹性下降,从而在肌肉伸缩的往复运动中产生皱纹,表现出皮肤老化的特征。紫外线还可能引发真皮的炎症反应,促使弹性蛋白酶、胶原酶等组织中溶解酶的释放,间接破坏真皮基质中维持皮肤张力、承受外作用力、保持外观丰盈饱满的弹性纤维、胶原纤维等重要物质,进而使皮肤逐渐松弛,在过度伸展后出现褶皱。

仲少敏等实验发现,与使用前相比较,在全面部使用含 EGCG 护肤品 8 周后,皮肤的角质层含水量有轻度的升高,皮肤脱落角质细胞下降 83.4%,差异达到极显著水平($P<0.001$);与皮肤皱纹和粗糙程度相关的参数从第 4 周开始得到改善,8 周后更为明显,差异有显著性($P<0.05$)。结果表明,含 EGCG 护肤品对皮肤早期老化所引起的皮肤干燥、角质细胞堆积、粗糙、细纹和皱纹有良好的改善作用[15]。

儿茶素类物质对胶原酶和弹性蛋白酶也有很好的抑制作用。赵伟康等以大鼠为试验对象,发现 100μg/mL 儿茶素对胶原酶的活性抑制力达到 98%,同时能与胶原蛋白和弹性纤维相互作用,从而减轻胶原蛋白被水解程度,保持皮肤的弹性和嫩度[16]。

4 茶多酚的修复美白作用

随着年龄的增长和受长时间高强度的日光照射,人体皮肤表面会产生不同程度的斑痕或色素沉着。事实上,皮肤色素沉积主要与皮肤内黑色素含量及脂褐素含量有关。

黑色素是皮肤黑色素细胞在光照下主动合成的天然紫外线吸收剂,具有保护皮肤免受射线伤害的作用。在黑色素细胞内,高尔基体负责合成黑素体,并以小泡的形式通过细胞突触向周围的角质形成细胞输送,共同完成黑素的周期性代谢。在向外运输的过程中,黑素体内的酪氨酸(Ty)在酪氨酸酶(TyR)及 O_2 的作用下发生氧化,逐步氧化成多巴、多巴醌等一系列中间体,并最终形成黑色素运至角质形成细胞,在外观上表现为肤色变黑。在这一生理过程中,紫外辐射会刺激黑素细胞的增殖、促进黑素细胞的功能表达、促进黑素体合成中关键酶 TyR 的合成、为氧化反应的中间产物继续氧化提供能量、促进黑素体向转运等,可通过多种作用途径加速皮肤的黑化[17]。

儿茶素类物质,可以从源头上直接吸收紫外线,减轻紫外线带来的一系列黑化效应。此外,儿茶素还能抑制黑色素的形成,一方面,大量的酚羟基结构络合酪氨酸酶中的 Cu^{2+},大大降低酶活性,降低了关键步骤的反应速率,减缓了黑色素的生成,如 EGCG 对体外培养的人表皮黑素细胞的酪氨酸酶活性及黑素合成均有抑制作用,且呈明显的剂量效应关系[18];另一方面,儿茶素多羟基结构使其具有很强的供氢能力,及时地还原多巴、多巴醌等氧化中间产物,使反应向还原态方向推进,从而有效遏制氧化进程,抑制皮肤斑痕及色素沉着。

影响皮肤黑化的另一个重要方面是脂褐素的形成与积累。阳光中的紫外线和其他辐射线的照射会使皮肤产生应激反应,导致自由基数目激增和活性增强,打破机体内正常的自由基生成与消除平衡。大量的活性自由基在短时间内极易攻击生物大分子,引发皮肤内丰富的多不饱和脂肪酸的脂质过氧化物链式反应,逐步产生 MDA。MDA 与蛋白质的氨基发生美拉德反应生成荧光物质脂褐素,在体表长期积聚产生斑痕。

儿茶素不仅可以消除自由基、减少自由基引发的脂质过氧化概率,还能够抑制超氧化物歧化酶(SOD)、过氧化氢酶(CAT)等的活性,也可以有效地调节谷胱甘肽过氧化物酶(GPx)等抗氧化酶的活性[19],从而有效地抑制脂质过氧化进程中的多步重要反应,减少 MDA 的生成,达到减少脂褐素目的。

国内已开发出以茶多酚和芦荟苷为主要成分用于祛除黄褐斑的复合制剂,经人体试食试验表明,试食后的黄褐斑颜色及面积的下降幅度与试食前和对照组相比,均有极显著性差异,且不产生新的黄褐斑,经安全毒理学评价和临床观察实验后表明为安全可靠,可以进一步开发成化妆品或保健品[20]。

5 茶多酚的收敛保湿作用

茶多酚是茶汤呈现出收敛性的重要物质,其涩味机理是茶多酚容易与口腔表皮蛋白质的氨基相结合,从而沉淀到口腔表层形成一层不透水层,给人以收敛的口感。当茶多酚应用于人体皮肤表面时,茶多酚也可与人体皮肤表皮层中的角质蛋白、真皮层的黏蛋白及黏多糖、细胞膜上的磷脂等分子,以疏水键和氢键发生类似的聚合反应,从而使人体皮肤产生一种收敛拉紧的感觉。关于茶多酚的收敛作用机理,Haslam 提出了多酚-蛋白质反应的多点疏水键氢键结合理论[21]。以 EGCG 与蛋白质结合为例,蛋白质分子中的疏水基团较集中的地方可以构成一种"疏水袋"的结构,EGCG 分子中的亲脂部分在疏水作用的牵引下进入"疏水袋"中,与蛋白质肽键(—NH—CO—)及侧链基团(—OH、—NH_2、—COOH)以氢键的形式加强结合,从而形成产生收敛性的茶多酚——蛋白质疏水结合。

这种复合作用,不仅仅是儿茶素杀菌、抗病毒、抑制酶活性作用的基础,所产生的收敛性在护肤品的开发中也大有用武之地。儿茶素成分和皮肤角蛋白的结合,一方面可以收缩毛孔,收敛、提拉松弛的皮肤,从而达到紧致肌肤、减少皱纹的功效;另一方面,收敛性还会刺激汗腺口的膨大、堵塞汗液的渗透,从而抑制排汗、减少油性皮肤皮脂腺的过度分泌,特别是在气温较高的夏季能起到防汗、抑脂、保持皮肤清洁的作用。

儿茶素与蛋白质的良好亲和性,有利于儿茶素在无水条件下牢固地附着于皮肤表面,从而能够长时间地发挥作用。儿茶素的多羟基结构,易于吸附空气中的水分,以保持皮表的含水量。有实验表明,98%茶多酚可显著提高皮肤的吸湿和保湿性[22]。茶多酚易与多糖、多元醇、脂质、蛋白质、多肽等形成分子复合物,在缓和收敛性的同时,可以增加稳定性,形成良好的保湿剂。此外,茶多酚还可以抑制透明质酸酶的活性,抑制保湿因子透明质酸的降解,对皮肤过敏与和发炎也有一定的预防作用。

目前市场上出现了利用茶叶功能成分来做宣传的爽肤水、乳液、眼霜、面膜等多款保湿补水类产品,大多数产品申称配方中含有绿茶提取物或茶叶精华素,但对其功能性的标注仍缺乏明确的科学阐述。

此外,儿茶素还可以抑制5-α还原酶活性[23]从而达到调节激素活性的作用,其中(—)-EGCG 和(—)-ECG 对 5-α 还原酶的 IC_{50} 约为 $10\mu mol/L$,从而可以抑制皮脂过度分泌。有研究表明:健康男性志愿者使用含3%绿茶提取物的乳液8周后,皮肤皮脂的生成量显著减少,可起到预防与治疗粉刺和痤疮的功效,有望进一步用于面部痤疮患者[24]。

6 含茶日用化妆品的开发展望

日用化妆品的产品种类丰富,能满足个人清洁、保养、美容、修饰等诸多需要,在全球范围内需求量大,消费人群广,蕴藏着巨大的经济效益。日用化妆品实现诸多功能的基础,在于产品配方成分在机体内发挥的生理活性作用。近年来,随着人们对天然型产品的渴求,向日用化妆品中添加植物来源功能性成分已然成为新时代的潮流。茶多酚(儿茶素)作为天然健康载体物质——茶叶中的主要功能性成分,能赋予日用型终端产品抗氧化、防辐射、保湿修复、延缓衰老等多重功能。

化妆品的液剂、乳剂、乳膏剂、固融体油膏剂和棒形剂、粉剂、啫喱、气雾剂等多种剂型[25],为茶多酚(儿茶素)及其衍生物的添加应用提供了条件。儿茶素类物质应用在日用化妆品中的研发,主要集中在日本、中国、美国等地,产品类型涉及外用软膏、牙膏、护唇膏、漱口水、发用染料、乳液、面霜、面膜等,例如韩国 NANA 绿茶水养美白化妆品、日本资生堂绿茶男士须后水、美国水之澳绿茶抗氧化面霜、法国 Lancome 绿茶清透控油、雅芳白茶抗氧化等系列[26]。

化妆品是一类由各类物料经科学调配而成的混合物,包括基质原料、辅助原料、功效性原料等[27]。茶多酚(儿茶素)除了作为功能性成分添加到化妆品中以外,其本身具有良好的抗氧化、抑菌、螯合金属离子等作用,也可以作为天然抗氧化剂、防腐剂、金属离子螯合剂等辅助成分应用,在护肤品中的应用前景广阔。

参考文献

[1] Zhong R, Zhen Z, Dao W, et al. Effect of tea catechins on regulation of antioxidant enzyme expression in H_2O_2-induced skeletal muscle cells of goat in vitro [J]. Agricultural and Food Chemistry, 2011, 59(20): 11338—11343.

[2] Liu M L, Wen J Q, Fan Y B. Potential protection of green tea polyphenols against 1800MHz electromagnetic radiation-Induced injury on rat cortical neurons [J]. Neurotoxicity Research, 20(3): 270—276.

[3] Haza A I, Morales P. Effects of (+)catechin and (−)epicatechin on heterocyclic amines-induced oxidative DNA damage[J]. Journal of Applied Toxicology, 2011, 31(1): 53—62.

[4] Masami S, Achinto S, Hirota F, et al. New cancer treatment strategy using combination of green tea catechins and anticancer drugs[J]. Cancer Science, 2011, 102(2): 317—323.

[5] Maurya P K, Rizvi S I. Protective role of tea catechins on erythrocytes subjected to oxidative stress during human aging[J]. Natural Product Research, 2009, 23(12): 1072—1079.

[6] Rainsa T M, Agarwalb S, Maki K C. Antiobesity effects of green tea catechins: A mechanistic review [J]. The Journal of Nutritional Biochemistry, 2011, 22(10): 1—7.

[7] 李大祥,鲜殊,杨卫. 茶叶的神经保护作用研究进展[J]. 茶叶科学,2011, 31(2): 79—86.

[8] Verschuren L, Wielinga P Y, Duyvenvoorde W V, et al. A dietary mixture containing fish oil, resveratrol, lycopene, catechins, and vitamins E and C reduces atherosclerosis in transgenic mice [J]. Nutr, 2011, 141(5): 863—869.

[9] 陈宗懋. 茶叶的益思美容功效[J]. 中国茶叶,2009, 31(12): 10—11.

[10] 廖勇. 皮肤光老化的分子机制[J]. 中国美容医学,2010, 19(3): 444—447.

[11] 刘玮,张怀亮. 皮肤科学和化妆品功效评价[M]. 北京:化学工业出版社, 2005.

[12] 徐向群. 茶叶提取物在化妆品上的应用[J]. 中国茶叶, 1993, 15(3): 31—32.

[13] 杨贤强. 茶多酚化学[M]. 上海：上海科学技术出版社，2003. 201-267.
[14] Jeon S E, Choi-Kwon S, Park K A, et al. Dietary supplementation of (+)-catechin protects against UVB-induced skin damage by modulating antioxidant enzyme activities [J]. Photodermatol Photo, 2003, 19(5)：235-241.
[15] 仲少敏，Lee H K, Yeon J H, et al. EGCG 护肤品对女性面部皮肤的抗老化作用[J]. 中国皮肤性病学杂志，2010，24(2)：183-185.
[16] 赵伟康，刘平. 固真饮延缓皮肤胶原衰老的实验研究[J]. 中国医药学报，1989，4(6)：23-24.
[17] 刘玮，张怀亮. 皮肤科学和化妆品功效评价[M]. 北京：化学工业出版社，2005.
[18] 岳学壮. 部分天然植物成分对皮肤微循环、黑素合成以及 VEGF 分泌的影响[D]. 南京：南京医科大学，2006.
[19] 孙培培. 茶多酚芦荟苷复合制剂安全毒理学评价及祛黄褐斑功效研究[D]. 杭州：浙江大学，2010.
[20] Haslam E, Lilley H T, Cai Y, et al. Traditional herbal medicines——the role of polyphenols [J]. Planta Medica, 1989, 55(1)：1-8.
[21] Wei X, Liu Y, Xiao J, et al. Protective effects of tea polysaccharides and polyphenols on skin[J]. J Agric Food Chem, 2009, 57(17)：7757-7762.
[22] 隋丽华，郭珉. 茶多酚皮肤药理学的研究进展[J]. 医药导报，2003，22(11)：807-809.
[23] Mahmood T, Akhtar N, Khan B A. Outcomes of 3% green tea emulsion on skin sebum production in male volunteers [J]. Bosnian Journal of Basic Medical Sciences, 2010, 10(3)：260-264.
[24] Yamamoto M, Kirita M, Sami M. Gene encoding methylated catechin synthase[P]. EP 20050806065. 2007.
[25] 黄静红，李德如，阎国富. 绿茶在皮肤美容中的应用研究进展[J]. 中国中西医结合皮肤性病学杂志，2008，7(3)：193-195.
[26] 裘炳毅. 化妆品化学与工艺技术大全[M]. 北京：中国轻工业出版社，1997.

附：基金项目：浙江省重大科技专项重点农业项目 2010C12023-2

苦瓜绿茶含片的生产工艺研究

梁 进 侯如燕 宛晓春

安徽农业大学农业部茶树生物学与茶叶加工重点实验室,中国 安徽

摘 要:以苦瓜为原料制备苦瓜浸提物,用苦瓜浸提物、速溶绿茶粉、木糖醇为主要原料,添加玉米淀粉及其他辅料,经混合、制粒、干燥、压片等工序,利用正交试验筛选出制备苦瓜绿茶含片的最佳配方和工艺。结果表明:苦瓜浸提物 0.3g、速溶绿茶粉 0.1g、填充剂(玉米淀粉:β—环状糊精=1.5:0.8)1.5g、木糖醇 3.0g,可制得口感好、具有苦瓜和绿茶特有风味,表面光滑美观且食用方便的新型苦瓜绿茶复合营养含片。

关键词:苦瓜;绿茶;含片;工艺

Research on Processing Technology of Bitter Gourd-green Tea Tablet

LIANG Jin HOU Ru-yan WAN Xiao-chun

Key Laboratory of Tea Tree Biology and Tea Processing, Ministry of Agriculture, Anhui Agricultural University, Auhui, China

Abstract: In this study, bitter gourd extracts was prepared from bitter gourd. Bitter gourd extracts, instant green tea powder and xylitol were used as the major raw materials with the addition of corn starch and other additives as supplementary materials to prepare bitter gourd-green tea tablet through a series of processing including mixing, granulation, drying and tablet. The optimal formula and processing parameters were explored by orthogonal experiments. Results showed that the optimal formula were bitter gourd extracts 0.3g, instant green tea powder 0.1g, filling agents (Corn starch : β-cyclodextrin=1.5:0.8) 1.5g and xylitol 3.0g, the new bitter gourd-green tea compound nutritional tablet could be obtained with good taste, special flavor of bitter gourd and green tea, smooth surface, edible facility.

Keywords: bitter gourd; green tea; tablet; processing

苦瓜是我国传统的一种药食同源植物,其味苦性寒,具有清暑涤热、明目解毒的功效[1]。苦瓜含有多糖类、蛋白质和多肽、维生素、氨基酸和矿物质等丰富的营养成分,其中又以 Vc 的含量居于瓜果蔬菜之首[2]。此外,苦瓜中还含有生物碱、三萜类、黄酮类、皂苷类等多种药用活性成分,具有较高的药用价值,有辅助降血糖、降血脂、抗氧化、增强免疫力等多种保健功能[3]。

绿茶是一种天然健康饮料,绿茶中富含酚类、咖啡因、茶色素、茶多糖、茶皂素和茶氨酸等功能性成分[4],对人体具有抗氧化[5]、抗突变[6]和防癌与抗癌[7]等保健功效。现代科学研究证实,绿茶所含有多酚和黄酮类等多种功效成分与人体健康密切相关[8-9],其研究越来越广泛,应用领域也不断扩大,在医药及保健食品等领域有着广阔的发展潜力。

苦瓜和绿茶均含有丰富的对人体有益的功能成分,具有较高的营养与保健功效。然而,目前市场上将二者结合起来开发的产品还较为少见。为充分利用我国绿茶及苦瓜的资源优势,最大限度地保留其有效成分,本实验将苦瓜浸提物与速溶绿茶粉两者配伍,研制成携带方便且口感良好的苦瓜绿茶含片,不仅改善苦瓜的风味与口感,具有一定的保健作用,还能为苦瓜茶的多元化开发提供一条新的途径。

1 材料与方法

1.1 材料与仪器

苦瓜:市购;速溶绿茶粉:南京融点食品科技有限公司;β-环状糊精:孟州市华兴生物化工有限责任公司;硬脂酸镁:天津市光复精细化工研究所;玉米淀粉:市购;木糖醇:郑州天乐化工产品有限公司;柠檬酸:湖南洞庭柠檬酸化学有限公司;滑石粉:上海聚千化工有限公司;乙醇等化学试剂:均为分析纯。

DK-S24 型电热恒温水浴锅:上海精宏实验设备有限公司;DHG-9146A 电热恒温鼓风干燥箱:上海精宏实验设备有限公司;FA1104A 型电子天平:上海精天电子仪器有限公司;DFT-200 型手提式高速中药粉碎机:温岭市大德中药机械有限公司;4#筛(0.63mm):上海嘉定粮油仪器有限公司;TDP-5 型单冲压片机:上海超亿制药机械设备有限公司。

1.2 实验方法

1.2.1 苦瓜浸提物的制备

苦瓜的预处理：先将从超市购买的苦瓜去瓤漂洗后切片并置于鼓风烘干机中于50℃烘干30min，再用手提式高速中药粉碎机将烘干后的苦瓜片粉碎，并过20目筛，所获得的苦瓜粉置于密封袋中备用。

苦瓜浸提工艺参数优化：苦瓜总皂甙是苦瓜中的主要活性物质，本研究主要采用一定浓度的乙醇溶液为浸提液对苦瓜中的皂甙类成分进行有效浸提，通过香草醛显色法[10]测定醇提液中皂甙类成分的含量。在预实验的基础上，以料液比、浸提温度、浸提时间3个主要影响因素，通过正交试验设计，确定苦瓜浸提物的最佳浸提方法。

苦瓜浸提物制备工艺流程：苦瓜→切片→烘干→粉碎→过筛→浸提→抽滤→浓缩→干燥→苦瓜浸提物。

1.2.2 苦瓜绿茶含片的制备

填充剂的选择：依据填充剂的种类及用量对苦瓜绿茶含片质量的影响，以口感和组织状态为考察指标，通过预实验确定β-环状糊精、玉米淀粉作为含片的填充剂。

润湿剂的选择：依据润湿剂的种类及体积分数对苦瓜绿茶含片质量的影响，通过预实验确定乙醇作为湿润剂，其体积分数优选范围为70%～95%。

甜味剂的选择：作为保健品的一种，在选择甜味剂的种类上有一定的要求，由于木糖醇所含的热量少，不被人体吸收，不会使人体血糖升高，其甜度也能达到消费者的口感要求，故优选木糖醇作为甜味剂。

苦瓜绿茶含片工艺优化：在预实验基础上，选出了对苦瓜绿茶含片有明显影响的苦瓜浸提物用量、速溶绿茶粉用量、填充剂（玉米淀粉：β-环状糊精＝1.5∶0.8)用量、木糖醇用量等4个因素，采用正交试验优化苦瓜绿茶含片的配方，以色泽、组织状态、口感、风味作为考察指标确定其最佳工艺。

生产工艺流程：

苦瓜浸提物与速溶茶粉→混合→制软材→制粒→干燥→整粒→压片→灭菌→包装

↑ ↑

填充剂 湿润剂

1.2.3 操作要点

苦瓜浸提物原料的制备:应选用市售新鲜苦瓜,用清水清洗干净后去瓤、漂洗,再切成片状并烘干成苦瓜片。为了降低高温对苦瓜中 Vc 等的营养损失,烘干温度应控制在 55℃ 以下为宜。为了缩短生产周期,提高干燥速度,便于工业化生产,本实验主要采用喷雾干燥法对浸提液进行处理,得苦瓜浸提物干粉。

原辅料混合:预先将木糖醇、β-环状糊精等经粉碎机粉碎(过 80 目筛),用电子天平准确称量原辅料,并按照配方配料要求,将原辅料进行均匀混合。

制软材:预先将柠檬酸溶解于湿润剂(乙醇)中,缓慢加入上述混合均匀的物料中,同时不断的均匀搅拌,调整物料湿度,制成软材。软材干湿度应适宜,即符合"手握成团,轻压即散"的标准。

制粒与干燥:以 4#筛作为造粒工具,将软材紧握成团,压过 4#筛,使软材变成小型颗粒状。将所制湿粒置于 55℃ 鼓风干燥箱中干燥,每隔 30min 翻动 1 次,以加快干燥速度,颗粒水分含量控制在 3%～5% 为宜。水分过高,取片时易变形,还影响压好片的干燥时间;水分低于 3% 时,不易压片或者压出的片剂质地过于疏松。

压片:将干燥好的颗粒按物料总质量的 0.25%(m/m)加入硬脂酸镁并混合均匀,然后用单冲压片机压制成片剂。

灭菌与包装:将压好的含片放在紫外线下照射 20～30min 进行灭菌,灭菌后检验,达到卫生标准即可包装。

1.2.4 感官评定

由食品加工专业人员组成的 10 人评定小组,分别以苦瓜绿茶含片的色泽、组织状态、口感、风味为考核指标确定其最佳工艺,评分标准见表1。

表1 苦瓜绿茶含片的感官评定标准

项 目	评 价	评 分
色泽(20)	淡焦糖色、有光泽、色泽自然、无麻点	20～18
	黄绿色、有光泽、基本无麻点	17～15
	淡黄色、色泽过淡、有麻点	14～12
	深褐色、色泽过浓	11～8

续表

项 目	评 价	评 分
组织状态(20)	形态完整、表面光滑、断面组织细腻紧密	20～18
	形态完整、表面微混、断面组织紧密	17～15
	形态基本完整、表面粗糙、断面不够紧密	14～11
	形态不很完整、有碎裂、表面粗糙、断面粉状	10～7
口感(30)	爽口、细腻、入口柔顺	30～26
	入口顺滑、无糊口、无粉粒感	25～21
	稍有糊口感、稍有粉粒感、稍觉粗糙	20～16
	较糊口、稍有粉粒感、感觉粗糙	15～10
风味(30)	有苦瓜和绿茶特有的风味、可口、酸甜苦味协调、且苦后回甜	30～25
	有苦瓜和绿茶特有的风味、可口、酸甜苦味适当	24～21
	较淡或较重的苦瓜味、口味正常、甜酸基本协调	20～17
	基本没有苦瓜味、稍有异味	16～10

2 结果与讨论

2.1 苦瓜粉浸提工艺的确定

以经粉碎过20目筛的苦瓜粉为原料,用索氏提取器在热水中浸提,通过研究不同的料水比、浸提时间及水温对浸提率的影响,采用$L_9(3^3)$正交表确定苦瓜粉的最佳浸提方案,结果见表2。

表2 苦瓜浸提工艺的优化

试验号	A 料水比	B 浸提时间(h)	C 浸提水温(℃)	吸光度值
1	1∶20	3	80	0.366
2	1∶20	4	90	0.592
3	1∶20	5	100	0.518
4	1∶30	3	90	0.487

续表

试验号	A 料水比	B 浸提时间(h)	C 浸提水温(℃)	吸光度值
5	1：30	4	100	0.415
6	1：30	5	80	0.308
7	1：40	3	100	0.318
8	1：40	4	80	0.350
9	1：40	5	90	0..303
K_1	0.492	0.390	0.341	
K_2	0.403	0.452	0.461	
K_3	0.324	0.376	0.417	
R	0.168	0.076	0.120	

由表 2 可知,对苦瓜浸提的影响程度为 A 料水比＞C 浸提水温＞B 浸提时间,即最佳浸提条件:$A_1B_2C_2$。当苦瓜经粉碎过 20 目筛、料水比 1：20、在温度 90℃的热水中浸提 4h 为最佳浸提条件。用 90℃的热水还可以除去苦瓜的生涩味和使产品有较好的焦糖味,丰富含片的口感和风味。

2.2 填充剂的选择

填充剂的加入不仅保证产品一定的体积大小,而且能够减少功效成分的剂量偏差,改善物料的压缩成型性等。实验表明,仅采用苦瓜提取物、速溶绿茶粉和木糖醇制作苦瓜绿茶含片,可以达到产品的功效性,但产品的嗜好性不足,口感较差,口味单调,苦瓜风味和绿茶风味均不明显;而且,由于苦瓜提取物有一定的吸湿性和较强的黏性,再加上木糖醇的黏性,在压片过程中易发生黏冲现象,不利于生产。因此,在配料中选择了速溶绿茶粉、玉米淀粉、β-环状糊精、柠檬酸、硬脂酸镁作为辅助原料。其中,β-环状糊精通过其对苦瓜提取物起到分散和包裹作用,极大缓解了压片过程中的粘冲现象,使得产品生产可以比较顺利地进行,并且对其苦味有一定的包埋和掩盖[11],达到保留有效成分和改善口感的双重作用。速溶绿茶粉中保健功能性成分较丰富,还是良好的天然防腐剂和着色剂,延长产品的保质期和美化产品色泽[12]。此外,柠檬酸的添加既丰富了产品的风味、口感,也对苦瓜浸提物带来的苦味起到了一定的掩盖作用,进一步提高了产品的嗜好性。

对不同填充剂进行试验并比较其结果(表 3),可知,选择玉米淀粉：β-环状糊精＝1.5∶0.8作为填充剂,对制粒压片有良好的效果。

表3 填充剂的选择

填充剂	组织状态	口 感
β-环状糊精	压片形态不完整,断面组织较松软,不易成型,碎裂现象严重	口感稍显粗糙,苦味过淡,有糊口感,有粉粒感
玉米淀粉	表面粗糙,形态不完整	口感粗糙,糊口严重,稍黏,粉粒感较重,苦味较重
玉米淀粉∶β-环状糊精=1.5∶0.8	形态完整,表面光滑,质地紧密厚实	苦味适中,基本无糊口感,无粉粒感

2.3 润湿剂的选择

在制粒中常用的润湿剂有蒸馏水和乙醇,由于物料水溶性成分较多,因此最好选择适当体积分数的乙醇溶液来克服发黏、结块、湿润不均匀、干燥后颗粒发硬的现象[13](表4)。

表4 润湿剂的选择

乙醇(%,V/V)	效 果
90	黏性适中,相互黏合,软材成团,颗粒完整
85	细粉少,紧贴颗粒,软材翻滚成浪,干燥后颗粒稍硬
80	颗粒不成型,黏性大
75	黏性大,结块,不易干燥

2.4 润滑剂的选择

总混时,取3份等重的颗粒,分别加入3%的硬脂酸镁、滑石粉,另一份不加任何润滑剂,充分混合均匀后,测定3种粉体的休止角,其结果如表5所示。

表5 休止角测定结果

润滑剂	硬脂酸镁	滑石粉	不 加
休止角	$(37\pm0.5)°$	$(41\pm0.5)°$	$(46\pm0.5)°$

从表5中可以看出,加入硬脂酸镁后粉体休止角最小,摩擦力小,流动性好,具有较好的润滑性。因此,选硬脂酸镁作为最合适的润滑剂。

2.5 苦瓜绿茶含片的最佳工艺确定

为了使产品有一个良好的口感,且具有一定的保健功能,在预实验基础上,选出了对苦瓜绿茶含片有明显影响的苦瓜浸提物用量、填充剂用量、速溶绿茶粉用量、木糖醇用量等4个因素,采用正交试验优化苦瓜绿茶含片的配方,以色泽、组织状态、口感、风味为考核指标确定其最佳工艺条件。选用L9(3^4)安排试验,结果见表6和表7。

表6 正交因素水平表

水 平	因 素			
	A 苦瓜浸提物	B 速溶绿茶粉	C 填充剂	D 木糖醇
1	0.2	0.1	1.0	1.0
2	0.3	0.2	1.5	2.0
3	0.4	0.3	2.0	3.0

表7 正交试验设计和结果

试验号	因 素				综合评分
	A 苦瓜浸提物(g)	B 速溶绿茶粉(g)	C 填充剂(g)	D 木糖醇(g)	
1	1	1	1	1	80
2	1	2	2	2	79
3	1	3	3	3	75
4	2	1	2	3	88
5	2	2	3	1	74
6	2	3	1	2	75
7	3	1	3	2	72
8	3	2	1	3	73
9	3	3	2	1	70
K_1	78.000	80.000	76.000	74.667	
K_2	79.000	75.333	79.000	75.333	
K_3	71.667	73.333	73.667	78.667	
R	7.333	6.667	5.333	4.000	

由表 7 可得:以色泽、组织状态、口感、风味为考核指标时,各因素对产品的影响度为 A>B>C>D,最佳组合条件为 $A_2B_1C_2D_3$,即苦瓜浸提物 0.3g、速溶绿茶粉 0.1g、填充剂(玉米淀粉:β-环状糊精=1.5:0.8)1.5g、木糖醇 3.0g,用此配方进行压片,片剂硬度适中,具有普洱茶特有的风味,甜度适中,无黏附感,且先苦后甘,此方为最佳配方。

2.6 干燥温度的选择

由于苦瓜浸提物和速溶绿茶粉都极易吸水,而较难失水,故采取高温干燥的方法,但温度不能过高,若干燥温度过高,木糖醇容易发生转化,影响其口感,而且易于溶解使颗粒结块,进而影响其溶解效果;若干燥温度过低,又会大大延长干燥时间,降低干燥效率,拉长生产周期,不利于大批量生产。以物料水分含量在 3%～5%为终点,采用不同的干燥温度进行实验时发现,干燥温度为 60℃～65℃,颗粒之间相互黏结成块,不利于干燥,并伴有少许焦糊味;干燥温度为 45℃～50℃,颗粒紧凑、沉实、适合压片,但由于温度为 50℃时,干燥至终点至少需要 4h,会大大降低干燥效率,因此本实验选择 55℃为最佳干燥温度。

3 结 论

本实验主要研究了苦瓜绿茶含片的加工工艺,围绕其加工过程存在的苦瓜浸提物制备工艺参数、原辅料添加量等问题进行研究。本实验中苦瓜浸提物的最佳浸提条件确定为:料液比为 1:20,温度为 90℃的热水中浸提 4h。在此条件下,所得的浸提物得率为最高。

苦瓜绿茶含片的研制受原辅料用量、填充剂用量、润湿剂浓度等多种因素的影响。本研究结果显示,以糊精、淀粉等作为填充剂时,其用量不宜过多,否则容易产生糊口感;当填充剂以玉米淀粉:β-环状糊精=1.5:0.8 进行混合,可有效地改善含片的口感;用木糖醇作为甜味剂,不仅能增进含片风味和口感,同时还增加了产品抗龋齿、改善肠胃的功能,尤其适合糖尿病人群的食用;以 90%乙醇作为润湿剂时,能防止物料结块,可得到质量好的颗粒。

本研究获得的制备苦瓜绿茶含片的最佳配方:苦瓜浸提物 0.3g、速溶绿茶粉 0.1g、填充剂(玉米淀粉:β-环状糊精=1.5:0.8)1.5g、木糖醇 3.0g,经 55℃干燥,制得具有苦瓜及绿茶所特有风味、口感好、表面光滑、色泽均匀、硬度适中的保健食品,适用于高血脂、高血糖、肥胖人群食用,具有广阔的市场开发前景。

参考文献

[1] 刘学俊,井瑞洁,于辉.苦瓜超微粉营养面包的制作工艺及配方研究[J].食品工业,2011,(11):13—15.

[2] 颜海燕,刘娅,王庆玲等.苦瓜脯的研制[J].食品研究与开发,2006,27(10):80—82.

[3] 陈敬鑫,张子沛,罗金凤等.苦瓜脯的研制[J].食品科学,2012,33(1):271—275.

[4] 宛晓春.茶叶生物化学[M].第3版.北京:中国农业出版社,2003.8—9.

[5] 陈金娥,丰慧君,张海容.红茶、绿茶、乌龙茶活性成分抗氧化性研究[J].食品科学,2009,30(3):62—66.

[6] 王刚,赵欣.两种白茶的抗突变和体外抗癌效果[J].食品科学,2009,30(11):243—245.

[7] 陈宗懋.两种白茶的抗突变和体外抗癌效果[J].茶叶科学,2009,29(3):173—190.

[8] 任廷远,安玉红,王华.绿茶功能性成分提取及保健作用的研究现状[J].食品与发酵科技,2009,45(5):15—18.

[9] 雷蕾.绿茶药理研究进展[J].海峡药学,2007,19(6):15—17.

[10] 毛清黎,施兆鹏,李玲等.超声波乙醇浸提快速测定苦瓜总皂甙的方法研究[J].食品科学,2006,27(12):622—626.

[11] 徐燕,邹沛.苦丁茶和苦瓜清凉饮料的研制[J].饮料工业,2007,10(3):17—20.

[12] 朱永兴.茶的功效及其运用[J].中国茶叶加工,2010(4):49—53.

[13] 范青生.保健食品工艺学[M].北京:中国医药科技出版社,2006.227—230.

径山抹茶及其在茶食品中的应用

吴茂棋　庞英华　吴步畅　施海根

浙江老茶缘茶叶研究中心，中国 浙江

摘　要：余杭径山茶是传统的历史名茶。据考证，南宋时期的径山茶是抹茶，经来径山学佛的日本高僧圆尔辩圆、南浦绍明等传至日本，所以径山抹茶应该是日本抹茶之父。为了促进径山茶业的转型升级，浙江老茶缘茶叶研究中心承担《径山抹茶及茶食品的研究和开发》项目（浙江省科技厅下达），完成了对南宋时期径山抹茶的原型、恢复、创新，及其在食品领域的应用技术研究，提出食品加工中抹茶或绿茶粉的保绿和抗氧化技术，获得中国发明专利一项，并已成功应用于"西湖茶点"，入选2012年中国杭州西湖国际博览会特许经营产品。

关键词：径山抹茶；茶食品；保绿；抗氧化

Study and Development of Jingshan Mattea for Foods

WU Mao-qi　PANG Ying-hua　WU Bu-chang　SHI Hai-geng

Essence Tea Research Centre of Zhejiang, Zhejiang, China

Abstract：Yuhang Jingshan tea is a traditional tea of China. In Chinese Southern Song dynasty, Jingshan tea was in the shape of Japanese matcha. The reason is Japanese monks who learn Buddhism in Jingshan mount in Chinese Tang dynasty and when they returned to Japan, they learned the technique for processing Jinshan tea and spread it in Japan. We created a technique for keeping green and antioxadation of Jingshan mattea which was mixed in foods. This technique got a paten of invention from China. Three kinds of tea cakes were selected as marketing foods of Hangzhou West-lake Expo. 2012.

Keywords：Jingshan mattea; tea foods; keeping green; antioxadation

1 南宋时的径山茶——抹茶

据考证,南宋时期的径山茶是用蒸青团茶或散茶碾磨而成的抹茶。蒸青团茶的工艺流程是采摘一芽一叶初展的芽叶,经拣芽→蒸芽→研茶→造茶→过黄等5道工序制成。蒸青散茶的工艺为鲜叶→蒸汽杀青→摊凉→揉捻→烘干,基本与现今径山茶的工艺技术相同。

抹茶的碾磨分团茶和散茶两种。

(1)团茶的研磨。用来研磨团茶的主要工具叫"碾"。流程是:炙茶→碾茶→罗茶。"炙茶"就是火中烤茶饼,至足够干燥。碾后要"罗",罗是用绢绑紧在竹圈上做成的筛。《茶录》云:"罗底用蜀东川鹅溪画绢之密者,投汤中揉洗以幂之"。画绢,是用未脱胶的桑蚕丝织成的不需精练的绢类丝织物,结构紧密,表面平洁,足可印证南宋时期的抹茶之"细",故能在茶汤中久浮不沉,沫饽咬盏,洁白细腻。

(2)散茶的研磨。散茶的研磨工具为"磨",宋·审安老人《茶具图赞》中被拟人化,称之谓"石转运"。《茶具图赞》成书于公元1269年,恰好与径山抹茶制法东传日本的年代相符,同时也能间接证明当时径山的抹茶制法已经是改团为散了。这种专用茶磨后来随径山抹茶制法东传日本,能把抹茶磨细至 $2\sim20\mu m$ ($2\mu m$ 相当于6250目),而且其显微外形为不规则撕裂状薄片,点茶后能长时间悬浮水中,故被日本茶界一直沿用至今,并规定只有用这种茶磨碾磨出来的绿茶粉才能叫"抹茶"。

2 径山抹茶的传承性创新研究

南宋径山抹茶的研究目的有二,一是古为今用,推陈出新,恢复和创新径山抹茶品牌;二是为重新找回"吃茶"的理念,将抹茶广泛应用于食品工业领域。为此,在创新理念上是在传承基础上创新,并不受"日本抹茶"概念所束缚,创新工艺既有历史依据,又体现现代技术进步。研究者认为,现今径山茶的工艺已基本传承了南宋后期的蒸青散茶(抹茶原料)做法,其流程简洁科学,有利于保持茶的真色、真香、真味,因此可以采纳,其工艺流程是:鲜叶→摊放→汽热杀青→散热散湿→揉捻→烘干→碾磨(茶磨)→密封冷藏。

(1)鲜叶采摘标准。一芽一二叶初展,用于茶食品的可放宽到一芽三叶。过去南宋时期追崇的"水芽"、"小芽"太嫩,日本采用一芽四五叶覆下茶,虽叶绿素含量高,但毕竟偏成熟。所以,一芽一二叶初展,乃至一芽三叶应该是较为合理的。从现代科学而言,原料细嫩,则不必再考虑工艺过程中梗、叶脉的分离技术问题,这样工艺更趋简洁和科学。

(2)摊放。这是现今径山茶工艺的成熟经验,对抹茶工艺则又是一种创新,摊放对发展茶香、促进蛋白质水解、促进脂型儿茶素向简单儿茶素的转变、减少苦涩味等都有着明显效果。

(3)杀青。南宋时期和日本抹茶都是蒸气杀青,其优点是能最大限度保持茶的原质原味,但我国明朝后就改蒸为炒了,而且国人也不再喜欢那种蒸青的青草味。所以,创新的径山抹茶采用锅炒杀青。

(4)揉捻。揉捻能使叶细胞破碎,有利于改善抹茶的显微结构,南宋时的蒸青散茶工艺也是要"揉捻"的,这是历史依据。

(5)烘干。日本碾茶(抹茶的原料茶)的烘焙有专用的烘房,设计先进而科学,说到底同南宋时期的炭火焙笼是同出一理。但是,日本碾茶的这种专用烘房造价很高,所以现今余杭径山茶仍然在沿用历史上的炭火焙笼这一古老而又不失科学的烘焙方法。

(6)碾磨。采用专用的茶磨,这是对我国古代抹茶文明的传承,而且舍此也没有其他技术能达到抹茶的细度和显微形状的要求。在操作上要求:环境清洁、干燥、低温,转速均匀、低速。但如用作食品添加剂,则也可采用机械超微粉碎,以降低成本。

创新径山抹茶、南宋径山散茶的径山抹茶和现代日本抹茶的对比如表1所示。

表1 创新后的径山抹茶与古今抹茶的比较

项 目	现代日本抹茶	南宋时的径山抹茶	创新径山抹茶
品种	薮北种	当地群体种	薮北、翠峰、鸠坑等
覆盖与否	覆盖20d	原生态不覆盖	原生态不覆盖
采摘标准	一芽四五叶展	单芽至一芽一叶初展	一芽一二叶初展
摊青与否	否	否	摊青
杀青方式	蒸汽杀青,轻蒸	蒸汽杀青,偏熟	汽热杀、炒杀均可
揉捻工艺	不经揉捻直接烘干	捣(团茶)、揉(散茶)	充分揉捻、揉透
烘焙工具	现代化专用烘焙炉	竹制炭火焙笼	竹制炭火焙笼
碾磨工具	专用茶磨	茶碾(团)、茶磨(散)	专用茶磨

综上所述,创新工艺的径山抹茶的特点是:①原料嫩度介于日本抹茶和南宋时的径山抹茶之间;②既传承了南宋径山抹茶工艺精华,又整合了现代科技成果,工艺流程设计简洁科学、有利于今后的机械化生产;③工艺简洁,生产成本大为降低,有利于进入广泛的食品工业领域。

3 径山抹茶在茶食品中的应用研究

3.1 抹茶食品的技术关键及研究现状

随着绿茶保健功效不断被开发和茶食品的迅速发展,普遍做法是把茶叶超微粉碎,然后添加到食品中去。但绿茶中的多酚类、叶绿素很容易被氧化,同时又是热敏性物质,在湿热条件下更容易被氧化变性,从而失去原有的绿色和原有的风味。

目前解决上述问题大致有 4 种方法:①β-环状糊精包埋法,②pH 值调节(碱性)法,③做在食品内部,④尽量避免高温处理或在高温处理后加入。但这些方法都无法真正经受住高温热处理的考验,也正是由于这个原因,以致目前的抹茶食品开发领域还大多停留于"低温处理"类产品,如:抹茶慕司、抹茶乳酪蛋糕、抹茶蛋糕卷、抹茶拿铁、抹茶冰糕、抹茶冰淇淋、抹茶凉面等等。如何避免抹茶(或绿茶粉)在食品热处理过程中叶绿素的失绿,避免茶多酚等热敏性成分的氧化变性,即加工中"绿茶粉的抗氧化保绿"问题已成为抹茶(或绿茶粉)食品开发必须直接面对的技术关键。

3.2 绿茶粉的抗氧化保绿技术研究

众所周知,无论是叶绿素中的镁离子被氢离子置换变成氢代叶绿素,还是多酚类等热敏性物质在湿热条件下产生一系列非酶性氧化,转变成茶黄素、茶红素、茶褐素等,从而失去抹茶(或绿茶粉)原有的绿色和风味,这些生化反应的发生都需要有空气(氧)和水的参加。所以,我们设法利用抹茶(或绿茶粉)颗粒超强的界面能和吸附性,选择某种性质致密并相对稳定的材料,通过被绿茶微粉的强力吸附,形成类似于超微胶囊的保护,从而实现它与空气、水分间的阻隔,达到水热加工条件下的抗氧化保绿目的。

在这一基本思想的指导下,经过比较,以各种食用油脂为最理想,如金华火腿就用涂油的方法抗氧化保鲜,铁锅用涂油的方法防锈(氧化)等等。而且,绿茶中的叶绿素和大多数香气成分的属性都是脂溶性的,溶解进油脂以后的叶绿素和香气成分在油脂的保护下更为稳定。当然,这一技术在实际应用上是首先要求对绿茶粉进行科学的保绿技术处理,处理时对绿茶粉的干燥度、细度、油脂的热稳性、环境条件、操作技术等都有相应的技术要求。现这一技术已获得中国发明专利,专利名称:绿茶粉的抗氧化保绿方法及其应用(专利号:200910308087)。

3.3 抗氧化保绿技术在茶食品中的应用

该项技术是以食用油脂作为保护剂,加工温度不能超过所用油品的沸点,不然也会失去其保护作用。该技术特别适用于加工各种蒸煮类含茶食品,如面条、水饺、面包、汤圆、年糕、冰皮月饼等,因温度只有100℃。同时也较适用于煎炸类含茶食品,因煎炸本身用的就是油脂。最受考验的是用于烘烤类含茶食品的加工,这是因为烘烤类食品的烘烤温度都很高,加之相关标准规定的"干燥失重率"要求,特别是在烘烤到快出炉时的食品温度很容易超过所用油品的沸点,这时的感官现象是茶的原有绿色会瞬间变黄。下面是几则应用实例。

3.3.1 抗氧化保绿技术在蒸煮类食品中的应用实例

(1)抹茶汤圆。

主料:糯米粉、径山抹茶(或绿茶粉)。馅料:自选(如:芝麻馅、糖馅、肉馅、海鲜馅等)。

做法:①用开水和糯米粉揉透,然后切一小团到沸水里煮至浮起,重新放回原来的糯米团里一起揉透(目的是有利于成团);②先将径山抹茶(或绿茶粉)作抗氧化保绿技术处理,抹茶的用量为2%~3%,然后加进前面的糯米团里,揉透至色泽均匀;③搓成适当大小的面团并压扁,包进馅料搓圆;④把水烧开后把生汤圆放进锅里,煮至浮起即可。

(2)抹茶冰皮月饼。

主料:糯米粉250g±15g、白糖20g、牛奶250mL、茶叶籽油30g、径山抹茶(或绿茶粉)12~15g。

馅料:豆沙或莲蓉250g(或其他馅料)。

做法:①先将径山抹茶(或绿茶粉)作抗氧化保绿技术处理;②将糯米粉用

文火慢慢炒制,至微微发黄;③牛奶、茶叶籽油、白糖置锅中用小火加热至沸关火,倒入经炒制的糯米粉250g,用筷子搅拌使其结块,待凉后揉成光滑面团,然后再加入经技术处理的径山抹茶,揉至光滑和色泽均匀,盖上保鲜膜静止半小时;④将面团擀成条,平均切成10等份,擀成中厚边薄的圆皮;⑤包馅、收口、压模成型、蒸制7min出笼。

上述产品色泽绿润,并能保持茶味、茶香纯正。

3.3.2 抗氧化保绿技术在烘烤类食品中的应用实例

(1)抹茶广式月饼。

皮料:月饼专用面粉200g、食用油45g、径山抹茶(或绿茶粉)10g、转化糖浆110g等。

馅料:可用于月饼的馅料很多,这里用豆沙600g,豆沙中也要加入经技术处理的茶粉,比例应不少于2%。

饰料:鸡蛋黄两个,加花生油半茶勺搅匀待用。

做法:①茶粉的抗氧化保绿技术处理;②转化糖浆+适量碱水+适量盐,搅匀后再加入食用油40g搅匀;③筛入低筋粉135g揉成团,覆盖保鲜膜松弛1h;④加入剩下的低筋粉、食用油和经技术处理的茶粉揉至色泽翠绿均匀和耳垂般柔软;⑤分割月饼皮(每份20g),然后搓圆压扁、包馅、压模、上烤;⑥烤箱温度预热到160℃,烘烤5min取出刷蛋黄液上色,再烘烤7min取出刷蛋黄液上色,最后再烘烤5~7min出炉冷却。

(2)西湖茶点。

西湖茶点是在"绿茶粉的抗氧化保绿方法及其应用"专利正式获批后推出的,并入选2012年中国杭州西湖国际博览会首批(3种)特许经营产品。这款"西湖茶点"由绿茶桃酥、绿茶椒桃片、绿茶四季饼和绿茶茶糕4个品种组成,全是烘烤类含茶食品,技术方法与含茶月饼雷同。

4 结论与展望

(1)抹茶是中国唐宋时期的主流茶类,所以,径山抹茶的传承性恢复和创新,对于打造传统的"中国抹茶"品牌来说,无疑是一个很好的铺垫。

(2)对于径山茶产业而言,还可将南宋式的龙凤团茶、抹茶以及其茶碾、茶磨等抹茶工具开发成旅游产品,并把一系列抹茶食品的开发作为径山茶产业新的增长点。

(3)"绿茶粉的抗氧化保绿方法及其应用"这一专利,对于一系列绿茶食品的开发无疑是一项突出的贡献,以此为契机进一步将抹茶及同类茶产品打入广泛的食品工业领域,拓展茶产业链,提高茶的增值空间,促进茶产业为造福人类健康作出更大的贡献。

茶—粮复合发酵体系茶叶发酵行为研究

李大伟　施海根　朱跃进　张海华　张士康

中华全国供销合作总社杭州茶叶研究院,中国 浙江

摘　要:在传统半固态发酵酿酒工艺基础上,建立茶—粮复合发酵体系,研究茶叶影响发酵过程的各种因素。相比纯粮发酵基质,茶——粮复合发酵体系酵母菌较早出现衰亡期,可能与茶多酚的抑菌作用有关;随着发酵的进行,发酵液酒精度逐渐增加,pH 值和可溶性固形物含量随发酵的进行而下降;还原糖的含量在 24h 前呈上升趋势;发酵液中儿茶素含量呈下降趋势,在 12～24h 下降趋势最为明显,36h 后趋于平稳。

关键词:茶—粮复合发酵;茶叶;酵母菌;儿茶素;pH 值;还原糖

The Behavior of Tea Fermentation in the Tea-cereals Composite Fermentation System

LI Da-wei　SHI Hai-gen　ZHU Yue-jin
ZHANG Hai-hua　ZHANG Shi-kang

Hangzhou Tea Research Institute, China Coop, Zhejiang, China

Abstract: In the brewing process on the basis of the traditional semi-solid state fermentation, the establishment of tea-cereals composite fermentation system, conducted study on the various factors of the tea fermentation process. The results showed that compared to pure cereals fermentation, the yeast in tea-cereals composite fermentation system appeared earlier in the decline phase, this may be the antibacterial around with tea polyphenols. As the fermentation progresses, the alcohol content gradually increased, and the concentration of soluble solids and pH value decreased during the fermentation. Reducing sugar had been increasing in 0～24h. The catechin content of the fermentation broth showed a downward trend, the downward trend is most obvious in 12～24h and stabilized after 36h.

Keywords: tea-cereals composite fermentation; tea; yeast; catechin; pH value; reducing sugar

在传统半固态发酵酿酒工艺基础上,建立茶－粮复合发酵体系,对茶叶影响发酵过程的各种因素进行了深入研究。主要包括茶叶对发酵过程中酵母生长的影响、对酒精度的影响、对发酵基质还原糖和可溶性固形物的影响以及儿茶素在发酵过程中含量变化情况,较为全面地阐明了茶叶在茶－粮复合发酵体系中的发酵行为,为茶叶参与传统酿酒提供了必要的基础性数据,为优化茶－粮复合酿酒工艺提供技术参考。

1 材料与方法

1.1 试验材料与仪器

炒青绿茶(儿茶素含量 13.9％,浙江省景宁县际头茶叶有限公司);东北优质大米(市售);酿酒曲:安琪增香高产酿酒曲、安琪耐高温酿酒高活性干酵母(安琪酵母股份有限公司);标样:EGCG、ECG、EGC、EC、GCG、GC、CG、C、caffeine(Sigma 公司);无水乙醇、斐林试剂、偏重亚硫酸钠、葡萄糖、浓硫酸、浓盐酸、草酸、30％过氧化氢溶液、抗坏血酸、碳酸氢钠、2,6－二氯靛酚、酒石酸、硝酸铝、亚硝酸钠、酚酞、淀粉指示剂等(分析纯,杭州华东医药股份有限公司)。

LHS-100CL 型恒温恒湿培养箱(上海一恒科学仪器有限公司);压力蒸汽灭菌锅 YXQ-SG41-280(上海医用核子仪器厂);3-K15 高速冷冻离心机(Sigma 公司);pH 测定仪(赛多利斯公司);UV-2102 PC 型紫外可见分光光度计(尤尼柯上海仪器有限公司);高效液相色谱仪(Waters 公司);DK-S26 型电热恒温水浴锅(上海精宏实验设备有限公司);ZHWY-2102C 型双层小容量全温度恒温培养振荡器(上海智城分析仪器制造有限公司);电子天平(梅特勒－托利多仪器有限公司);糖度计(上海光学仪器进出口有限公司);酒精计(余姚市方桥实验仪表厂)。

1.2 试验方法

茶汁浸提:用纯净水浸提,茶水比例为 1∶70(w/V),90℃热水浸提 10min,用 200 目滤布粗滤得到第 1 道汁,重复第 1 道汁工序 3 次,目的是使茶叶中的有效成分和化学成分尽可能多地进入茶汤。将 3 道汁混合得茶叶萃取汁,灭菌备用。

酒坯制备:各取 100g 大米,将大米淘洗后,分装 500mL 三角瓶中,30℃浸泡 24h,浸泡过程中换 1 次水,淋去水分,分摊在铺有纱布的蒸屉上,用 100℃蒸汽蒸 30min,取出冷却,各添加 100mL 无菌水,添加酒曲 0.4%(按大米干重计算),30℃糖化 30h,进行糖化。糖化完全后,形成酒坯。

酵母活化:酿酒酵母采用安琪耐高温高活性干酵母,在使用前需进行活化,即在 35℃～40℃含糖 5%的温水中加入活性干酵母,使其质量分数为 10%,小心混匀后保温静置,每隔 10min 轻轻搅拌一下,进行 20～30min(不得超过 30min)的活化。

接种发酵:将茶汁与糖化后的酒坯按一定比例混合,并补充 5%的蔗糖。将活化好的酵母加入茶—粮混合介质中,加入恒温培养箱中进行发酵。发酵之初发酵容器内留有少量空气以利酵母生长,以后保持无氧发酵,发酵温度控制在 20℃～25℃。当发酵液的可溶性固形物含量低于 5%时,停止发酵。

1.3 检测方法

可溶性固形物(SSC):手持糖量计法;还原糖[1]:直接滴定法;酒精度[2]:比重计法测定;pH 测定:采用 PHS-3TC 型 pH 计;酵母生长曲线:600nm 下,分光光度计测定吸光度 A,茶浸提液做空白;茶多酚及儿茶素测定[3]:GB/T 8313—2008。

儿茶素色谱分析条件:Waters 高效液相色谱仪,色谱柱为大连依利特 hypersil ODS2,250mm×4.5mm,2487 双波长紫外检测器:$\lambda=278$nm。样品采用梯度洗脱,洗脱液由 A、B 两相组成,流动相 A:分别将 90mL 乙腈、20mL 乙酸、2mL EDTA 加入 1000mL 容量瓶中,用水定容至刻度,摇匀,溶液需过 0.45μm 膜;流动相 B:分别将 800mL 乙腈、20mL 乙酸、2mL EDTA 加入 1000mL 容量瓶中,用水定容至刻度,摇匀,溶液需过 0.45μm 膜。洗脱条件:100% A 相保持 10min,15min 内由 100% A 相 68% A 相、32% B 相 68% A 相、32% B 相保持 10min 100% A 相由 0 到 100%,柱温 35℃,流动相流速 1.0mL/min。

2 结果与分析

2.1 不同发酵基质中酵母生长曲线

由图 1 所示,对不同发酵基质的酵母生长状况实施动态监测,在监测的 8d

(192h)内可以看出,随着发酵的进行,纯粮发酵基质中酵母总数先快速增加,以后增加缓慢直至稳定,发酵开始的48h内酵母生长速度最快,48h后增加趋缓,而后逐步稳定。而在茶—粮复合发酵基质中,酵母的生长曲线却呈现典型的迟滞期(0~12h)、对数生长期(12~36h)和衰亡期(36~144h)[4]。其原因可能是在发酵初期,酵母为适应茶—粮复合培养基质,需要合成必需的酶类或某些中间代谢物,以适应新环境,故呈现生长延滞现象;对数生长期是酵母适应了新环境后,进行快速繁殖,酵母菌的细胞数呈几何级数增加;而在衰亡期中部分酵母衰老死亡,并伴随细胞自溶,细胞数下降。

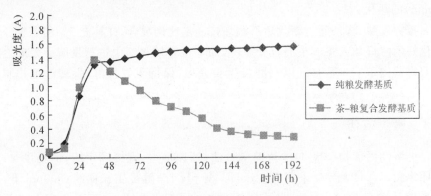

图1 不同发酵基质中酵母生长曲线

2.2 不同发酵基质中pH值随发酵时间变化

图2表明,在不同发酵基质发酵过程中,发酵液的pH值随发酵时间的变化

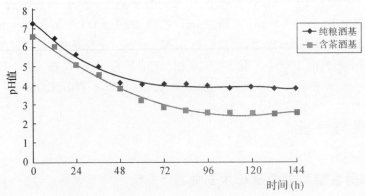

图2 不同发酵基质中pH值随发酵时间的变化

情况。可以看出,两种不同发酵基质中发酵液 pH 值变化动态趋势基本一致,主要表现为:随着发酵时间的延长,发酵液 pH 值逐渐下降,在 0~60h 均呈现较快下降趋势,纯粮发酵基质发酵液 pH 值由 7.26 降至 4.08,茶-粮复合发酵基质 pH 值由 6.52 下降至 3.19;此后发酵速度趋缓,72h 后基本趋于稳定,纯粮发酵基质发酵醪液最终 pH 值为 3.81,茶-粮复合发酵基质发酵醪液最终 pH 值为 2.85。从整体 pH 值变化趋势而言,两种不同发酵基质 pH 值变化规律基本与酵母的生长曲线相符合。在迟滞期,酵母的代谢较缓,因此有机酸的增加不明显;12h 后,酵母生长逐渐加速,代谢非常活跃,此时有机酸浓度迅速增高,pH 值变化明显;酵母对数生长期后,酵母部分死亡,生长停止,pH 值也趋于稳定[5]。

2.3 不同发酵基质中酒精度、还原糖及可溶性固形物含量变化

从图 3 至图 5 可以看出,无论发酵基质有无区别,发酵前期,酒精度增加和可溶性固形物含量降低的速度很快;发酵后期,酒精度、还原糖和可溶性固形物 3 种指标都趋于稳定。这可能是由于发酵的后期酒精的抑制作用以及活酵母数减少的缘故。图 3 和图 4 显示,不同发酵基质进行发酵时,酒精度的变化曲线之间有较明显的区别,还原糖的变化曲线之间也有一定区别,发酵初期茶-粮复合发酵基质的产酒及降糖能力均低于纯粮发酵基质。另外,图 4 显示,茶-粮复合发酵基质中,还原糖含量变化曲线呈先升高后降低的趋势,这可能是由于向发酵基质补充糖源后蔗糖并没有立即转化为还原糖,而是当酵母生长旺盛后,酵母菌分泌蔗糖酶将蔗糖分解成葡萄糖和果糖进一步发酵的缘故。

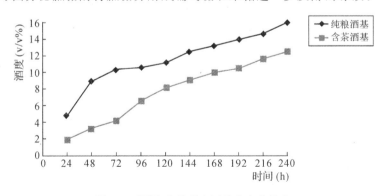

图 3 不同发酵基质中酒精度变化趋势

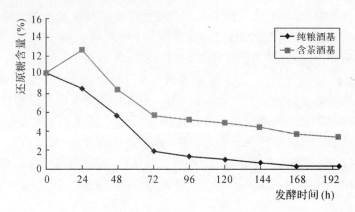

图4 不同发酵基质中还原糖变化情况

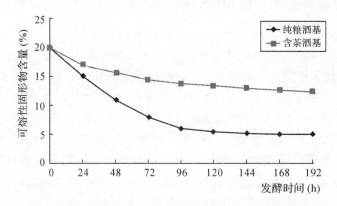

图5 不同发酵基质中可溶性固形物随发酵时间的变化趋势

2.4 茶-粮复合发酵基质发酵过程中儿茶素的变化

对茶-粮复合发酵基质中儿茶素各单体的检测中可以发现:总体上儿茶素各单体含量均呈下降趋势,在12～24h下降趋势尤为明显(图6)。在0～24h,儿茶素总量由0.965mg/mL下降至0.565mg/mL,下降了41.5%,此后下降不明显。截至144h,EGCG、EGC、EC、ECG和C含量分别由最初的0.48mg/mL、0.25mg/mL、0.07mg/mL、0.14mg/mL和0.025mg/mL下降到0.18mg/mL、0.10mg/mL、0.04mg/mL、0.02mg/mL和0.025mg/mL(表1)。儿茶素含量下降可能是由两种作用引起:一是被酵母细胞吸收利用,可能在菌体中不断积累;另一主要原因可能是酸性环境中酚类物质自身的络合反应[5]。24h后发酵液的pH值下降较快,60h后发酵液pH值已在3.0以下,这时儿茶素的减少主要是由于它在酸性条件下的分解所致。

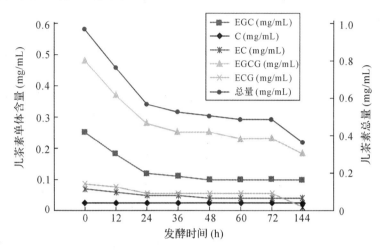

图 6 发酵过程儿茶素含量随发酵时间的变化情况

表 1 发酵液中儿茶素含量随发酵时间变化情况　　　　　　　　　mg/mL

发酵时间(h)	EGC	C	EC	EGCG	ECG	总　量
0	0.25	0.025	0.07	0.48	0.14	0.965
12	0.18	0.025	0.06	0.37	0.13	0.765
24	0.12	0.025	0.05	0.28	0.09	0.565
36	0.11	0.025	0.05	0.25	0.09	0.525
48	0.1	0.025	0.04	0.25	0.09	0.505
60	0.1	0.025	0.04	0.23	0.09	0.485
72	0.1	0.025	0.04	0.23	0.09	0.485
144	0.1	0.025	0.04	0.18	0.02	0.365

3　结　论

(1)在茶－粮复合发酵基质中,酵母的生长曲线呈现典型的迟滞期(0~12h)、对数生长期(24~36h)和衰亡期(36~144h)。相比纯粮发酵基质而言,茶粮复合发酵基质较早出现衰亡期(36~144h),在衰亡期中部分酵母衰老死亡,并伴随细胞自溶,细胞数下降。可能是由于茶叶中含有较高含量的茶多酚,对酵母菌生长产生了一定程度的抑制作用,这也可能是导致发酵液酒精度提高缓慢的主要原因之一。

(2)茶-粮复合发酵基质中发酵液的pH值变化趋势与纯粮发酵基质基本一致。随着发酵时间的延长,发酵液pH值逐渐下降,72h后基本趋于稳定(pH值2.85),茶叶的加入对发酵液pH值变化影响不大,只是茶-粮复合发酵基质初始pH值较纯粮发酵基质低,伴随发酵产酸,发酵液最终pH值维持在较低水平。

(3)茶-粮复合发酵基质酒精度和可溶性固形物变化趋势基本与纯粮基质一致,但茶叶的加入为发酵液引入了较高浓度的茶多酚及其他茶叶生物活性物质,对酵母菌会产生一定程度的抑制,导致酵母菌对发酵性糖的利用度下降,最终发酵酒精度较低,还原糖含量较高。另外,值得说明的是,在茶-粮复合发酵基质中,还原糖含量变化曲线呈现先升高后降低的趋势,这可能是因为向发酵基质补充糖源后蔗糖并没有立即转化为还原糖,而是当酵母生长旺盛后,才开始对蔗糖的利用。

(4)对茶-粮复合发酵基质中儿茶素各单体的检测中可以发现:总体上儿茶素各单体均呈下降趋势,在12~24h下降趋势最为明显,36h后趋于平稳。儿茶素的含量下降可能是由两种作用引起:一是被酵母细胞吸收利用,可能在菌体中不断积累。从儿茶素总量和儿茶素各单体的变化统一来看,在0~24h,pH值为4.0~4.5,可以看出,此pH范围儿茶素较稳定,基本上不被酸所降解,而在24h后pH值降低,儿茶素仍会部分与酸络合更易被透膜吸收[6]。儿茶素下降的另一主要原因可能是酸性环境中酚类物质自身的络合反应。24h后发酵液的pH值下降较快,60h后发酵液pH值已在3.0以下,这时儿茶素的减少主要是其酸性条件下的分解。

参考文献

[1] 穆华荣,于淑萍. 食品分析[M]. 北京:化学工业出版社,2001.4-56.

[2] GB/T 13662—2000 黄酒[S].

[3] GB/T 8313—2008 茶叶中茶多酚及儿茶素的检测[S].

[4] 陈思妤,萧熙佩. 酵母生物化学[M]. 山东:山东科学技术出版社,1990.

[5] 陈暄,屠幼英,童启庆等. 绿茶活性成分在酵母发酵过程中的代谢动力学研究[J]. 茶叶科学,2002,22(1):66-69.

[6] Nakagawa K. The absorption and metabolism of catechins of green tea[C]. Proceeding of the 15th meeting of Japanese tea society,2000:41-44.

脱咖啡因花草袋泡绿茶配方及其抗肠癌活性研究初探

涂云飞 杨秀芳 张士康 朱跃进 王盈峰 孔俊豪

中华全国供销合作总社杭州茶叶研究院,中国 浙江

摘 要:为改善茶叶香气与滋味偏淡的脱咖啡因绿茶产品,本实验通过二次通用回归旋转设计方法,结合感官评定与羟自由基体外评定手段,从常见植物花草中筛选获得了较为理想的与之配伍的花草配方,从而使产品的口感风味更具特色;同时对花草袋泡绿茶70%乙醇提取物进行了体外抗肠癌活性初步评价。

关键词:脱咖啡因绿茶;花草;配方;抗癌活性

Optimization the Formula of Herbal Green Tea Bags and Preliminary Evaluation the Anti-colon Cancer Activity in-vitro

TU Yun-fei YANG Xiu-fang ZHANG Shi-kang
ZHU Yue-jin WANG Ying-feng KONG Jun-hao

Hangzhou Tea Research Institute, All-China Federation of Supply and Marketing Co-operatives, Zhejiang, China

Abstract: In this paper, we optimized the formula of herbal green tea bags by both methods of quadratic general rotate regression design and the index of scavenging the hydroxyl free radicals; and furthermore we preliminarily evaluated the anti-colon cancer activity *in-vitro*. The results indicated that decaffeinated green tea mixed with herbals not only overcame the low flora, but also improved the antioxidant ability with scavenging hydrogen free radical. The results also showed that 70% ethanol extraction of the herbal green tea bags has anticancer effect in the inhibition aspects with proliferation, cell cycle in G_0/G_1 phase and tumor metastasis.

Keywords: decaffeinated green tea; herbal; formula; anticancer bioactivity

绿茶具有生津止渴、降火明目、去腻减肥、止痢除湿、预防心脑血管病、癌症等药理作用[1]。绿茶中咖啡因含量为2%～5%，不适宜易失眠者、小孩、孕妇以及老人[2-3]饮用。脱咖啡因绿茶的生产可以满足特定人群的需求，但采用超临界 CO_2 脱除绿茶中咖啡因[4-5]的同时也将茶叶中的内质成分萃取出，致使茶叶香气与滋味偏淡。

植物中的花草如桂花、甘草及绞股蓝等不仅具有良好的保健作用，而且口感风味较佳。本文通过流动注射化学发光体外评价羟自由基清除能力为指标对产品的配方进行了优化，并在细胞水平上对脱咖啡因花草袋泡绿茶的抗结直肠癌活性方面进行了相关评价。

1 材料与方法

1.1 材料与试剂

桂花、绞股蓝、甘草购自本地中药材店。脱咖啡因蒸青茶为本院基地产品。RKO 结肠癌细胞系（浙江省生物治疗重点实验室惠赠）；IFFM-E 型流动注射化学发光分析仪（西安瑞迈分析仪器有限公司）；Waters 高效液相色谱仪〔配备1525泵，2487紫外可见检测器，717-plus 自动进样器，Breeze 控制软件，Hypersil ODS-C18（4.6mm×250mm）色谱柱〕；紫外－可见分光光度计（尤尼柯上海仪器有限公司）；色谱级乙腈（美国 TEDIA 公司）。福林酚（上海源聚生物科技有限公司），其余试剂均为分析纯。

1.2 配方筛选方案

袋泡茶以 2g 为计量单位。采用三因子二次通用旋转回归设计，在单因素实验的基础上，将绿茶和桂花作为一个整体 A，记为因子 Z_1，将绿茶记为因子 Z_2；将绞股蓝与甘草作为一整体 B，同时以绞股蓝作为因子 Z_3。各水平值设置如表1所示。

表1 因素编码表

因子	$-\gamma$	-1	0	1	γ
Z_1(%, w/w)	50	56.08	65	73.92	80
Z_2(%, w/w)	50	60.14	75	89.86	100
Z_3(%, w/w)	20	32.16	50	67.84	80

实验实施序号如表 2 所示,并对实验所获取羟自由基清除能力以统计软件对其进行统计回归分析,优化最佳花草袋泡茶拼配配方。

表 2　三因子二次通用旋转设计实施序号

序　号	Z_1	Z_2	Z_3	清除率(%)	感官评审
1	−1	−1	−1	64.36	浅黄绿,桂花香,滋味甜涩,爽口,利喉
2	−1	−1	1	61.13	黄绿,桂花香,微涩,利喉,绞股蓝甜香
3	−1	1	−1	49.54	黄绿,青草气,涩,利喉,弱甜
4	−1	1	1	51.36	黄绿,青草气,甘草甜,利喉
5	1	−1	−1	67.99	黄绿,桂花香,微涩,爽口,甘甜
6	1	−1	1	65.82	黄绿,青草气,涩,甘,绞股蓝香
7	1	1	−1	59.81	黄,青草气,微涩,利喉
8	1	1	1	40.68	浅黄,青草气,桂花香,涩较重,利喉
9	−γ	0	0	50.56	黄绿,淡桂花香,甜爽,利喉
10	γ	0	0	66.17	黄绿,桂花香,酸涩,利喉
11	0	−γ	0	71.44	黄绿,桂花香,回甘
12	0	γ	0	48.14	黄,青草气,酸涩弱,甘爽
13	0	0	−γ	60.66	淡黄,桂花香,绞股蓝香,甜,爽口
14	0	0	γ	61.09	黄亮,青草气,绞股蓝香,甜,爽口
15	0	0	0	58.05	黄亮,桂花香,绞股蓝香,甜,爽口
16	0	0	0	57.70	同 15
17	0	0	0	59.35	同 15
18	0	0	0	56.07	同 15
19	0	0	0	58.05	同 15
20	0	0	0	63.18	同 15

注:各试验实施过程中所测定的羟自由清除能力为本实验室自建方法[6]。

1.3　茶样制备

将拼配的花草茶依据 GB/T8303—2002《茶 磨碎试样的制备及其干物质含量测定》进行样品的制备。

1.4 感官评定及理化指标测定

称取2g样品,置于150mL审评杯中,用沸水注满,加盖,冲泡5min后将茶汤沥入评茶碗中,评定香气、汤色和滋味。茶汤冷却后转移并用去离子水定容于200mL容量瓶中待用;茶多酚、儿茶素和咖啡因测定方法依据GB/T8313—2008《茶叶中茶多酚和儿茶素类含量的检测方法》进行;羟自由基清除能力采用流动注射化学发光法测定。

1.5 细胞增殖存活率

将拼配的花草茶以70%的乙醇水溶液提取、过滤、浓缩后真空干燥,以去离子纯净水配制成28μg/mL,置4℃冰箱中待用。取对数生长期的RKO接种于96孔板中(5×10^4 个/mL),培养24h后,加入100、200、300、400mg/L的提取液,于含5% CO_2 的空气中,37℃培养96h后,吸去培养液,用PBS洗3次后,加入100μL的MTT(500mg/kg)培养4h后,吸去此液并加入150μL DMSO溶解沉淀,于酶标仪上测定570nm的光密度值[7]。

1.6 DNA-Ladder法检测细胞凋亡

取对数生长期细胞接种于12孔板中,24h后采用0、100、200、300、400μg/mL处理48h并收集细胞,70%乙醇固定2h,500×g离心5min,去掉上清液,PBS洗2次;加入400μL裂解液,酚:氯仿:异戊醇(25:24:1)处理后,用无水乙醇沉淀DNA,并溶解于TE中;最后进行1%琼脂糖凝胶电泳100V,30min,凝胶成像系统中拍照[7]。

1.7 流式细胞仪检测细胞周期

细胞培养后,收集 1×10^6 个细胞,加入1mL 70%冷乙醇4℃过夜后,加入PI最终质量浓度为50μg/mL,RNase最终质量浓度为20μg/mL,常温避光保存1h后上机检测[7]。

1.8 细胞迁移率测定

取对数生长期的细胞接种于12孔板中,待板铺满至80%时,吸出培养液,并加入不含胎牛血清蛋白的高糖DMEM培养液培养24h后,用200μL的黄色移液枪头划出交叉形细线,进行培养,于48h进行光学显微镜下观察拍照[8]。

2 结果与分析

2.1 拼配袋泡茶中茶质量分数对茶汤多酚浓度的影响

为考察不同茶质量分数对茶汤中酚类等成分的影响,将1、1.5、2.0g茶置于150mL评审杯中,注满沸水冲泡,分别以A、B、C代表,重复2次,并将第1次冲泡(以1表示)后茶汤滤入评审碗中,进行第2次冲泡(以2表示)。

结果(表3)表明,茶汤中咖啡因及茶多酚的浸出浓度相对稳定,但儿茶素各组分及总量在不同冲泡过程中,变化幅度较大,可能一方面由于儿茶素的浸出率低于多酚类中其他组分[9],同时儿茶素的浸出受外界因素影响较大,并且在茶汤中儿茶素的质量浓度也偏低。另外,第2次冲泡中,茶汤中亦有可观的茶多酚生物活性浓度[10-13]。按人们的饮用习惯,袋泡茶只冲泡1次,即会弃去,故应尽可能的选用合适的茶叶质量分数,达到效益最大化。

表3 儿茶素、咖啡因及茶多酚在茶汤中的浓度   mg/L

编号	EGC	Caffeine	EGCG	EC	ECG	儿茶素总量	茶多酚
A1	22.39±3.32	1.85±0.11	5.65±3.80	13.43±3.41	1.65±1.51	43.11±5.40	308.89±6.29
A2	10.10±4.07	1.69±0.00	3.77±2.48	11.45±4.48	1.55±1.27	26.86±12.30	296.67±14.14
B1	32.97±27.59	2.62±0.01	26.13±31.71	17.03±6.12	7.02±8.13	83.14±73.56	494.45±77.00
B2	48.79±2.56	2.44±0.02	43.89±19.92	21.26±2.10	12.11±6.67	126.04±21.93	462.23±9.43
C1	54.45±3.59	3.00±0.30	28.69±8.56	23.44±2.84	6.91±3.95	113.49±11.75	571.11±188.56
C2	66.67±35.84	3.15±0.40	40.98±29.49	24.46±7.62	9.96±7.70	142.06±80.65	553.33±97.43

2.2 二次通用旋转实验设计分析

按表4对应的回归系数,其所对应的回归方程即为:$y=58.79+2.5Z_1-7.11Z_2-1.61Z_3-1.09Z_1 \times Z_2-2.48Z_1 \times Z_3-1.49Z_2 \times Z_3-0.58Z_1 \times Z_1-0.07Z_2 \times Z_2+0.31Z_3 \times Z_3$。

表5说明,二次通用旋转回归方程的失拟性不显著,说明模型是适合的,拟合不足以被否定。其中Z_0、Z_1、Z_2、$Z_2 \times Z_3$回归系数通过t检验达到显著性,而二次项中3个因素均不显著,因此不构成响应面。

表 4　因素回归系数及显著性检验

因素	Z_0	Z_1	Z_2	Z_3	$Z_1 \times Z_2$	$Z_1 \times Z_3$	$Z_2 \times Z_3$	$Z_1 \times Z_1$	$Z_2 \times Z_2$	$Z_3 \times Z_3$
回归系数	58.79	2.5	−7.11	−1.61	−1.09	−2.48	−1.49	−0.58	−0.07	0.31
t 检验值	34.97	1.49	4.23	0.96	0.65	1.48	0.89	0.34	0.04	0.18
显著性	***		*			*				

注：*、*** 分别代表显著与极显著性。

表 5　方差来源及回归显著性统计分析

方差来源	回归显著性
总平方和	1072.82
剩余平方和	177.59
回归平方和	895.23
误差平方和	29.22
失拟平方和	148.38
拟合度检验（F 值）	5.08（$P=0.05$，不显著）
回归方程检验（F 值）	5.60（$P \leqslant 0.05$，高度显著）

2.2.1　主效应分析

由于试验各因素均经无量纲线性编码处理，且一次项与二次项回归系数间都是不相关的，因此，可以由回归系数绝对值的大小直接比较各因素一次项对羟自由基清除率的影响，故 $Z_2 > Z_1 > Z_3$。虽然 Z_2（绿茶）在本次实验的模型中作为负效应难以分析，但是由于 Z_2 作为 Z_1（绿茶与桂花）中的因素，而 Z_1 为正效应，可以间接说明桂花的体外羟自由基清除能力强于绿茶。

2.2.2　单因素效应分析

将回归模型方程中的 3 个因素中的 2 个固定在 0 水平上，可得到单因素模型方程：

Z_1：$y_1 = 58.79 + 2.5 Z_1 - 0.58 Z_1 \times Z_1$，

Z_2：$y_2 = 58.79 - 7.11 Z_2 - 0.07 Z_2 \times Z_2$，

Z_3：$y_3 = 58.79 - 1.61 Z_3 + 0.31 Z_3 \times Z_3$

依次对单因素模型方程取不同水平值，求得 y 值，如图 1 所示。

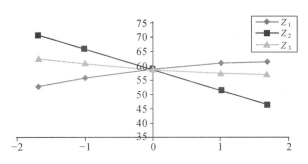

图 1　单因素与羟自由基清除率回归方程曲线

由图 1 所示，Z_2 的变异系数最大，说明桂花与茶的变化对羟自由基的清除率影响显著。而 Z_1 与 Z_3 的变异系数接近，但方向相反。故桂花与茶整体对羟自由基的贡献率为正效应，在水平 1 时，接近最大水平；同时，随着茶质量分数的降低，拼配茶的抗氧化性也随着增大。而绞股蓝的质量分数增加，羟自由基清除率则降低，这与上述的分析一致。

2.2.3　单因素边际效应分析

各因素在不同水平下的边际效应方程为：

$(y_1)' = 2.5 - 1.16 Z_1$，
$(y_2)' = -7.11 - 0.14 Z_2$，
$(y_3)' = -1.61 + 0.62 Z_3$。

取不同水平值作图。如图 2 所示，各因素对羟自由基清除率的边际影响，Z_1 的边际效应为下降趋势，与 Z_3 的变化趋势相反，同时，Z_2 的边际效应较小。

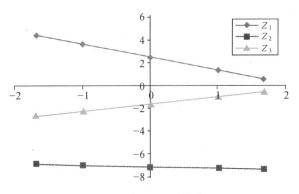

图 2　单因素边际效应

2.2.4 两因素间的交互效应

Z_2 与 Z_3 间的交互作用相对较为明显,所得交互方程为

$$y_{(Z2\times Z3)}=58.79-7.11Z_2-1.61Z_3-1.49Z_2\times Z_3-0.07Z_2\times Z_2+0.31Z_3\times Z_3$$

对 Z_2 与 Z_3 分别求偏导得 Z_2 为 -50.79, Z_3 为 2.59,而这一极值点已超出设定范围,因此,符合单因素分析,即在 A 组分中,随着茶叶量的降低和桂花量的上升,羟自由基的清除率会上升。

各配方(表2)的羟自由基清除率以第 11 号最大,第 5 号达到了 67.99%,排在第 2 位,第 10 号达到了 66.17%,排在第 3 位。并且 5 与 10 号中茶叶所占的份额较大,茶的苦涩味偏大了点,而 11 号不仅抗氧化能力强,而且汤色黄绿,桂花香显,滋味回甘。各花草茶成分中单位多酚清除羟自由基的能力不同,造成第 3、4、8、9、12 号明显低于其他花草茶的拼配组合,特别是第 8 号,只有 40.68% 的羟自由基清除率。同时在 3、4、6、7、8、12 号实验茶汤中,明显带有青草味;虽然第 10 号拼配茶中茶的份额占到了组分质量分数的 60%(高于 3、4、6、7、8、12 号配方中茶质量分数),但青草气不显,而且桂花香显,汤色黄绿,饮之利喉。因此,花草的风味改善了脱咖啡因蒸青绿茶的不良滋味。

2.3 70%乙醇提取物对抗结直肠癌细胞活性评价

(1)70%乙醇提取物对细胞增殖活性的影响。脱咖啡花草拼配绿茶醇提取物对细胞生长抑制作用结果如图 3 所示,图中可见,70%乙醇提取产物对结肠癌细胞培养 96h 后的增殖有明显的体外剂量依赖性抑制作用,当浓度增加至 $200\mu g/mL$ 时,结肠癌细胞增殖率即得到了显著性的抑制,抑制率为 38.3%($n=3$, $P=0.0013<0.01$, t-test)。

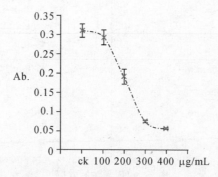

图 3 不同质量分数的 70%乙醇提取物对结肠癌细胞体外增殖抑制效果

（2）70％乙醇提取物对细胞迁移率与周期的影响。侵袭转移是恶性肿瘤临床治疗危及患者生命和预后复发的主要原因。本研究通过划痕实验初步揭示了醇提产物可以有效抑制结肠癌细胞的体外迁移能力。结果如图4(a)～(e)所

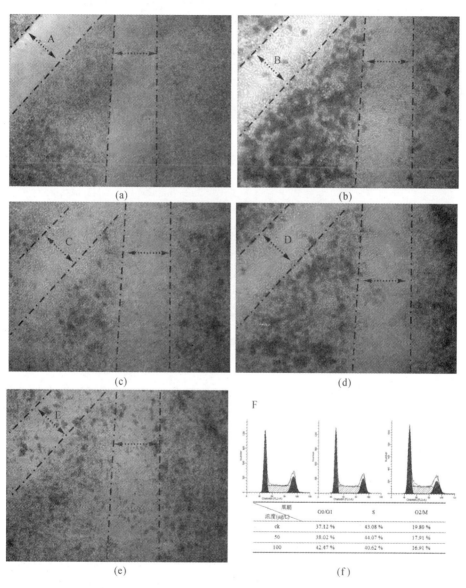

图4 不同质量分数的70％乙醇提取物对结肠癌细胞体外迁移与细胞周期的影响

示,其中,(a)为划痕原始状态,(b)为未加醇提取物,(c)为培养液中含有 100g/mL 醇提取物,(d)为培养液中含有 200g/mL 醇提取物,(e)为培养液中含有 400g/mL 醇提取物,其中图(b)~(e)的处理时间为 48h;(f)为细胞醇提产物处理 96h 后的分析结果。与对照组相比,醇提产物处理组随着质量分数的增加,癌细胞迁移也受到相应的抑制作用。为进一步确定醇提产物诱导细胞周期分布的影响,通过流式细胞仪定量比较了醇提产物处理后体外周期的分布情况,结果如图 5 所示,100μg/mL 的醇提物即可将癌细胞阻止于细胞周期的 G0/G1 期。

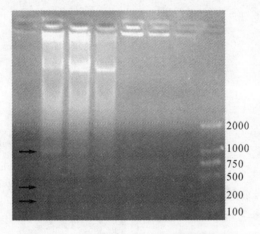

图 5　不同质量分数的 70％乙醇提取物诱导结肠癌细胞 DNA 片断断裂情况
(从左至右浓度依次为 100、200、400g/mL)

(3)70％乙醇提取物对细胞凋亡影响。本实验过程中以不同醇提产物处理如图 5 所示,虽然有 DNA 片断出现,但并未出现 180~200bp 的 DNA 产物或其整数倍凋亡信息[7],且无规律可循,推测细胞增殖抑制或漂浮可能是由坏死引起[14],具体结果有待进一步实验确定。

3　讨　论

本实验所获配方产品迎合了当今人们对营养、健康、天然等要素聚集的绿色食品追求。产品醇提取物亦具有体外抗肿瘤活性。并且近年来,随着社会发展带来的社会竞争力加大、环境污染以及人口老龄化等问题,使得癌症的发病率近年来呈快速增长趋势[15-16],此产品将有可能为癌症患者在药物治疗过程中作为辅助产品而起增效作用。

综上所述,本实验通过二次通用旋转回归设计方案,有效地将植物中的花草与口感滋味欠佳的脱咖啡因绿茶进行了配方筛选与优化,使得拼配的花草茶产品风味特色,并从生物活性角度初步阐述了其抗肿瘤活性,故此产品的开发和产业化应用将有利于各植物成分相互补长,有利于长期得不到有效利用的夏秋茶资源获得应有的利用。

参考文献

[1] Brahma N S, Sharmila S, Rakesh K S. Green tea catechin, epigallocatechin-3-gallate (EGCG): Mechanisms, perspectives and clinical application [J]. Biochemical Pharmacology, 2011, 82: 1807−1821.

[2] 最新研究发现:咖啡因增加流产风险[J]. 中华中医药学刊, 2008, 26(5): 1036.

[3] 咖啡因会影响青少年晚间睡眠[N]. 中华医学信息导报.

[4] 周海滨, 宋新生. 超临界 CO_2 脱除茶叶咖啡因的工艺优化研究[J]. 食品科学, 2003, 24(5): 88−91.

[5] 朱跃进, 周卫龙, 毛志方等. 超临界 CO_2 脱咖啡因绿茶加工对茶内含成分影响的研究[J]. 中国茶叶加工, 2009(2): 19−23.

[6] 涂云飞, 毛志方, 杨秀芳. 茶叶抗氧化能力测定的流动注射化学发光法研究[J]. 中国茶叶加工, 2009(1): 42−45.

[7] 明艳林, 郑志忠, 陈良华等. 人参皂苷 IH901 抗肝癌生长及其侵袭转移的作用[J]. 中国科学: 生命科学, 2011, 41(3): 219−225.

[8] 陈花. 炎症、氧化应激与动脉粥样硬化的关系及辛伐他汀抗炎、抗氧化作用的研究[D]. 西安: 中国人民解放军第四军医大学西京医院, 2008.

[9] 杨秀芳, 涂云飞, 孔俊豪. 茶树次生代谢产物在铁观音茶冲泡过程中的溶出规律初探[J]. 福建茶叶, 2010(5): 30−35.

[10] Fujiki H, Suganuma M, Imai K, et al. Green tea: Cancer preventive beverage and /or drug [J]. Cancer Letter, 2002, 188: 9−13.

[11] Nakachi K, Matsuyama S, Miyake S, et al. Preventive effects of drinking green tea on cancer and cardiovascular disease: Epidemiological evidence for multiple targeting prevention [J]. BioFactors, 2000, 13: 49−54.

[12] Nakachi K, Suemasu K, Suga K, et al. Influence of drinking green tea on breast cancer malignancy among Japanese patients, Jpn [J]. Journal

of Cancer Research, 1998, 89: 254—261.

[13] Bettuzzi S, Brausi M, Rizzi F, et al. Chemoprevention of human prostate cancer by oral administration of green tea catechins in volunteers with high-grade prostate intraepithelial neoplasia: A preliminary report from a one-year proof-of-principle study [J]. Cancer Research, 2006, 66: 1234—1240.

[14] Guido M, Isabelle J. Apoptosis, oncosis, and necrosis—an overview of cell death [J]. American Journal of Pathology, 1995, 146(1): 3—15.

[15] Ahmedin J, DVM, Freddie B, et al. Elizabeth ward, david forman. Global cancer statistics [J]. CA: A Cancer Journal for Clinicians, 2011, 61: 69—90.

[16] 靳宝峰, 夏晴, 张学敏. 军事医学科学院肿瘤研究进展[J]. 中国科学: 生命科学, 2011, 41(10): 807—811.

夏秋茶鲜叶发酵茶酒的品质形成规律

黄友谊　肖　平　张嫣嫣　倪　超
潘　欣　乔如颖　黄莹捷　常银凤
园艺植物生物学教育部重点实验室，
华中农业大学园艺林学学院茶学专业，中国 湖北

摘　要：夏秋茶树鲜叶经杀青后直接制成茶水，接种酵母发酵成茶酒，分析了茶酒品质形成规律。结果表明发酵时间过短或过长均不利于茶酒感官品质的形成，以发酵第4d至第8d之间茶酒的感官品质最好。总糖与可溶性固形物的变化趋势基本一致，其中总糖在茶酒发酵前7d下降约达92.83%，至发酵第12d后大部分被消耗完。茶多酚、黄酮、游离氨基酸、pH值均是前期（特别是发酵第1d）下降迅速，后期相对稳定。酒精度和总酸均是先增高后降低，酒精度在发酵第6d达到最大值，约为1.60%，总酸在发酵第7d达到最大值，约为3.74g/L。了解茶酒发酵过程中品质形成规律，有助于阐明茶酒品质形成机理，并为优化茶酒发酵工艺提供基础。

关键词：鲜叶；茶酒；发酵；品质形成

Quality Formation of Law in Tea Wine Fermented by Fresh Leaves (*Camellia sinensis*)

HUANG You-yi　XIAO Ping　ZHANG Yan-yan　NI Chao　PAN Xin
QIAO Ru-ying　HUANG Ying-jie　CHANG Ying-feng
Key Lab of Horticultural Plant Biology, Ministry of Education;
College of Horticulture and Forestry Sciences,
Huazhong Agricultural University, Hubei, China

Abstract: When fresh leaves were fixed, crushed, and inoculated yeast for fermentation, the quality formation of law in tea wine was studied. The results showed that the sensory quality of tea wine was best during the fourth to the eighth day of fermentation time, and too short or too long fermentation time is not conducive to it. The trend of the total sugar and soluble solids in tea wine was same. The total sugar in tea wine decreased about 92.83% at the seventh day, and mostly was consumed after the twelfth day. Polyphenols, flavonoids, free amino acids, and pH values all decreased rapidly at early stage of fermentation especially the first day, and are relatively stable at the latter. Alcohol and total acid both increased firstly and then slightly lowered. The maximum content of alcohol was about 1.60% at the sixth day, and the maximum content of total acid is approximately 3.74g/L at the seventh day. To understand the quality formation of law in tea wine during fermentation process will promote to clarify the mechanism of tea wine quality formation and provide a basis for optimization of tea wine fermentation technology.

Keywords：fresh leaves；tea wine；fermentation；quality formation

茶酒是一种具有保健功能的特殊风味饮品[1]，目前主要以浸提调配和酵母发酵两种途径制成[2-4]。对发酵型茶酒已有工艺技术优化等方面的研究报道[5-13]，但还不清楚发酵型茶酒品质形成机理[2-3]。现有茶酒均是以制好的成茶为原料来生产的，还未见以茶树鲜叶直接为原料来发酵生产的[2-3]。当前我国夏秋茶鲜叶浪费严重，迫切需要有效开发利用[14]。为此我们开展了夏秋茶树鲜叶发酵茶酒的研究，分析了茶酒品质形成规律，以探明茶酒品质形成机理，为茶酒发酵工艺优化提供依据，并为夏秋茶树鲜叶的开发利用提供有效途径。

1 材料与方法

(1)材料。茶树鲜叶为一芽四五叶，2011年7—10月采摘于华中农业大学茶学教学实习基地，茶树品种为福鼎大白茶有性系群体品种。白砂糖为市售食用产品，酿酒活性干酵母为安琪酵母有限公司生产。

(2)主要仪器、设备。可见分光光度计、酒精计、糖度计、旋转蒸发器、循环水式真空泵、手提式灭菌锅、离心机、pH计、培养箱、超净工作台等。

1.3 试验方法

(1)茶酒发酵工艺流程及操作要点。茶鲜叶→杀青处理→浸提茶汤→分装并密封→发酵→过滤→茶酒
↑
白砂糖、酵母活化液

茶树鲜叶经龙井锅160℃～180℃杀青约7min后(杀青叶含水率为50%～60%),用打汁机打碎。按1g杀青叶加60mL无菌水的比例将无菌水与打碎的杀青叶混匀,以65℃水浴浸提10min,用脱脂棉过滤并定容,得茶水,加入0.1kg/L白砂糖,溶解冷却备用。酵母菌种活化,在5%质量分数的蔗糖水中加入50g/L酵母菌粉,摇匀,在37℃保温静置培养活化15～30min。按2.5%体积分数接种量向茶水中加入酵母菌种活化液,摇匀,分装于无菌玻璃瓶中(玻璃瓶体积为300mL,装200mL发酵液),加盖密封,室温发酵。茶酒发酵周期定为20d,前10d每天取样,后10d隔天取样,及时审评各试样,并及时测定各理化指标。所有试验均重复3次。

(2)茶酒感官审评。茶酒感官审评包括外观、香气、口感和整体风格4项,茶酒感官评分标准如表1所示。

表1　茶酒感官审评标准

项　目	标　准	分　值
外观(20)	澄清透明,具有本品应有的色泽	18～20
	澄清,无夹杂物,有原茶汤色	12～17
	微浑,失光或氧化变色	<12
香气(30)	酒香宜人,有茶香,协调	27～30
	酒香偏淡,稍有茶香	23～26
	酒香单薄,无茶香,有不良气味	19～22
	无香味,不良气味突出	<22
口感(40)	酒体丰满,醇厚协调,茶味适中,酸甜适口	37～40
	酒体单薄,口味好,有茶味,稍苦涩	33～36
	无不良口味,无茶味,苦涩	29～32
	略酸,较甜腻	25～28
	酸,涩,苦,平淡,有异味	<25
风格(10)	典型完美,风格独特	9～10
	稍有典型性,风格良好	6～8
	无典型性	<5

(4)理化指标测定方法。酒精度:酒精计法(GB/T 15038—2006)。

pH值与总酸量:pH值用pH计测定,总酸用酸碱滴定法(GB/T 12456—90)。

可溶性固形物(SSC):手持式折光仪(0%~32%)。
茶多酚:酒石酸亚铁比色法(GB/T 21733—2002)。
可溶性氨基酸:茚三酮法[15]。
黄酮:三氯化铝比色法[15]。
总糖:蒽酮比色法[15]。

(5)数据处理分析。所有数据采用SPSS软件(PASW Statistics 18.0)进行统计分析。单向方差分析法用于方差分析,处理间的多重比较采用Duncan法。$P<0.05$,处理间有显著性差异,用小写字母表示;$P<0.01$,处理间有极显著性差异,用大写字母表示。数据的表示方式均为平均值±标准差,图中数据均为试验的平均值,误差线表示试验的标准偏差。

2 结果与分析

2.1 茶酒发酵过程中感官品质的变化

对不同发酵时间的茶酒样品进行了感官审评,从表2可知,茶酒发酵第1d产生大量二氧化碳,形成刺鼻感;发酵第3d产生浓郁的酒香,并有麻舌感,未出现苦涩味;发酵至第12d,苦涩味明显;随着发酵时间加长,香气变化不大,但苦涩味加重,酸味不明显。对不同发酵时间的感官得分进行差异显著性检验,可知不同发酵时间的茶酒感官品质之间整体差异显著,以发酵第4d至第8d之间的感官品质最好(表2)。可见茶酒要求有一定的发酵时间,发酵时间过短或过长均不利于茶酒感官品质的形成。

表2 茶酒发酵过程中的感官品质

发酵时间(d)	感官评价	得 分
0	甜茶味,茶清香,无酒味,黄绿	58.3±1.5 Jj
1	甜茶味,香气略刺鼻,黄绿	72.3±0.6 Gg
2	入口略刺,较麻舌,略有酒香,浅黄绿	77.0±1.0 Ff
3	酒香浓郁、刺鼻,麻舌,浅黄绿	86.0±0.0 Cc
4	酒香浓郁、刺鼻,麻舌,浅黄绿	87.7±0.6 ABCab
5	酒香浓郁、刺鼻,麻舌,气泡多而均匀,浅黄绿	89.0±0.0 Aa

续表

发酵时间(d)	感官评价	得 分
6	酒香浓郁、刺鼻、麻舌、浅黄绿	88.3±0.6 Aba
7	酒香浓郁、刺鼻、麻舌、浅黄绿	88.3±0.6 Aba
8	酒香浓郁、刺鼻、麻舌、浅黄绿	88.3±0.6 Aba
9	酒香浓郁、刺鼻、麻舌、有苦味、浅黄	86.3±0.6 BCbc
10	酒香浓郁、刺鼻、麻舌、有苦味、浅黄	83.7±0.6 Dd
12	较苦、麻舌、米酒味、浅橙黄	80.7±0.6 Ee
14	较苦、麻舌、米酒味、浅橙黄	82.0±1.0 Dee
16	苦、麻舌、米酒味、浅橙黄	70.3±0.6 Hh
18	苦、麻舌、米酒味、浅橙黄	67.7±1.5 Ii
20	苦、麻舌、米酒味、浅橙黄	67.7±1.5 Ii

注：表中同一列相同字母表示经 Duncan 法检验在 0.01 或 0.05 水平差异不显著。A，b：$P<0.05$；A，B：$P<0.01$。

2.2 茶酒发酵过程中糖浓度、可溶性固形物及酒精度的变化

对茶酒不同发酵过程中的总糖、可溶性固形物和酒精度进行了检测，从图1可知，总糖在茶酒发酵前7d下降显著，尤以发酵前两天下降迅速；至发酵第2d糖含量已由发酵开始的128.25g/L下降到58.84g/L，约降低了54.12%；至发酵7d，糖含量为9.20g/L，约下降了92.83%；至发酵第12d后，糖含量在2g/L以下，绝大部分已被消耗。可溶性固形物(SSC)含量的变化，从图2可知，与总糖含量的变化趋势一致，前期下降迅速，后期缓慢减少并逐渐趋于稳定。酒精含量是茶酒一项重要的质量指标，本试验以酒精度来表示。从图2可知，在茶酒发酵前期酒精度逐步增加，至发酵第6d酒精度达到最大值，约为1.60%；在发酵第6d后，茶酒的酒精度略有下降；在发酵前12d，不同发酵时间之间茶酒的酒精度差异极显著；在发酵第12d后酒精度保持不变，维持在1.30%。

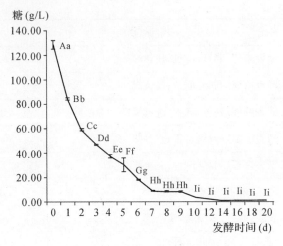

图 1　茶酒发酵过程中总糖含量的变化

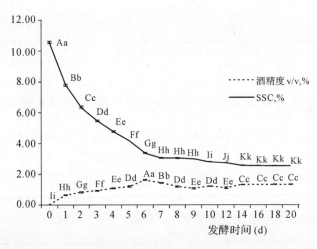

图 2　茶酒发酵过程中酒精度与可溶性固形物的变化

2.3　茶酒发酵过程中茶叶常规成分的变化

茶酒融合了茶和酒的特点,了解茶叶常规成分在发酵过程中的变化规律,有助于了解茶酒的品质与功能,为此分析了茶酒发酵过程中茶多酚、黄酮、游离氨基酸的含量变化。从图 3 可知,茶多酚在茶酒发酵前两天下降迅速,由 1.72g/L 下降到 1.28g/L,约下降 25.58%,有极显著差异;发酵第 2d 后,茶多酚含量在 1.31~1.37g/L 之间波动,相对稳定。从图 4 可知,黄酮在茶酒发酵

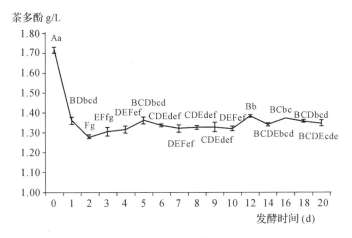

图 3 茶酒发酵过程中茶多酚含量的变化

第 1d 下降迅速,由 51.20mg/L 下降到 39.36mg/L,约下降 23.12%,有极显著差异;发酵第 1d 后,黄酮含量在 37.46～39.36mg/L 之间波动,相对稳定。从图 5 可知,游离氨基酸在茶酒发酵第 1d 下降迅速,由 0.14g/L 下降到 0.05g/L,约下降 64.28%,有极显著差异;发酵第 1d 至第 5d 之间氨基酸含量缓慢下降,但发酵第 5d 后缓慢上升,整体在 0.04～0.06g/L 波动。

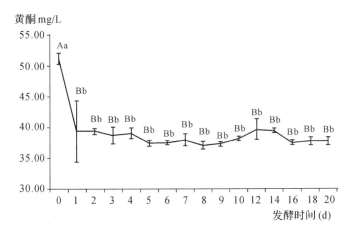

图 4 茶酒发酵过程中黄酮含量的变化

2.4 茶酒发酵过程中总酸与 pH 值的变化

茶酒在发酵过程中会产生酸,引起 pH 值下降。从图 6 可知,pH 值在发酵

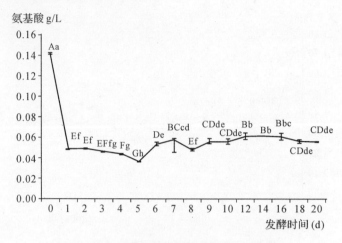

图 5 茶酒发酵过程中游离氨基酸含量的变化

第1d下降迅速,由初始的4.59下降到3.37,发酵第1d至第5d之间下降趋缓,但相互间差异极显著;发酵第5d后,pH值略有上升,但大体维持在2.9左右。从图6还可知,总酸含量在茶酒发酵前7d是逐步增加的,至发酵第7d达最高,约为3.74g/L;在发酵第7d后,总酸含量整体呈现缓慢降低的趋势。

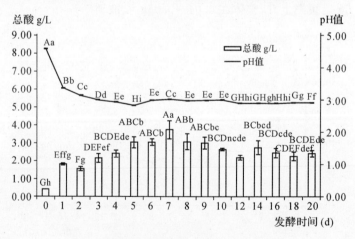

图 6 茶酒发酵过程中总酸与pH值的变化

3 讨 论

对发酵型茶酒已有的文献报道除主要针对发酵工艺外[11,13,16-20]，还涉及原料、香气成分、澄清剂等方面[10,12,21]，但未见有茶酒品质形成规律方面的报道[2-3]。而探明茶酒品质形成规律，对阐明茶酒品质形成机理和优化茶酒发酵工艺无疑具有重要作用。

3.1 茶酒主发酵时间

酵母发酵过程都有一个启动期，一般为 14～16h，以后便进入主发酵，时间在 3～4d。根据酒精含量变化，结合糖含量及感官品质，可以确定主发酵时间。根据本试验的条件，茶酒主发酵过程时间较短，3d 即产生浓郁酒香，茶酒发酵至第 6d 酒精度不再升高；发酵至第 7d，糖含量在 10g/L 以下，口感适宜，无明显苦涩及酸馊味，因此利用夏秋叶发酵茶酒的主发酵时间以 5～7d 较适宜，这与周丹丹[19]的研究结果一致，但李建芳[18]报道的绿茶酒发酵时间在 12d 左右，邱新平[11]报道的茶酒为 20℃发酵 10d。造成这种主发酵时间差异的主要原因，可能是各发酵茶酒要求的酒精度不一样，加糖量也不一样，并且发酵温度也不同。

3.2 茶叶常规成分对茶酒品质的影响

茶饮料存在一个普遍问题就是产品质量不稳定，主要表现在颜色加重、氧化感增加、茶香变淡，严重者有失光、沉淀等现象[22]。在茶酒发酵中，邱新平等专门进行了茶酒澄清剂筛选，以解决茶酒产品稳定性问题[12]。目前还缺乏对茶酒发酵过程中有效成分转化的研究，在本研究中游离氨基酸、茶多酚、黄酮在茶酒发酵第 1d 即大量减少，氨基酸可能被作为氮源消耗，茶多酚、黄酮可能在光和酵母分泌的酶等作用下氧化、水解、聚合或缩合为其他物质。茶多酚的减少，能有效减少茶饮料的沉淀，有利于茶酒后期储存和延长货架期。然而茶多酚是茶叶最主要的功能活性成分，茶多酚的转化途径和转化生成的成分及其保健功效等，还有待于进一步深入研究。此外酒类存在后发酵过程，在后发酵过程中会越陈越香，而茶酒目前还缺乏后发酵研究，包括后熟工艺及储藏条件等。茶叶常规成分在茶酒后发酵过程中的稳定性如何，也还有待研究。

3.3 利用夏秋鲜叶开发茶酒前景广阔

目前所有关于茶酒的研究报道，均是利用成茶浸提茶水发酵而成的[2-3]。

我国夏秋茶资源浪费严重[14]，为提高夏秋茶资源的利用率，本研究选择以夏秋茶鲜叶为原料，经杀青后直接打碎浸提制备茶水来发酵茶酒，减少了成茶的加工工序，而且夏秋茶鲜叶原料成本低廉，有利于开发生产茶酒，将为夏秋鲜叶资源的开发利用提供有效途径。

致谢

本项目由中央高校基本科研业务费专项资金（2009PY023）、华中农业大学自主科技创新基金（2009SC013）资助。

参考文献

[1] 王家林，王煜，吕丽丽. 茶酒的保健作用[J]. 食品研究与开发，2011(8)：133－136.

[2] 卫春会，罗惠波，豆永强. 我国茶酒生产现状及发展[J]. 酿酒科技，2007(10)：126－127，129.

[3] 徐洁昕. 茶酒的酿酒技术研究进展[J]. 酿酒科技，2010(10)：87－89，92.

[4] 周剑丽，谌永前，吴广黔等. 配制型茶酒稳定性研究[J]. 酿酒科技，2011(2)：29－31.

[5] 胡峰，向祖祥，余成光. 浓香型保健茶酒的研制[J]. 酿酒科技，1994(5)：91.

[6] 钟宜春. 蜂蜜酒、茶酒试制总结[J]. 食品科学，1988(11)：23－26.

[7] 韩珍琼，魏明. 浓香型保健茶酒的研制[J]. 饮料工业，2005(2)：19－21，34.

[8] 陈立杰，赖萍，谭书明等. 干型茶酒的加工技术[J]. 山地农业生物学报，2008(4)：371－376.

[9] 卫春会，罗惠波，黄治国等. 液态发酵茶酒的研制[J]. 酿酒科技，2008(11)：97－99.

[10] 张帅，董基，陈少扬. 发酵型铁观音茶酒的研制[J]. 食品工业科技，2008(10)：159－161.

[11] 邱新平，李立祥，蒋其忠等. 发酵型茶酒酿造工艺参数研究[J]. 食品科学，2010(16)：300－304.

[12] 邱新平，李立祥，倪媛等. 发酵型茶酒澄清剂的筛选[J]. 茶叶科学，2011，31(6)：537－545.

[13] 彭小东，唐维媛，张义明. 茶酒的生产工艺研究[J]. 中国酿造，2011(9)：185－187.

[14] 李永章. 深度开发夏秋茶,充分利用茶资源[J] 茶叶,2007:46-47.

[15] 钟萝. 茶叶品质理化分析[M]. 上海:上海科技出版社,1989.

[16] 徐亚军,赵龙飞. 微生物发酵生产绿茶酒的工艺研究[J]. 酿酒科技, 2008(1):96-97,101.

[17] 关琛,牛广财,李志江. 绿茶酒生产工艺的研究[J]. 农产品加工:学刊, 2010(6):44-46.

[18] 李建芳,周颖,周枫等. 绿茶酒液态发酵工艺参数的优化研究[J]. 茶叶科学,2011,31(4):313-318.

[19] 周丹丹,高逢敬,李延云. 发酵型茶叶酒生产工艺的研究[J]. 酿酒科技, 2010(6):72-74.

[20] 王煜,王家林. 发酵型绿茶酒的研究[J]. 中国酿造,2010(9):163-167.

[21] 邱新平,李立祥,赵常锐等. 发酵型茶酒香气成分的 GC-MS 初步分析[J]. 酿酒科技,2011(9):100-102,106.

[22] Mondal T K, Bhattacharya A, Laxmikumaran M, et al. Recent advances of tea (*Camellia Sinensis*) Biotechnology [J]. Plant Cell, Tissue and Organ Culture,2004(3):195-254.

十两茶水提物降糖作用及机制研究

黄春桃[1]　杜万红[1,2]　刘仲华[3]　吴浩人[4]　刘小阳[2]　施兆鹏[3]

1. 湖南师范大学第二附属医院,中国 湖南；2. 解放军第163医院,中国 湖南；
3. 湖南省天然产物工程技术研究中心,中国 湖南
4. 湖南省茶业有限公司,中国 湖南

摘　要：目的：本文研究湖南安化黑茶的主要品种十两茶水提物的体外降糖作用,初步探讨其作用机制。方法：通过体外蛋白质非酶糖基化试验、葡萄糖苷酶试验、醛糖还原酶试验比较十两茶与茯砖茶-9、茯砖茶-11、红茶、绿茶体外降糖作用。结果：该5种茶叶水提物在体外均能不同程度地抑制蛋白质非酶糖基化产物形成、葡萄糖苷酶以及醛糖还原酶活性,其中以十两茶作用为优。结论：十两茶水提物对糖尿病相关酶具有抑制作用,具有一定的降糖和防止糖尿病并发症作用。

关键词：黑茶；十两茶；葡萄糖苷酶；蛋白质非酶糖基化；醛糖还原酶

Research on the Anti-diabetic Effects of Shi-liang Tea Extract

HUANG Chun-tao[1]　　DU Wan-hong[1,2]　　LIU Zhong-hua[3]
WU Hao-ren[4]　　LIU Xiao-yang[2]　　SHI Zhao-peng[3]

1. The Second Affiliated Hospital of Hunan Normal University, Hunan, China；
2. The 163rd Hospital of PLA, Hunan, China； 3. The Hunan Engineering and Technology Center for Natural Products, Changsha, China；
4. Hunan Tea Company Limited, Hunan, China

Abstract：Aims：The aims of the study were to study the anti-diabetic effects of shi-liang tea, a major Anhua deep-black tea, in vitro. Methods：The glucosidase test, non-enzymatic glycation of proteins test and aldose reductase test were done to investigate the anti-diabetic activities of three kinds of shi-liang tea, fuzhuan tea-9,11 and black tea as well as green tea. Results：The five tested tea exerted the inhibiting effects on glucosidase activity, non-enzymatic glycation of proteins and aldose reductase activity at different extent. Among them, the anti-diabetic effects of shi-liang tea were better than those of others. Conclusion: shi-liang tea exerts a strong lowering-glucose and preventing diabete complications effect in vitro.

Keywords: deep-black tea; shi-liang tea; glucosidase; non-enzymatic glycation of proteins; aldose reductase

糖尿病是严重危害人类健康的主要疾病之一,是人类致死、致残的主要原因。糖尿病的发病率逐年增加,其中 2 型糖尿病的发病率占糖尿病发病率的 90%以上[1]。2 型糖尿病的特征有肥胖、高血糖、高胰岛素血症。糖尿病导致血管并发症的原因与体内糖基化终末产物增加、糖代谢紊乱密切相关。因此,治疗糖尿病不仅要降糖,而且要能减少糖尿病并发症。目前,人们已经把目标转移到一些降糖植物上,把它作为糖尿病的辅助治疗。经过前人的初步研究证明,茶叶具有明显的降糖效果。

作为我国的主要饮品,茶具有抗衰老、调血脂和降糖等功效[2-3]。研究报道,多种品种的茶叶如绿茶、乌龙茶、普洱茶等的降血糖作用与茶叶中富含多酚类化合物和多糖类化合物有关[4-5]。十两茶属于湖南黑茶之花卷茶中一种,产于湖南安化一带。我们前期研究发现,在高胆固醇大鼠模型,十两茶水提物能显著降低血脂水平,并明显改善血管内皮依赖性舒张功能[6]。为了确证十两茶水提物是否具有降糖及改善糖尿病并发症作用,本实验使用体外多种酶学试验(蛋白质非酶糖基化、葡萄糖苷酶和醛糖还原酶试验)观察了十两茶水提物降糖作用并探讨其机制。同时,选取茯砖茶、绿茶和红茶作为对照。

1 实验材料

(1)受试药物。5 种不同茶叶水提物(茯砖茶-9、茯砖茶-11、红茶、绿茶和十两茶)由湖南农业大学茶学重点实验室提供。茯砖茶-9、茯砖茶-11、红茶、绿茶和十两茶提取的得率分别为 21%、20%、22%、23%和 25%。实验临用时用双蒸水配制成相应浓度,现配现用。

(2)试验动物。雄性 SD 大鼠,体质量 230~270g,购自长沙市开福区东创实验动物技术服务部,动物合格证号:SCXK(湘)2009—0012。

(3)其他试剂。牛血清白蛋白(BSA)、阿卡波糖、果糖、葡萄糖、DL-甘油醛、NaN_3 和盐酸氨基胍购自美国 Sigma 公司,还原型辅酶 II(NADPH-Na_4)为

Roch产品,葡萄糖氧化酶法测试盒为中生北控生物科技公司产品,其余试剂均为国产分析纯。

(4)主要仪器。LS-55型荧光光谱仪(美国铂金埃尔默仪器有限公司)。

2 实验方法

2.1 葡萄糖苷酶试验[7]

大鼠禁食24h后,颈椎脱臼处死。立即取出小肠,用4℃生理盐水冲洗。翻转小肠后再漂洗,将洗净的小肠置于冰台上,用滤纸掭干小肠上的黏液,用载玻片轻刮下小肠黏膜,称重。将刮取物约按1:10加入4℃ 0.2mol/L磷酸缓冲液(pH 6.8),在冰浴上用玻璃匀浆器匀浆,4℃离心(4000r/min)30min。取上清液(即酶液)分装于-20℃保存备用。反应体积为500μL,其中对照组为缓冲液300μL、酶液100μL和蔗糖100μL,空白组为缓冲液400μL和蔗糖100μL,样品组为缓冲液300μL、酶液100μL、蔗糖100μL和受试茶溶液100μL。为排除茶本身的颜色对试验结果的干扰,另设样品对照组为缓冲液300μL、蔗糖100μL和受试茶溶液100μL。于37℃保温15min,加入1mL 1mol/L Na_2CO_3 终止液,用葡萄糖氧化酶法试剂盒测定产生的葡萄糖量。阳性对照药为阿卡波糖。按以下公式计算药物对葡萄糖苷酶的抑制百分率:

$$抑制百分率(\%) = \frac{1-给药组葡萄糖含量}{对照组葡萄糖含量} \times 100\%$$

2.2 蛋白质非酶糖基化试验[8]

(1)蛋白质糖基化溶液体系的建立。样品组:在800μL 10g/L BSA溶液中加入100μL 100mmol/L NaN_3 溶液,然后加入1600μL 500mmol/L的葡萄糖溶液,再依次加入终浓度为0.01~1mg/mL的各种受试茶水溶液800μL,并用20mmol/L碳酸缓冲溶液(pH 9.0)将溶液稀释至4mL。对照组(两组,1A组置于55℃恒温箱,2A组置于4℃冰箱,避光孵育10d后取样):以同体积的碳酸缓冲溶液(pH 9.0)代替样品,以氨基胍为阳性对照。体系配制好后,置于55℃的恒温箱或者4℃冰箱(2A组)避光孵育10d后取样进行荧光分析。

(2)受试茶水溶液对蛋白质非酶糖基化(ACEs)形成的抑制活性测定。取1.5mL糖基化溶液用碳酸缓冲液稀释至3mL,用LS-55型荧光光谱仪检测。激发波长为340nm,发射波长为420nm,狭缝为1.5nm,扫描发射光谱,读

420nm 的荧光强度得出样品管 A(F)值。应用以下公式计算化合物对 BSA 糖基化反应的抑制作用,用抑制百分率(IR)表示,然后以抑制率对抑制剂浓度的对数作图。

$$IR = 100\% - \frac{\text{样品管 A(F)} - \text{对照管 2A(F)}}{\text{对照管 1A(F)} - \text{对照管 2A(F)}} \times 100\%$$

2.3 醛糖还原酶(AR)试验[9]

(1) AR 的提取。取 SD 大鼠晶状体 100 个,约 40g,剪碎后用 5mmol/L、pH 7.4 的 PBS 溶解,匀浆,15000r/min 离心 15min;取上清,加入 40% 硫酸铵溶液,间歇搅拌 30min;再 15000r/min 离心 15min;取上清,加入 75% 硫酸铵溶液,间歇搅拌 30min 再 10000r/min 离心 15min。用 PBS 溶解沉淀,用 10 倍体积的 50mmol/L、pH 7.4 的 PBS 透析液透析,每 4h 换 1 次透析液,透析 3 次。以上操作均需在 4℃下进行。透析后的提取物−70℃冻存。

(2) AR 活性的测定。酶促反应体系总体积为 200μL,各反应液的终浓度分别为:底物 DL-甘油醛 2mmol/L,辅酶Ⅱ NADPH 0.2mmol/L,Li_2SO_4 为 400mmol/L,β-巯基乙醇为 5mmol/L,缓冲液 PBS 的 pH = 6.2,120mmol/L,加入 40μL 醛糖还原酶提取物。反应温度为 37℃,反应时间为 3min,于 340nm 测定吸光度值(OD)。酶活性单位定义:37℃时使反应体系吸光度值变化 0.001 单位为 1 个酶活性单位。公式为:

$$\text{酶活性单位} = \Delta AOD \times 1000,$$

$$\text{AR 抑制率}(\%) = \frac{1 - (\Delta Abs \text{样品} - \Delta Abs \text{空白})}{\Delta Abs \text{对照} - \Delta Abs \text{空白}} \times 100\%。$$

3 试验结果

3.1 受试茶水溶液对葡萄糖苷酶的抑制

受试茶水溶液抑制葡萄糖苷酶形成的 IC50 如表 1 所示。5 种茶提取物均具有不同程度的体外抑制葡萄糖苷酶活性作用,其中绿茶作用最强,为(0.44±0.07)g/L;十两茶次之,为(0.76±0.38)g/L。阳性对照药阿卡波糖对葡萄糖苷酶活动的抑制率为(0.0028±0.0004)g/L。

表 1 受试茶水溶液抑制葡萄糖苷酶的 IC50（均数±标准差）

受试品	IC50(g/L)
红 茶	2.92±0.12
绿 茶	0.44±0.07
茯砖茶－9	2.00±0.27
茯砖茶－11	1.53±0.04
十两茶	0.76±0.38
阿卡波糖	0.0028±0.0004

3.2 受试茶水溶液对 ACEs 形成的抑制

受试茶水溶液抑制 ACEs 形成的 50％抑制浓度（IC50）如表 2 所示。5 种茶提取物均具有不同程度的体外抑制蛋白质非酶糖基化作用，其中茯砖茶－9 作用最强，为(4.9±8.6)mg/L；十两茶次之，为(27.0±10.0)mg/L。阳性对照药氨基胍对 ACEs 的抑制率为(91.0±10.0)mg/L。

表 2 受试茶水溶液抑制 ACEs 形成的 IC50（均数±标准差）

受试品	IC50(mg/L)
红 茶	108.7±21.7
绿 茶	39.5±16.6
茯砖茶－9	4.9±8.6
茯砖茶－11	47.5±9.2
十两茶	27.0±10.0
氨基胍	91.0±10.0

3.3 受试茶水溶液对醛糖还原酶的抑制

受试茶水溶液抑制醛糖还原酶的 IC50 如表 3 所示。5 种茶提取物均具有不同程度的体外抑制醛糖还原酶活性作用，其中绿茶作用最强，为(1.30±0.09)mg/L；十两茶次之，为(3.00±0.90)mg/L。阳性对照药槲皮素对醛糖还原酶活性的抑制率为(1.53±0.25)mg/L。

表3 受试茶水溶液抑制醛糖还原酶的 IC50（均数±标准差）

受试品	IC50(mg/L)
红茶	5.80±0.50
绿茶	1.30±0.09
茯砖茶－9	11.70±0.60
茯砖茶－11	4.70±0.20
十两茶	3.00±0.90
槲皮素	1.53±0.25

4 讨 论

关于茶叶提取物降糖作用,国内外已有大量的研究。在临床前动物实验研究,大多数研究者都认为无论是绿茶还是黑茶,其提取物都具有一定的降血糖效果,能够降低空腹血糖和血浆中胰岛素的浓度,增加葡萄糖载体的表达,对于降低和恢复糖尿病大鼠血浆中葡萄糖的水平具有重要的作用。在临床实验研究,不少研究亦报道茶叶提取物能够改善健康人体内胰岛素的敏感性和口服葡萄糖的耐受量,显著降低2型糖尿病患者血糖水平,具有明显的降血糖效果[10-11]。但茶叶提取物降糖作用的机制尚不完全清楚。α-葡萄糖苷酶能使淀粉类分解为葡萄糖,促进肠道内葡萄糖的吸收,进而导致高血糖。α-葡萄糖苷酶抑制剂能竞争性抑制位于小肠的各种 α-葡萄糖苷酶,使淀粉类分解为葡萄糖的速度减慢,从而减缓肠道内葡萄糖的吸收,降低餐后高血糖。本研究显示十两茶能显著抑制 α-葡萄糖苷酶活性,提示其降糖作用可能与抑制 α-葡萄糖苷酶活性,减少葡萄糖吸收有关。

蛋白质非酶糖基化是指在无酶催化条件下,葡萄糖分子游离醛或酮基与蛋白质氨基通过亲核加成反应形成糖基化蛋白的过程。近年大量研究证明,在高血糖状态下蛋白质发生的非酶糖基化在糖尿病并发症的发病中起重要作用[12]。糖基化终末产物(AGEs)形成后引起蛋白质分子间广泛交联,致使蛋白质结构、机械强度、溶解性和配位结合等性质均发生改变。体内多种蛋白质糖基化可从多个方面影响机体,如引起血管通透性增大、血管基底膜增厚和细胞外基质积聚等。AGEs 与其细胞表面受体结合,通过趋化和活化单核巨噬细胞,激活转录因子 NF-KB,促进细胞因子和组织因子的释放,灭活 NO 和产生氧自由基等途径,参与糖尿病慢性并发症的发生和发展。血清和组织中,晚期 AGEs 水平与糖尿病慢性并发症的程度明显相关,阻止糖基化反应可减少 AGEs 形成,减少并发症。本研究显示,十两茶能显著抑制蛋白质非酶糖基化,提示其可能能减少 AGEs 生成,改善糖尿病并发症。

醛糖还原酶是多元醇通路的关键限速酶。当血糖浓度维持在正常生理水平时,它并不激活,它对葡萄糖的亲和力较低,此时葡萄糖很少转化为山梨醇。在高血糖状况下,磷酸己糖激酶被饱和,这时醛糖还原酶激活,促使体内的葡萄糖转化为山梨醇。山梨醇本身由于极性强不易通过细胞膜,在细胞内会形成蓄积,使细胞膜的通透性发生改变,并使细胞中 Na^+-K^+-ATP 酶活性下降,造成肌醇丧失,导致细胞代谢与功能的损害。由于眼睛和神经细胞等组织内醛糖还原酶的含量较高,糖尿病病人体内高血糖的环境使这一通路很容易被打开,造成对这些组织的病理损害,如糖尿病、白内障、神经病变、肾脏病变、视网膜病变、动脉粥样硬化等糖尿病并发症[13]。因此,抑制醛糖还原酶能有效地改善糖尿病并发症的发生发展。本研究显示,这几种茶水提均有抑制醛糖还原酶活性的作用,以绿茶和十两茶活性最强,提示茶水提取物对糖尿病并发症有改善作用。

综上所述,十两茶提取物在体外能抑制葡萄糖苷酶,减少蛋白质非酶糖基化和抑制醛糖还原酶,提示其具有降糖和改善糖尿病并发症作用。然而,其作用需在糖尿病动物模型上进一步确证。

参考文献

[1] Ali D, Kunzel C. Diabetes mellitus: Update and relevance for dentistry[J]. Dent Today, 2011, 30(12): 45—46, 48—50.

[2] 黄春桃,杜万红. 茶叶的药理作用研究新进展[J]. 中南药学, 2010, 8(9): 692—695.

[3] 邹盛勤. 茶叶的药用成分、药理作用及应用研究进展[J]. 中国茶叶加工, 2004(3): 35—37.

[4] 丁仁凤,何普明,揭国良. 茶多糖和茶多酚的降血糖作用研究[J]. 茶叶科学, 2005, 25(3): 219—224.

[5] 杨冯,赵国华. 茶多酚降血糖研究进展[J]. 食品工业科技, 2009, 30(6): 324—327.

[6] 杜万红,刘仲华,施玲等. 十两茶提取物抗高脂诱导家兔动脉粥样硬化形成涉及调节 ADMA/NO 通路[J]. 中国临床药理学与治疗学, 2008, 13(7): 747—752.

[7] 张海风,黄亚林,胡萨萨. 五种中药对两种不同来源的葡萄糖苷酶活性的

抑制作用比较[J]. 中药材, 2008, 31(7): 1024-1027.

[8] Sun Y X, Hayakwa S, Izumori K. Modification of ovalbumin with a rare ketohexose through the millard reaction: Effect on protein structure and gel properties [J]. Agric. FoodChem, 2004, 52(5): 1293-1299.

[9] 彭剑, 顾性初, 龚炳永. 醛糖还原酶抑制剂的微量筛选模型的建立[J]. 中国医药工业杂志, 1998(5): 221-223.

[10] Hosoda K, Wang M F, Liao M L, et al. Ntihyperglycemic effect of oolong tea in type 2 diabetes [J]. Diabetes Care, 2003, 26(6): 1714-1718.

[11] Ryu O H, Lee J, Lee K W, et al. Effects of green tea consumption on inflammation, insulin resistance and pulse wave velocity in type 2 diabetes patients [J]. Diabetes Res Clin Pract, 2006, 71(3): 356-358.

[12] Puddu A, Viviani G L. Advanced glycation endproducts and diabetes. Beyond vascular complications [J]. Endocr Metab Immune Disord Drug Targets, 2011, 11(2): 132-140.

[13] Ramana K V, Yadav U C, Calhoun W J, et al. Current prospective of aldose reductase inhibition in the therapy of allergic airway inflammation in asthma [J]. Curr Mol Med, 2011, 11(7): 599-608.

茶多酚和茶黄素通过 MAPK 途径诱导癌细胞凋亡

李 伟[1]　屠幼英[2]

1. 浙江树人大学茶文化研究与发展中心，中国 浙江；
2. 浙江大学农业与生物技术学院茶学系，中国 浙江

摘 要：茶多酚和茶黄素的抑制癌症细胞的作用是茶叶预防癌症研究的重点。茶多酚及其氧化产物可抑制癌细胞增殖和诱导凋亡。大量的证据显示茶多酚可抑制癌症发展和阻断癌细胞生长的信号通路，尤其是丝裂原活化蛋白激酶途径是作用的分子靶点。本文对茶多酚和茶黄素通过丝裂原活化蛋白激酶途径对癌细胞的作用机理做一综述。

关键词：癌症；丝裂原活化蛋白激酶；茶多酚；茶黄素

Tea Polyphenols and Theaflavins Induce Apoptosis of Cancer Cells through MAPK Signaling Pathways

LI Wei[1]　TU You-ying[2]

1. Department of Tea Science, Zhejiang Shuren University, Zhejiang, China;
2. Department of Tea Science, Zhejiang University, Zhejiang, China

Abstract: The biological activities and mechanisms of tea polyphenols and their polymeric have been attractive issues in cancer research. The inhibition of tea polyphenols on cancer cells resulting in decreased cell proliferation and increased apoptosis. There are tremendous evidences that tea polyphenols suppress tumor promotion by inhibiting enzyme activities and blocking signal transduction pathways. Specifically, the mitogen activated protein kinases (MAPK) pathways has been implicated as an important target molecular for cancer prevention and therapy. The purpose of this review is to discuss the relationship between tea polyphenols and MAPK signalling pathways in cancer.

Keywords: cancer; MAPK; signalling pathways; tea polyphenols

癌症的发病机理相当复杂,受控于多种因素和多个基因,而这些基因发挥功能涉及许多酶、生长因子、转录因子和信号转导因子等。如何控制这些酶和因子的活性是防癌和抗癌的关键[1]。近年来科学家们对茶多酚和茶黄素的抗癌作用和机理进行大量研究,并取得较多进展[2-3]。

丝裂原活化蛋白激酶(Mitogen-activated protein kinase,MAPK)是一类细胞内丝氨酸/苏氨酸蛋白激酶,是一类高度保守的癌基因产物,其特点是其丝氨酸/苏氨酸和酪氨酸须同时被磷酸化,才能获得活性。MAPK 可被丝裂原活化蛋白激酶激酶激酶激酶(MAPKK)激活,后者再被丝裂原活化蛋白激酶(MAPKKK)激活,由此构成一个三级的磷酸化级联反应,激活的 MAPK 可激活下游的多种靶蛋白,包括一系列转录因子和蛋白激酶。哺乳动物的 MAPK 信号传导系统主要有 3 条通路:生长因子刺激细胞增殖时是 ERK 途径,应激时的 p38 和 JNK 途径,JNK 又称为应激活化蛋白激酶(Stress-activated protein kinase,SAPK)[4]。MAPK 途径的简图如图 1 所示。

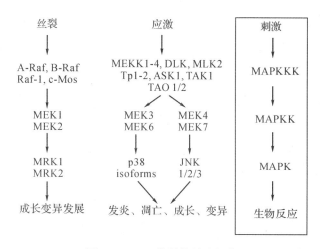

图 1　MAPK 信号转导途径简图

细胞凋亡是在特定时空发生的、受机体严密调控的细胞"自杀"现象[5]。在肿瘤细胞中,活化的 p38 可增强 c-myc 表达、磷酸化 p53、参与 Fas/FasL 介导的凋亡[6];可增强 TNF-α 表达[7];作用于半胱天冬酶(Cysteine-requiring Aspartate Portease,Caspase)家族的上游而诱导肿瘤细胞凋亡[8-9]。某些作用于 p38 通路的化疗药物,也是通过诱导凋亡,产生抗肿瘤作用[10]。c-jun 氨基末端激酶(c-jun N-terminal minase,JNK)信号转导通路参与多种凋亡反应,现阶段研究表明,JNK 通路介导肿瘤细胞凋亡的机制是通过磷酸化 Bcl-2 和 Bcl-xl,促进线粒体释放细胞色素 C,进而激活 Caspase 级联反应,最终作用于 Caspase-3,导致细胞凋亡。细胞外调节蛋白激酶(Extracellular regulated protein kinases,ERK)通路是目前研究最为透彻的丝裂原活化蛋白激酶通路,其激酶可被依赖于高浓度 Ca^{2+} 的 Ras 激活,而其底物蛋白就包括 c-myc 等。大量研究表明,ERK 介导的肿瘤细胞正是通过 c-myc 起作用。同时,ERK 诱导肿瘤细胞凋亡的机制也包括:下调 Bcl-2、上调 Bax,作用于 Caspase 家族等。

  Chen 等[11]比较了茶多酚和茶黄素对 NIH3T3 细胞中的 PKC 活性抑制作用,20μmol/L TF3 可以强烈抑制由 12-O-十四烷酰佛波醇-13-醋酸酯(12-O-tetradecanoylphorbol 13-acetate,TPA)提高的 NIH3T3 膜结构上蛋白激酶 C(Protein kinase C,PKC)的活性,抑制率为 94.5%,同样条件下 EGCG 抑制率为 9.4%。Chung 等[12]在体外利用小鼠表皮 JB6 细胞层和一个突变型 H-ras 基因构建模拟的癌细胞,并研究了茶黄素对这些细胞的作用。结果表明,茶黄素表现出对 30.76ras 红细胞生长的抑制以及对 AP-1 细胞的抑制。在已加入茶黄素的细胞中加入过氧化氢,不能阻止茶黄素对 AP-1 作用的抑制。Tsai 等人研究发现,EGCG 和 TF3 可抑制由脂多糖激活的鼠巨噬 RAW264.7 细胞中 NO 的产生,TF3 比 EGCG 强,而其他多酚类没有这种作用。通过蛋白质印迹法和逆转录聚合酶链式反应(RT-PCR)分析证实了 TF3 可降低 130-kda 蛋白的产生[13]。

  茶多酚和茶黄素可以促进各种肿瘤细胞的凋亡,抑制肿瘤的增殖和扩散。Yang 等[14]研究发现,用 EGC、EGCG 和茶黄素处理人支气管上皮细胞 33BES 和 21BES 时发现,24h 后 EGC、EGCG 和 TF3 的半抑制浓度(The half maximal inhibitory concentration,IC_{50})均为 23μmol/L。经 TF3 处理的细胞死亡较早,形态学观察表明,细胞形状不规则,且出现细胞质微粒,表现出毒性效应,而 EGC 和 EGCG 诱导细胞凋亡则出现滞后现象,未表现任何明显的形态学变化,跟 H_2O_2 诱导的细胞凋亡相似。以上实验若加入外源过氧化氢酶则会阻止 EGC 和 EGCG 诱导的细胞凋亡,而对 TF3 诱导细胞凋亡无作用,综合以上结果表明 EGC 和 EGCG 诱导细胞凋亡可能与 H_2O_2 有关,而 TF3 则表现出不同的诱导机制。Giovanna 等研究绿茶提取物、红茶提取物以及红葡萄酒对偶氮甲烷诱导(Azoxymethane,AOM)的小鼠(16 周)肠道肿瘤时发现,在喂食红茶提取物和葡萄酒的实验组中患腺癌的小鼠数量明显低于对照组(对照组、红茶提取物、绿茶提取物以及红葡萄酒的实验组患腺癌的小鼠分别为 86%、59%、90% 和 50%),诱导细胞凋亡的形态学特征是细胞发生浓缩,与相邻细胞失去联系且铬氨盐浓度升高,形成圆形或椭圆形的核碎片[15]。

茶叶中的多酚类物质通过受体(膜受体和核受体)结合引发细胞内的一系列生物化学反应,直至基因表达。精细的细胞间通讯网络控制着细胞的生长、分裂、分化和死亡及各种其他生命过程。MAPK 是一组可被多种信号激活的丝/苏氨酸激酶,经双重磷酸化激活后可参与细胞的多种生物活性,如调节基因转录、诱导细胞凋亡、调节细胞周期等。MAPK 对细胞凋亡的诱导作用,是近年来研究的重点。尤其是对肿瘤细胞凋亡的诱导作用,更是人们关注的焦点。现在已经发现 p38、ERK5、ERK 和 JNK 4 个亚族,其中 ERK、JNK、p38 3 条通路与肿瘤的关系密切。绿茶多酚可通过影响 p38 蛋白,下调影响血管功能的窖蛋白的基因表达[16]。Bhattachaya 等[17]研究了红茶中的多酚对人恶性黑色素瘤 A375 细胞的影响,茶黄素和茶红素可以通过 JNK 和 p38 通路诱导细胞凋亡。

对于癌症细胞来说,激活的 JNK 可进一步使核内 c-jun 等转录因子活性增强。而 JNK 促进细胞凋亡的机制有:一是上调促凋亡蛋白的表达;JNK 通过使 AP-1 活性增强进一步促进 p53、Bax、Fasl 和 TNF 等促凋亡蛋白的表达,二是作用于线粒体,如 Bax、Bak 等促使细胞色素 C 释放入细胞质,细胞色素 C 和 caspase 9 结合,最终作用于 caspase 3,激活的 caspase 3 与凋亡底物结合引起细胞凋亡。p53 是 JNK 作用的底物之一,JNK 磷酸化 p53 并抑制泛素蛋白介导的降解,从而稳定 p53 蛋白,JNK 通过调节 p53 的半衰期来控制 p53 的表达水平。Gupta 等[18]揭示了 EGCG 显著升高 LNCap 细胞的活性并导致凋亡。EGCG 是 p53 蛋白转录调控的有力诱导剂,并且也是通过 p53 信号通路来诱导 A549 细胞凋亡的[19]。

现已经证明茶多酚和茶黄素可通过 MAPK 家族转导并调控细胞凋亡信号,而 c-jun 氨基末端激酶信号通路在其中起着十分重要的作用。然而对茶多酚类诱导细胞凋亡的机制研究还远远不够,特别是细胞信号转导机理知之甚少,这将是以后茶多酚类对癌症作用的研究热点。

参考文献

[1] Grivennikov S I, Greten F R, Karin M. Immunity, inflammation, and cancer [J]. Cell, 2010, 140(6): 883-899.

[2] Liang Y C, Chen Y C, Lin Y L, et al. Suppression of extracellular signals and cell proliferation by the black tea polyphenol, theaflavin-3, 3′-digallate [J]. Carcinogenesis, 1999, 20(4): 733-736.

[3] Saeki K, Sano M, Miyase T, et al. Apoptosis-inducing activity of polyphenol compounds derived from tea catechins in human histiolytic lympho-

ma U937 cells [J]. Bioscience Biotechnology and Biochemistry, 1999, 63(3): 585—587.

[4] Pearson G, Robinson F, Beers G T, et al. Mitogen-activated protein (MAP) kinase pathways: Regulation and physiological functions [J]. Endocrine Reviews, 2001, 22(2): 153—183.

[5] Williams G T. Programmed cell death: Apoptosis and oncogenesis [J]. Cell, 1991, 65(7): 1097—1098.

[6] Shen G, Xu C, Chen C, et al. p53-Independent G1 cell cycle arrest of human colon carcinoma cells HT-29 by sulforaphane is associated with induction of p21CIP1 and inhibition of expression of cyclin D1 [J]. Cancer Chemotherapy and Pharmacology, 2006, 57(3): 317—327.

[7] Weitsman G E, Ravid A, Liberman U A, et al. The role of p38 MAP kinase in the synergistic cytotoxic action of calcitriol and TNF-alpha in human breast cancer cells [J]. The Journal of Steroid Biochemistry Moecularl Biology, 2004, 89/90(1/2/3/4/5): 361—364.

[8] Jameel N M, Thirunavukkarasu C, Wu T, et al. p38-MAPK-and caspase-3-mediated superoxide-induced apoptosis of rat hepatic stellate cells: Reversal by retinoic acid [J]. Journal of Cell Physiology, 2009, 218(1): 157—166.

[9] Zou W, Zeng J, Zhuo M, et al. Involvement of caspase-3 and p38 mitogen-activated protein kinase in cobalt chloride-induced apoptosis in PC12 cells [J]. Journal of Neuroscience Research, 2002, 67(6): 837—843.

[10] Kang S J, Kim B M, Lee Y J, et al. Titanium dioxide nanoparticles induce apoptosis through the JNK/p38-caspase-8-Bid pathway in phytohemagglutinin-stimulated human lymphocytes[J]. Biochemical and Biophysical Research Communications, 2009, 386(4): 682—687.

[11] Chen Y C, Liang Y C, Lin-Shiau S Y, et al. Inhibition of TPA-induced protein kinase C and transcription activator protein-1 binding activities by theaflavin-3,3′-digallate from black tea in NIH3T3 cells [J]. Journal of Agricultural and Food Chemistry, 1999, 47(4): 1416—1421.

[12] Chung J Y, Huang C S, Meng X F, et al. Inhibition of activator protein 1 activity and cell growth by purified green tea and black tea polyphenols in H-ras-transformed cells: Structure-activity relationship and mecha-

nisms involved [J]. Cancer Research, 1999, 59(18): 4610—4617.

[13] Chan M M, Fong D, Ho C T, et al. Inhibition of inducible nitric oxide synthase gene expression and enzyme activity by epigallocatechin gallate, a natural product from green tea [J]. Biochemical Pharmacology, 1997, 54(12): 1281—1286.

[14] Yang G Y, Liao J, Li C, et al. Effect of black and green tea polyphenols on c-jun phosphorylation and H_2O_2 production in transformed and non-transformed human bronchial cell lines: Possible mechanisms of cell growth inhibition and apoptosis induction [J]. Carcinogenesis, 2000, 21(11): 2035—2039.

[15] Caderni G, De F C, Luceri C, et al. Effects of black tea, green tea and wine extracts on intestinal carcinogenesis induced by azoxymethane in F344 rats [J]. Carcinogenesis, 2000, 21(11): 1965—1969.

[16] Li Y, Ying C, Zuo X, et al. Green tea polyphenols down-regulate caveolin-1 expression via ERK1/2 and p38MAPK in endothelial cells [J]. J of Nutritional Biochemistry, 2009, 20(12): 1021—1027.

[17] Bhattacharya U, Halder B, Mukhopadhyay S, et al. Role of oxidation-triggered activation of JNK and p38 MAPK in black tea polyphenols induced apoptotic death of A375 cells [J]. Cancer Science, 2009, 100(10): 1971—1978.

[18] Gupta S, Ahmad N, Nieminen A L, et al. Growth inhibition, cell-cycle dysregulation, and induction of apoptosis by green tea constituent (—)-epigallocatechin-3-gallate in androgen-sensitive and androgen-insensitive human prostate carcinoma cells [J]. Toxicology and Applied Pharmacology, 2000, 164(1): 82—90.

[19] Yamauchi R, Sasaki K, Yoshida K. Identification of epigallocatechin-3-gallate in green tea polyphenols as a potent inducer of p53-dependent apoptosis in the human lung cancer cell line A549 [J]. Toxicology in Vitro, 2009, 23(5): 834—839.

十两茶提取物对小鼠免疫功能的影响

杜万红[1]　彭世喜[1]　刘仲华[2]　段家怀[1]　吴浩人[3]
施　玲[2]　周重旺[3]　霍　治[4]　施兆鹏[2]

1. 湖南师范大学第二附属医院干部病房,中国 湖南;2. 湖南省天然产物工程技术研究中心,中国 湖南;3. 湖南省茶业有限公司,中国 湖南;
4. 中南大学基础医学院医学免疫学系,中国 湖南

摘　要:目的:观察十两茶提取物对小鼠免疫功能的影响。方法:利用昆明种小鼠模型,设立不同剂量的十两茶提取物给药组以及对照组,分别在给药后的第 7、14、28d 检测小鼠血清中补体成分含量、相关细胞因子含量以及免疫球蛋白含量,检测腹腔巨噬细胞吞噬功能,检测 T 细胞、B 细胞增殖活性,检测 NK 细胞的数量及杀伤活性,并做 T 细胞亚群分析。结果:给药后 14d,给药组小鼠脾脏中 NK 细胞的数量下调,但其细胞杀伤活性没有受到影响;给药后 28d,给药组小鼠 NK 细胞的杀伤活性显著高于对照组;而血清补体含量、IgG 含量、细胞因子含量以及腹腔巨噬细胞吞噬功能,T、B 细胞增殖活性以及 $CD4^+$ T 细胞/$CD8^+$ T 细胞比值等指标的差别没有统计学意义。结论:十两茶提取物可显著促进小鼠 NK 细胞的杀伤活性,在短期内对小鼠的适应性免疫功能没有明显影响。

关键词:十两茶;免疫功能;NK 细胞

Effects of the Extract of Shi-liang Tea on Immune Function in Mice

DU Wan-hong[1]　PENG Shi-xi[1]　LIU Zhong-hua[2]
DUAN Jia-huai[1]　WU Hao-ren[3]　SHI Ling[2]
ZHOU Zhong-wang[3]　HUO Zhi[4]　SHI Zhao-peng[2]

1. Cadres Ward of the Second Affiliated Hospital of Hunan Normal University, Hunan, China; 2. The Hunan Engineering and Technology Center for Natural Products, Hunan, China; 3. Hunan Tea Company Limited, Hunan, China; 4. Department of Immunology, College of Basic Medicine, Central South University, Hunan, China

Abstract: Objectives: To observe the effects of the extract of Shi-liang tea on immune function in mice. Methods: Kunming mouse model were fed to establish different doses of Shi-liang tea extract group as well as control group. After administration, on day 7, 14 and 28, detection of C3, IgG and related cytokines in the serum were performed. We also analysis the activity of peritoneal macrophages, the ability of T/B cell proliferation, the number and cytotoxic activity of NK cells and the subsets of T cells. Results: The extract of Shi-liang tea can significantly promote the NK cell killing activity. But the levels of serum C3, IgG and cytokines as well as macrophage activity, T/B cell proliferation and $CD4^+/CD8^+$ T-cell ratio were no significant impacted by the extract of Shi-liang tea on day 28. Conclusions: The extract of Shi-liang tea can significantly enhance the killing activity. But it has no significant effect on adaptive immune function in mice in short term.

Keywords: Shi-liang tea; immunologic function; NK cell

茶叶作为一种传统饮料在我国已有上千年的历史,东汉的《神农本草》、唐代陈藏器的《本草拾遗》、明代顾元庆的《茶谱》等史书均详细记载了茶叶的药用价值。近几十年来,茶叶生物活性成分的预防和保健作用成为医学界的研究热点,相关文献数以千计[1-2]。茶叶的成分很复杂,目前至少已鉴定出450种以上的有机成分和15种以上的无机元素,其中茶多酚是其主要功能性成分[3]。现代医学已经证明茶多酚及其氧化产物具有抗肿瘤、抗衰老、抗辐射、清除人体自由基、降血糖、调血脂等重要功效[4]。

十两茶是湖南黑茶花卷茶系列产品中的一种,分一两茶、十两茶、百两茶和千两茶 4 种,其品质优次,以一两茶最好,依次排列。百两茶与千两茶有近 200 年历史,一两茶和十两茶皆为新产品。我们的前期研究发现,十两茶水溶液提取物中茶多酚的含量较高(52.1%),并且通过不同的动物模型证实十两茶提取物具有调血脂、保护血管内皮和抗动脉粥样硬化的效应[5-6]。为了进一步明确十两茶保健功能,探讨十两茶水溶液提取物对心血管系统以外的生物学效应,我们应用小鼠模型,通过对相关指标的比较,初步证实了十两茶提取物对小鼠免疫功能的影响。

1 材料与方法

(1)药品与试剂。十两茶提取物由湖南农业大学天然产物研究中心及湖南省茶业股份有限公司白沙溪茶厂提供。将十两茶原料处理成碎状,按每克茶叶加 12mL 沸水进行提取,沸水浴 1h,精密过滤,冷却,经低温真空浓缩后冷冻干燥,得十两茶水溶液提取物粉末。高效液相色谱(HPLC)检测其中儿茶素含量 22.43%,紫外分光光度法(Folin-Denis 法)检测其茶多酚含量 52.1%。盐酸左旋咪唑(Sigma 公司)、MTT(sigma 公司);小鼠血清补体(C3)、血清抗体(IgG)、血清细胞因子(IL-2、IFN-γ、IL-4、IL-10)ELISA 检测试剂盒(R&D 公司);小鼠 CD3、CD4、CD8、NK1.1 荧光抗体(BD 公司);其余试剂均为国产分析纯。

(2)实验动物和分组。昆明种小鼠(雌性,4 周龄,18~20g)按体重随机分为纯水对照组、阳性药物对照组和实验组,其中实验组又分高、中、低 3 种剂量组,每日一次分别灌胃给予十两茶水溶液提取物 100、200、400mg/kg。纯水对照组每日一次以等体积灭菌单蒸水灌胃,阳性药物对照组每日一次灌胃给予盐酸左旋咪唑25mg/kg,相关药剂均以灭菌单蒸水溶解。每组 12 只动物,分别在给药后的第 7、14、28d 进行样本采集,每次每组采集 4 只小鼠。通过眼球摘除法采血,-70℃保存备用,同时无菌采集腹腔巨噬细胞、胸腺及脾脏。

(3)免疫功能指标检测。每只动物共计检测 8 类免疫学指标,具体包括:①腹腔巨噬细胞吞噬功能测定,中性红比色法;②血清细胞因子(IL-2、IFN-γ、IL-4、IL-10)含量测定,ELISA 法;③血清补体成分 C3 含量测定,ELISA 法;④血清抗体 IgG 含量测定,ELISA 法;⑤脾脏 B 细胞和胸腺 T 细胞增殖活性测定,MTT 法;⑥脾脏 NK 细胞计数,流式细胞术;⑦NK 细胞杀伤活性测定,MTT 法;⑧胸腺 T 细胞亚群分析,流式细胞术。各项检测均设复孔。

(4)统计学分析。所有数据均以均数±标准差表示,用 SPSS 12.0 进行统计学处理,组间差异采用方差分析。双侧 $P<0.05$ 认为有显著性差异。

2 结　果

（1）十两茶提取物对小鼠腹腔巨噬细胞吞噬功能的影响。在给药后7、14、28d，十两茶提取物各浓度组与纯水对照组、阳性药物对照组之间的腹腔巨噬细胞吞噬功能均无显著性差别。（数据未列出）。

（2）十两茶提取物对小鼠血清补体成分C3含量的影响。在给药后各个时间点，各组之间的血清补体成分C3浓度均无显著性差别。（数据未列出）。

（3）十两茶提取物对小鼠血清抗体IgG含量的影响。在给药后各个时间点，各组之间的血清抗体IgG含量均无显著性差别。（数据未列出）。

（4）十两茶提取物对小鼠血清细胞因子含量的影响。给药14d后，十两茶中浓度组较纯水对照组血清IL-2含量低（表1）；给药28d后，十两茶高浓度组较纯水对照组血清IL-4含量低（表2）；在给药后各个时间点，十两茶提取物各浓度组与纯水对照组、阳性药物对照组之间的血清IFN-γ、IL-10含量无显著性差别。（数据未列出）。

表1　十两茶提取物对小鼠血清含量的影响（$\bar{x}\pm s$, $n=4$）

组　别	血清IL-2含量（μg/mL）		
	给药7d后	给药14d后	给药28d后
纯水对照组	40.86±11.908	48.64±6.003	30.13±0.672
100mg/kg 十两茶提取物组	47.29±5.564	48.46±9.439	29.19±20.917
200mg/kg 十两茶提取物组	30.80±21.782	19.13±14.702*	28.12±16.458
400mg/kg 十两茶提取物组	40.01±12.970	36.09±8.776	33.69±10.572
左旋咪唑对照组（25mg/kg）	30.89±9.273	45.10±2.516	22.90±10.864

注：与纯水对照组比较 * $P<0.05$。

表2　十两茶提取物对小鼠血清IL-4含量的影响（$\bar{x}\pm s$, $n=4$）

组　别	血清IL-4含量（μg/mL）		
	给药7d后	给药14d后	给药28d后
纯水对照组	40.32±15.025	46.12±24.298	56.12±4.087
100mg/kg 十两茶提取物组	27.83±4.84	35.83±2.469	57.74±7.643
200mg/kg 十两茶提取物组	34.49±5.218	43.36±11.442	38.49±15.471
400mg/kg 十两茶提取物组	33.65±8.933	38.9±2.851	26.87±15.561*
左旋咪唑对照组（25mg/kg）	29.48±7.392	51.33±6.481	46.46±3.86

注：与纯水对照组比较 * $P<0.05$。

(5)十两茶提取物对小鼠 T、B 细胞增殖活性的影响。与纯水对照组相比较,十两茶提取物各浓度组在给药后各个时间点的 T、B 细胞增殖活性无显著性差别。(数据未列出)。

(6)十两茶提取物对小鼠脾脏 NK 细胞数量和杀伤活性的影响。在给药后 7d,3 个浓度十两茶提取物组的脾脏 NK 细胞数量都较纯水对照组低;给药后 14d,低浓度组和中浓度组的脾脏 NK 细胞比例仍低于纯水对照组;给药 28d 后这种差别消失(表3)。虽然给药后 7d 和 14d 十两茶提取物组脾脏 NK 细胞数量较纯水对照组低,但其杀伤活性没有显著性差别;而给药后 28d 时,3 个浓度十两茶提取物组的脾脏 NK 细胞杀伤活性均明显高于纯水对照组,与阳性药物左旋咪唑对照组表现一致(表 4)。

表3 十两茶提取物对小鼠脾脏 NK 细胞数量的影响($x\pm s$, $n=4$)

组 别	NK 细胞数量(%)		
	给药 7d 后	给药 14d 后	给药 28d 后
纯水对照组	5.87±0.47	5.74±1.45	6.21±1.67
100mg/kg 十两茶提取物组	3.16±1.37*	2.99±0.13*	4.39±1.02
200mg/kg 十两茶提取物组	3.78±0.75*	3.56±1.36*	3.16±0.24
400mg/kg 十两茶提取物组	3.38±0.26*	5.43±1.32	4.64±0.20
左旋咪唑对照组(25mg/kg)	2.88±0.20*	4.00±0.21	4.23±0.83

注:与纯水对照组比较 * $P<0.05$。

表4 十两茶提取物对小鼠脾脏 NK 细胞杀伤活性的影响($x\pm s$, $n=4$)

组 别	NK 细胞杀伤活性(杀伤百分率)		
	给药 7d 后	给药 14d 后	给药 28d 后
纯水对照组	62.47±0.608	62.739±2.235	54.5±1.68
100mg/kg 十两茶提取物组	59.255±0.803	61.683±1.747	63.727±1.055*
200mg/kg 十两茶提取物组	59.856±2.302	58.779±3.349	63.659±2.441*
400mg/kg 十两茶提取物组	65.264±0.393	60.528±3.767	64.295±1.44*
左旋咪唑对照组(25mg/kg)	61.178±2.583	61.733±1.738	60.75±2.408*

注:与纯水对照组比较 * $P<0.05$。

(7)十两茶提取物对小鼠胸腺 $CD4^+$ T 细胞/$CD8^+$ T 细胞比值的影响。在给药后各个时间点,各组之间的胸腺 $CD4^+$ T 细胞/$CD8^+$ T 细胞比值无显著性差别。(数据未列出)。

3 讨 论

大量已有的研究证实了茶叶提取物的生物保健作用,而免疫系统能够帮助机体清除异己物质从而维持内环境的稳定,那么,茶叶的有效成分是否会通过影响免疫系统来调节机体的功能状态,成为了一个值得探讨的科学问题。一般将人或哺乳动物的免疫系统划分为固有免疫和适应性免疫两个部分,两者之间各有侧重又互为补充。固有免疫的组成部分主要包括补体、细胞因子、吞噬细胞、NK 细胞等;而适应性免疫主要是指由 T 细胞、B 细胞所介导的细胞免疫和体液免疫,其效应物主要是抗体和效应 T 细胞。通过对不同免疫指标的检测,可以分别判断固有免疫和适应性免疫功能的变化情况。本次研究主要发现:①十两茶提取物在给药 28d 后可显著促进 NK 细胞的杀伤活性;②十两茶提取物在给药 14d 内,可下调小鼠脾脏中 NK 细胞的数量,但其细胞杀伤活性没有受到影响;③十两茶提取物对小鼠的腹腔巨噬细胞吞噬功能、血清补体含量、血清 IgG 含量、血清细胞因子(包括 IL-2、IFN-γ、IL-4、IL-10)含量,T、B 细胞增殖活性以及 $CD4^+$ T 细胞/$CD8^+$ T 细胞比值没有明显的影响。

这些结果提示我们,十两茶提取物对小鼠的适应性免疫功能没有显著影响,而对固有免疫的重要组成细胞——NK 细胞的杀伤功能有增强效应。NK 细胞是一种天然杀伤细胞,在机体的抗感染、抗肿瘤过程中都发挥重要的作用[7-8],而且 NK 细胞分布广、作用快,活化不需要抗原的预先刺激。十两茶通过促进 NK 细胞的杀伤,将有利于机体清除病原体,清除被病毒感染的细胞和突变的细胞,从这个角度而言,也可以说十两茶具有保健、抗癌的效应。此外,我们的结果虽然没有显示十两茶提取物对适应性免疫功能有显著影响,但是饮茶一般是一种长时间的行为,而动物模型的给药时间仅有 28d,茶叶对机体的远期效果或许尚未显示出来,因此,有必要通过延长给药的时间,深入探讨十两茶提取物对免疫系统的远期影响。

综上所述,十两茶提取物可显著促进小鼠 NK 细胞的杀伤活性,但在短期内对小鼠的适应性免疫功能没有明显影响。

参考文献

[1] Zavcri N T. Green tea and its polyphenolic catechins:Medicinial uses in cancer and noncancer applications[J]. Life Science,2006,78:2073—2080.

[2] LIU M-J, CHEN J-P. Advances on research of extraction of tea polyphe-

nols from green tea leaves[J]. Progress in Modern Biomedicine, 2006, 6(7): 70—72.

[3] Stangl V, Lorenz M, Stangl K. The role of tea and tea flavonoids in cardiovascular health [J]. Mol Nutr Food Res, 2006, 50(2): 218—228.

[4] YAN S-L, YANG X-Q. Eeffective of tea polyphenols[J]. Forestry Economy, 2005, 1: 70.

[5] ZHANG Y-M, CHEN X-Y, XU L. The effect of tea polyphenols on expression of tumor blood vessels related factors in tumor tissues and vital organs of implanted breast cancer[J]. Progress in Modern Biomedicine, 2007, 7(10): 1441—1445.

[6] DU W-H, LIU Z-H, SHI Ling, et al. Involvement of ADMA/NO pathway in antiatherosclerotic properties of the extract of shi-liang cha in hypercholesterolemic diet-fed rabbits[J]. Chinese Journal of Clinical Pharmacology and Therapeutics, 2008, 13(7): 747—752.

[7] Bauer S V, Gron J, Wu A S, et al. Activation of NK cells and T cells by NKG2D, a receptor for stress inducible MICA [J]. Science, 1999, 285: 727.

[8] Eleme K, Taner S B, Onfelt B, et al. Cell surface organization of stress-inducible proteins ULBP and MICA that stimulate human NK cells and T cells via NKG2D [J]. J EXP Med, 2004, 199: 1005.

附:基金项目:湖南省科技厅一般项目(2008FJ3056)

茶叶籽油中反式脂肪酸和苯并(α)芘含量的测定

朱晋萱[1,2]　朱跃进[1]　张士康[1]　刘国艳[2]　金青哲[2]

1. 中华全国供销合作总社杭州茶叶研究院,中国 浙江;
2. 江南大学食品学院,中国 江苏

摘　要:测定了15个茶叶籽自制毛油样品中反式脂肪酸(TFA)和苯并(α)芘含量。结果显示,茶叶籽浸提毛油中反式脂肪酸总量很低,其中又以t-C18:1及t-C18:2为主,未检测到t-C18:3类的反式脂肪酸;大部分样品中苯并(α)芘含量也远低于欧盟2μg/kg的标准,但有些样品含量明显较高;冷榨毛油中两种有害成分均未检测到,热榨毛油中仅反式脂肪酸明显升高,但仍未超过各类标准限定值。

关键词:茶叶籽油;反式脂肪酸;苯并(α)芘;检测

Detection of Trans Fatty Acids and Benzo(α) pyrene in Tea Seeds Oil

ZHU Jin-xuan[1,2]　ZHU Yue-jin[1]　ZHANG Shi-kang[1]
LIU Guo-yan[2]　JIN Qing-zhe[2]

1. Hangzhou Tea Research Institute, China Coop, Zhejiang, China;
2. School of Food Science and Techology, Jiangnan University, Jiangsu, China

Abstract:Contents of trans fatty acids and benzo(α)pyrene in tea seeds oil are detected of 15 samples that conclude solvent extraction crude oil from 13 provinces, cold pressing crude oil and hot pressing crude oil that made by HunanXiangtan sample. The results show that, the content of trans fatty acid in solvent extraction crude tea seeds oil is low and less than 0.05%. The type of t-C18:1 and t-C18:2 are more than others, type of t-C18:3 has not been detected. And benzo(α)pyrene content is far lower than EU standards (2μg/kg). But some samples are obviously too high. Both the two kinds of harmful substance has not been detected in cold pressing crude oil. Only TFA content increased significantly in hot pressing crude oil, but still under the standards.

Keywords: tea seeds oil; trans fatty acids; benzo(α)pyrene; detection

从茶叶籽中提取的油脂为茶叶籽油。茶叶籽是山茶科山茶属植物茶树(*Camellia sinensis* O. Kuntze)的种子,卫生部2009年第18号公告中正式批准茶叶籽油为新资源食品[1-2]。

茶叶籽油是一种高品质木本植物食用油,其开发对保障国家食用油供应意义重大[3-5]。其脂肪酸组成比例合理,更含有多种具有独特功能性成分的脂肪伴随物[6-7]。但对其中某些存在食用安全隐患的物质,如反式脂肪酸(TFA)及苯并(α)芘的含量测定则未见报道。反式脂肪酸对健康的影响主要在心血管方面,摄入过量会对人体健康产生不利影响[8]。多环芳烃是一类广泛存在于环境、食品及生物体内的污染物。其化学性质稳定,不易水解,最突出的特性是致癌性[9]。其中,苯并(α)芘的污染最广,致癌性最强。本文测定了来自全国13个省的茶叶籽浸提毛油及冷榨、热榨毛油共15个样品中的反式脂肪酸和苯并(α)芘含量,以期为茶叶籽油标准的制订及其产业化发展提供更全面的参考。

1 材料与方法

1.1 材料与仪器

(1)茶叶籽油样品。茶叶籽原料均为2011年11—12月收集自全国13个省的当年本地产新鲜茶叶籽,产地如下:云南普洱、贵州湄潭、江苏无锡、湖北随州、广西三江、四川雅安、湖南湘潭、江西赣州、浙江龙游、福建福安、河南信阳、安徽黄山、陕西汉中。将收集的新鲜茶叶籽在45℃下烘干至水分含量约7%后粉碎,干茶叶籽粉用六号溶剂在常温下进行超声波浸提,得到茶叶籽毛油。冷榨毛油及热榨毛油均为采自湖南湘潭的茶叶籽制取。冷榨油在常温下压榨制得,热榨油是将干燥好的茶叶籽在180℃～200℃下炒籽15min后进行压榨制得。

(2)试剂。顺、反异构的C18:1(46902-U,46903)、C18:2(47791)和C18:3(47792)脂肪酸甲酯混合标准品,均购自美国Sigma公司;苯并(α)芘标准品,J&K百灵威,CAS编号:50-32-8;正己烷、乙腈、四氢呋喃、甲苯均为色谱纯,美国J&K公司;正己烷、氢氧化钾、甲醇、一水合硫酸氢钠、无水硫酸钠、层析用中性氧化铝均为化学纯,国药试剂公司。

(3)仪器与设备。仪器:Waters1525 高效液相色谱仪,美国 Waters 公司;安捷伦 HPLC 1200 FLD 检测器(1321A),美国安捷伦公司;Trace GC Ultra 气相色谱仪,Al 3000 自动进样器,美国 TRACE 公司;液相色谱柱 LC-PAH supelcosil 15cm×4.6mm×5μm,美国 Sigma 公司;CP-Sil88(100m×0.25mm×0.2μm)毛细管柱,美国 Varian 公司。

设备:AR2140 电子分析天平,梅特勒－托利多仪器上海有限公司;RJ-TDL-50A 台式低速大容量离心机,无锡瑞江分析仪器有限公司;ZK-82BB 型电热真空干燥箱,上海实验仪器厂有限公司;HN 超声波发生器,无锡市华能超声电子有限公司;DHG-9076A 型电热恒温鼓风干燥箱,上海浦东荣丰科学仪器有限公司;Anke TGL-16G 台式离心机,上海安亭科学仪器厂;WK-800A 高速药物粉碎机,青州市精诚机械有限公司;荣华 HH-2 数显恒温水浴锅,江苏省金坛市荣华仪器制造有限公司;XW-80A 微型漩涡混合仪,上海沪西分析仪器厂有限公司;申科旋转蒸发仪,SHB-3 循环水真空泵,上海申生科技有限公司。

1.2 方　法

(1)反式脂肪酸测定。脂肪酸甲酯的制备:称取 70mg 左右的油样至具塞试管中,取 4mL 正己烷溶解试样,加 250μL 2mol/L 氢氧化钾甲醇溶液,剧烈震摇 2min,加 1g 一水合硫酸氢钾,剧烈震摇 2min,将上层液 10000r/min 离心 10min 冷藏备用。

色谱分析:载气 N_2(99.999%),燃烧气 H_2(99.999%)和空气;进样口温度 250℃,检测器温度 250℃,H_2 压力 60kPa,空气压力 50kPa,柱前压 220kPa;进样量 1.0μL,分流比 1:50。升温程序:120℃保留 3min,8℃/min 升温至 175℃保留 28min,3℃/min 升温至 215℃保留 20min。采用归一化法定量。

(2)苯并(α)芘测定。样品净化:称取 0.4g 样品,精确到 0.001g,用 2mL 正己烷溶解;向带聚四氟乙烯的层析柱中加入一半高度的正己烷,快速称取 22g 减活氧化铝于小烧杯中,立即转移到层析柱中,使其均匀沉淀,再转入约 8.5g 无水硫酸钠;向层析柱中加入样品溶液,用正己烷清洗内壁;向层析柱中加入 80mL 正己烷洗脱,流速为 1mL/min,弃去前 20mL,用圆底烧瓶收集后 60mL 洗脱液;将收集的洗脱液 65℃旋蒸至 0.5～1mL,转移至预先称量好的玻璃样品瓶中,N_2 吹干,用正己烷清洗圆底烧瓶两次,每次 1mL,继续吹干;用 100μL 丙酮溶解,漩涡混合;过膜,HPLC 分析。

分析测定:进样量 10μL,流动相为乙腈:水＝88:12(体积比),流速

1.0mL/min,荧光检测器:发生波长406nm(狭缝10nm),激发波长384nm(狭缝10nm);依据不同浓度的苯并(α)芘标品溶液的峰面积绘制标准曲线,依据标准曲线和样品的峰面积计算样品中的苯并(α)芘含量。

2 结果与讨论

2.1 茶叶籽油中反式脂肪酸含量

因茶叶籽油所含不饱和脂肪酸均集中在C18∶1－C18∶3,尤其是C18∶1－C18∶2[7],故仅选择该范围内的脂肪酸甲酯标品进行分析。在本试验的色谱条件下,顺、反异构的C18∶1、C18∶2、C18∶3脂肪酸甲酯标准品均实现了较好的分离,如图1。

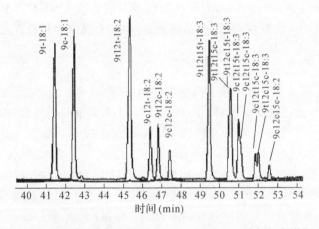

图1 脂肪酸甲酯混合标准品C18∶1－C18∶3区域气相色谱图

样品按上述方法进行气相色谱分析,依据标准品保留时间定性,采用归一化法计算样品提取油脂中反式脂肪酸含量,所有样品均未检测到C18∶3类的反式脂肪酸,结果见表1。

由表1可以看出,全国13省浸提茶叶籽毛油样品中反式脂肪酸含量很少。常温浸提毛油因无加热操作,其中反式脂肪酸总含量均低于0.05%,C18∶2类反式脂肪酸含量较C18∶1为高。因总量较少,并且可相互转化,t-C18∶1及t-C18∶2类别下的各种反式脂肪酸并不全部出现,相对而言出现较多的是9t12t-C18∶2,含量也稍高些。对该13个样品中反式脂肪酸总量进行统计学分析:13个样品TFA含量均值为0.0371%,在95%置信区间下,TFA含量

0.0308%～0.0434%。通过 SPSS 对全国 13 省茶叶籽样品的反式脂肪酸总量进行单样本 T 检验,无差异组:湖南、贵州、陕西、云南、四川、湖北、广西、福建、浙江、河南、江苏($P=0.070>0.05$),偏低组:江西、安徽($P=0.024<0.05$)。由此可见,茶叶籽油中反式脂肪酸含量是低于一定水平的。

表 1　茶叶籽油中反式脂肪酸含量

样　品	t-C18:1	t-C18:2				ΣTFA(%)
	9t-18:1	9t12t-18:2	9c12t-18:2	9t12c-18:2	Σt-C18:2	
云南普洱	0.0104	0.0302	0	0	0.0302	0.0406
贵州湄潭	0.0125	0.0228	0.0104	0	0.0332	0.0457
江苏无锡	0	0.0376	0	0.0105	0.0481	0.0481
湖北随州	0.0119	0.026	0	0	0.0260	0.0379
广西三江	0.0171	0	0.0194	0	0.0194	0.0365
四川雅安	0	0.027	0	0.0124	0.0394	0.0394
湖南湘潭	0	0.0291	0	0.0168	0.0459	0.0459
江西赣州	0	0	0.0177	0	0.0177	0.0177
浙江龙游	0.0116	0	0	0.0169	0.0169	0.0285
福建福安	0	0.0358	0	0	0.0358	0.0358
河南信阳	0.0151	0.0326	0	0	0.0326	0.0477
安徽黄山	0	0	0	0.0162	0.0162	0.0162
陕西汉中	0.0138	0.0152	0	0.0133	0.0285	0.0423
冷　榨	0	0	0	0	0	0
热　榨	0.0209	0.1080	0	0	0.1080	0.1289

压榨毛油样品数据则反映了常温冷榨与热榨的较大区别,冷榨毛油中未检测到反式脂肪酸,而热榨毛油中则出现超过 0.1% 的反式脂肪酸。由此可见,加热操作对茶叶籽油中反式脂肪酸的含量有较为明显的影响。总的来说,茶叶籽毛油中反式脂肪酸含量都维持在低于 0.2% 的水平,且仅在 180℃～200℃ 下炒籽 15min 后升至 0.12%,该含量显著低于欧盟 5% 和美国 2% 的限定值。

2.2 茶叶籽油中苯并(α)芘含量

将浓度为 0.4、0.8、4、8、20、40μg/L 的苯并(a)芘标准溶液按照最佳色谱分析条件分别进样 10μL，测定相应的峰面积。以苯并(a)芘浓度(μg/L)为横坐标，对应峰面积为纵坐标，制作标准曲线。试验结果表明，苯并(a)芘含量与峰面积的线性范围为 0.4～40μg/L，线性回归方程：$y=85975x+24108$，相关系数 $R^2=0.9992$，标准曲线见图 2。

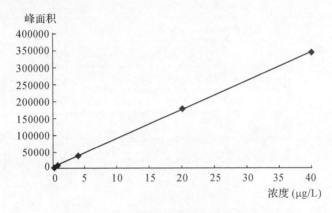

图 2 苯并(α)芘标准曲线

全国 13 个省茶叶籽浸提毛油中苯并(α)芘含量测定结果如图 3 所示，本次处理中冷榨毛油和热榨毛油中均未检测到苯并(α)芘。由此可见，大部分常温浸提茶叶籽毛油中苯并(α)芘含量是低于欧盟食用油规定限值 2μg/kg 的，有些甚至未检出(来自贵州和陕西的样品)。但也可看到，来自云南、浙江、福建的样品含量偏高。其中，云南和浙江的样品高于欧盟标准，但仍低于国家标准 10μg/kg；来自福建的样品中苯并(α)芘含量高达 37.17μg/kg。

结合本次实验中自榨毛油的检测结果，可以得出：①由于苯并(α)芘是一种非极性物质，茶叶籽本身并不含有，一般为环境污染带入，故大部分油中其含量较低，有些甚至为痕量或未被检出。同时也说明茶树的生长环境状况对茶叶籽油中苯并(α)芘含量高低有一定影响，这点可以通过比较浙江、云南样品与除福建之外的其他地区样品得出。②虽然加热对茶叶籽油中苯并(α)芘含量有影响，但本实验中炒籽后榨取的毛油依然未检出苯并(α)芘，说明本实验的热处理方式对茶叶籽油的安全性无不利影响。因此对比福建样品中很高的苯并(α)芘含量，以及收集到的福建茶叶籽外观(颜色明显较其他样品为深)，基本可以确

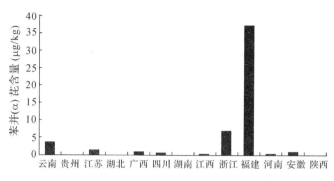

图 3 全国 13 个浸提毛油样品中苯并(α)芘含量

定该茶叶籽样品在本次实验收集前已经进行了较强烈的加热干燥,导致其中苯并(α)芘远高于其他样品。

3 结 论

本研究测定了收集到的全国 13 个省茶叶籽自制浸提毛油以及两个压榨毛油样品中反式脂肪酸和苯并(α)芘含量。结果显示,浸提毛油中反式脂肪酸总量为 0.01%~0.05%,常温冷榨毛油中未检测到反式脂肪酸,热榨毛油中其含量则有较为明显的升高,为 0.1289%,但仍低于欧盟等标准规定的限定值。大部分样品中苯并(α)芘含量低于欧盟等标准规定的限定值,常温冷榨毛油和热榨毛油均未检测到苯并(α)芘,但有 3 个样品中苯并(α)芘含量明显高于其他样品,可能为环境污染问题或过度加热操作造成,但明显后者的影响更大。总的来说,茶叶籽油的提取方法和是否进行加热操作以及加热的程度对茶叶籽毛油中反式脂肪酸和苯并(α)芘含量有明显影响,尤其是加热操作的影响更为显著。

参考文献

[1] 曹国锋,邬冰,钟守贤. 茶叶籽油、油茶籽油与茶树油的区别[J]. 中国油脂,2008,33(8):17-20.
[2] 马跃青,张正竹. 茶叶籽综合利用研究进展[J]. 中国油脂,2010,35(9):66-69.
[3] 中国茶叶籽油产业高峰论坛在京举行[J]. 中国茶叶,2011,33(1):23-24.
[4] 王瑞元. 充分利用茶叶籽制油 为国家增产食用油脂[J]. 农业机械·粮油加工,2011(1):20-21.

[5] 王兴国. 大力发展茶叶籽油产业[J]. 农业机械·粮油加工, 2011(1): 24.
[6] 李桂华. 茶叶籽油的营养与健康[J]. 农业机械·粮油加工, 2011(1): 25-26.
[7] 朱晋萱, 朱跃进, 张士康等. 茶叶籽油的脂肪伴随物成分分析初报[J]. 中国茶叶加工, 2011(4): 47-50.
[8] 金青哲, 王兴国, 曹万新等. 反式脂肪酸安全问题辨析[J]. 中国油脂, 2011, 36(1): 5-8.
[9] Ibanez R, Agudo A, Berenguer A, et al. Dietary intake of polycyclic aromatic hydrocarbons in a Spanish population [J]. J Food Protec, 2005, 68(10): 2190-2195.

附：基金项目：十二五国家科技支撑计划课题《传统优势特产资源生态高值利用技术研究与产品开发 2012BAD36B06》；十二五国家科技支撑计划项目《油料油脂危害因子新型高效检测技术及产品研发 2012BAK08B03》（涉及油脂安全的，主要是脂肪氧化物、甾醇氧化物、苯并芘、塑化剂等的检测和控制）；十二五国家科技支撑计划项目《高含油油料加工关键技术和装备研究与示范 2011BAD02B03》。

花卷茶提取物对高胆固醇血症大鼠血脂和内皮功能的影响

杜万红[1]　刘仲华[2]　施　玲[4]　周重旺[5]
郭立新[1]　刘小阳[1]　姜德建[3]　施兆鹏[2,4]

1. 解放军163医院内四科,中国 湖南;2. 湖南省天然产物工程技术研究中心,中国 湖南;3. 中南大学药学院药理学系,中国 湖南;4. 湖南农业大学茶学教育部重点实验室,中国 湖南;5. 湖南省茶业有限公司,中国 湖南

摘　要:目的:研究花卷茶提取物对高胆固醇血症大鼠血脂和内皮功能的影响及机制。方法:雄性SD大鼠高脂饲养4周诱发高胆固醇血症,其中实验组在高脂饲料饲养2周后开始每日灌胃给予不同剂量的花卷茶系列之十两茶水提物(50mg/kg、100mg/kg、200mg/kg),持续2周。实验结束后,颈动脉取血测定血脂、NO、非对称二甲基精氨酸(ADMA)和丙二醛(MDA)含量,分离胸主动脉检测血管内皮依赖性舒张功能。结果:十两茶提取物能呈剂量依赖性降低高脂血症大鼠血清总胆固醇和低密度脂蛋白及甘油三酯水平。十两茶提取物能显著改善高脂所诱导的血管内皮舒张功能障碍,降低血浆ADMA和MDA含量以及增加NO水平,且呈剂量依赖性。结论:十两茶提取物具有降低血脂和保护血管内皮功能作用,其作用与抑制脂质过氧化、调节ADMA/NO系统有关。

关键词:花卷茶;十两茶;高胆固醇血症;内皮功能;非对称二甲基精氨酸

Effects of the Extract of Hua-juan Tea on Blood Lipids and Endothelial Function in Hypercholesterolemic Rats

DU Wan-hong[1]　　LIU Zhong-hua[2*]　　SHI Ling[4]　　ZHOU Zhong-wang[5]
GUO Li-xin[1]　　LIU Xiao-yang[1]　　JIANG De-jiang[3]　　SHI Zhao-peng[2,4]

1. The Fourth Department of Internal Medicine, 163 Hospital of the PLA, Hunan, China; 2. Hunan Engineering and Technology Center for Natural Products, Hunan, China; 3. School of Pharmaceutical Sciences, Central South University, Hunan, China; 4. Key Laboratory of Tea Science, Ministry of Education, Hunan, China; 5. Hunan Tea Company Limited, Hunan, China

Abstract: Objectives: To investigate the effects of the extract of Hua-juan tea on blood lipids and endothelial function in hypercholesterolemic rats. Methods: Male SD rats were fed with food containing cholesterol for 4 weeks to induce hypercholesterolemia. For drug treatment, the animals were orally treated with different doses of the extract of Shi-liang tea (50, 100 or 200mg/kg per day) during the last two-week period of the experiment. At the end of the experiment, blood samples were collected to measure the plasma levels of nitric oxide (NO), asymmetric dimethylarginine (ADMA) and malondialdohyde (MDA), and the thoracic aortas were isolated to determine endothelium-dependent vasodilator response. Results: The extract of Shi-liang tea could dose-dependently decrease plasma levels of total cholesterol, low-density lipoprotein and triglyceride in hypercholesterolemic rats. Moreover, such treatment could improve endothelium-dependent vasodilator response and decrease plasma levels of ADMA and MDA as well as increase NO content. Conclusions: The extract of Shi-liang tea lowers blood lipids and protects endothelial function, and its effects may be related to inhibiting peroxidation and modulating ADMA/NO system.

Keywords: Hua-juan tea; Shi-liang tea; hypercholesterolemia; endothelial function; nitric oxide; asymmetric dimethylarginine

高脂血症被认为是动脉粥样硬化(AS)的独立危险因素之一。血管内皮功能不全是AS血管早期的一个重要病变特征,也是AS发生的始动环节[1]。大量动物实验和临床研究显示,高脂血症动物和患者均伴有血管内皮功能不全。

血中低密度脂蛋白(LDL)的浓度提升及氧化增加是导致高脂血症诱导内皮损伤的关键因素。因此,降脂和保护内皮功能被认为是防治 AS 的关键手段。

NO 是由 NO 合酶(NOS)催化 L-精氨酸转化而来,在维持血液循环稳定和调节心血管作用方面起关键作用。近年研究证实,内源性 NOS 抑制物——非对称二甲基精氨酸(ADMA)能抑制 NOS 和减少 NO 的生成。高脂血症所致血管内皮功能不全与血中 ADMA 水平升高密切相关,而且 ADMA 能促进氧化应激、诱导炎症反应等,被认为是一个新的致 AS 分子[2]。

茶叶是我国的主要饮品,含有多种多酚类化合物如儿茶素和茶多酚,具有抗氧化、调血脂和抗 AS 等功效[3]。研究发现,茶叶的单体成分表没食子儿茶素没食子酸酯(EGCG)能显著降低 LDL 诱导的大鼠血浆 ADMA 水平增加和改善血管内皮功能[4]。我国有六大茶类,其成分与功效不尽相同,目前研究比较多的为绿茶、普洱茶等。黑茶如花卷、茯茶为湖南茶叶的主要品种,花卷茶系紧压茶,产于湖南安化一带。十两茶是湖南黑茶花卷茶系列产品中的一种,分一两茶、十两茶、百两茶和千两茶四种,其品质优次,以一两茶最好,依次排列。百两茶与千两茶有近 200 年历史,一两茶和十两茶皆为新产品。关于湖南黑茶的心血管保护作用及机制的研究尚不多见。本实验拟采用高脂血症大鼠模型,观察花卷茶中的十两茶提取物对血脂和血管内皮功能的影响并进一步探讨其作用是否与调节 ADMA/NO 系统有关。

1 材料与方法

(1)药品与试剂。十两茶提取物由湖南农业大学天然产物研究中心及湖南省茶业股份有限公司白沙溪茶厂提供,将十两茶原料处理成碎状,按 1g 茶叶加 12mL 沸水进行提取,沸水浴 1h,精密过滤,冷却,经低温真空浓缩后冷冻干燥,得十两茶水提取物粉末。高效液相色谱(HPLC)[5]检测其中儿茶素含量 22.43%;紫外分光光度法(Folin-Denis 法)[6]检测其茶多酚含量 52.1%。

胆固醇(北京奥博星生物技术责任有限公司)、猪胆盐(上海源聚生物科技有限公司)、丙硫氧嘧啶(上海复星朝晖药业有限公司)、NO 和丙二醛(MDA)试剂盒(南京建成生物科技研究所)、辛伐他汀(Sigma 公司),其他试剂均为分析纯。

(2)实验动物与分组。雄性 SD 大鼠(180～200g)按体重分为正常对照组(10 只)和高胆固醇组(50 只)。高脂饲料构成为基础饲料中加入 10%猪油、2%胆固醇、0.5%猪胆盐和 0.2%丙硫氧嘧啶。于实验第 14d,尾尖取血,测定各组血脂水平。然后再将高胆固醇组按血脂水平随机分为 5 组：高胆固醇模型组，3 个不同剂量十两茶组（每日分别灌胃给予十两茶提取物 50、100、200mg/kg），阳性药物辛伐他汀组（每日灌胃给予辛伐他汀 10mg/kg）。连续给药 2 周，正常组和高胆固醇模型组灌以等体积双蒸水。末次给药后禁食 12h，颈动脉取血测定血脂、NO、ADMA 和 MDA 含量，分离胸主动脉检测血管内皮依赖性舒张功能。

（3）血脂、MDA 和 NO 测定。血浆甘油三酯（TG）、总胆固醇（TC）、高密度脂蛋白（HDL）和 LDL 采用全自动生化分析仪测定；硫代巴比妥酸法测定血浆 MDA 含量，反映脂质过氧化水平；硝酸还原酶法测定血浆 NO_2^- 水平，反映 NO 含量。

（4）ADMA 测定。取血浆 0.5mL 加 5-磺基水杨酸沉淀蛋白，混匀，离心 10min 后取上清液，HPLC 测定血浆中 ADMA 的含量（LC-10Advp；色谱柱：Nova-Pak C_{18}）。取 10μL 样品或标准品加入 100μL 衍生试剂（邻苯二甲醛与硼酸缓冲液及 β-巯基乙醇的混合物），混匀，室温下反应 3min 后进样，然后用线性梯度洗脱的方式将样品从色谱柱中洗脱［流动相 A 为 0.05mol/L 乙酸钠（pH 6.8）－甲醇－四氢呋喃＝82：17：1，流动相 B 为 0.05mol/L 乙酸钠－甲醇－四氢呋喃＝22：77：1，流速为 1mL/min］，经预柱处理后，用 RF-10Axl 型荧光检测器对其及内标 ADMA 进行检测（激发波长：338nm；吸收波长：425nm）。

（5）离体血管内皮依赖性舒张功能检测。分离和剪取大鼠胸主动脉段，置于充以 95％ O_2 和 5％ CO_2 混合气体的改良克氏液（NaCl 118mol/L、KCl 4.7mol/L、$MgSO_4$ 1.2mol/L、$NaHCO_3$ 25mol/L、$KHPO_4$ 1.2mol/L、$CaCl_2$ 2.5mol/L、Glucose 11mol/L）中。小心剪去血管周围结缔组织，清除血管内血液，剪成约 3～4mm 的血管环，一端固定于浴槽，另一端连接张力换能器，通过 Power Lab 生物记录系统连续记录血管张力变化。血管环置于充满 95％ O_2 和 5％ CO_2 混合气体的改良克氏液的水浴槽中（37℃），每隔 15min 更换克氏液一次。调节静息张力为 2g，平衡 60min，加用 KCl（60mmol/L）收缩血管，冲洗 2～3 次后，重新平衡血管环，用去氧肾上腺素（10^{-6}mol/L）收缩血管，待稳定后观察累积浓度的乙酰胆碱（ACh，$3×10^{-9}$～10^{-6}mol/L）诱导血管内皮依赖性舒张。

（6）统计学处理。所有数据均以均数±标准误（$\bar{x}±s$）表示。用 SPSS 进行统计学处理，组间差异采用方差分析。双侧 $P<0.05$ 认为有显著性差异。

2 结　果

（1）十两茶提取物对血脂的影响。高脂饲料喂养大鼠 4 周后，血清 TC、LDL 和 TG 水平明显升高（$P<0.05$ 或 $P<0.01$，与正常对照组比较），而 HDL

无显著变化。灌胃给予低剂量十两茶提取物(50mg/kg)2周对血脂无显著影响,但中、高剂量十两茶提取物(100、200mg/kg)和阳性药辛伐他汀(10mg/kg)均能显著降低高脂饲养所致的 TC 和 LDL 及 TG 水平的增加,但对 HDL 水平无显著影响(表1)。

表1 十两茶提取物对高脂血症大鼠血脂的影响(均数±标准差,$n=10$)

组别	TC(mg/dL)	HDL(mmol/L)	LDL(mmol/L)	TG(mmol/L)
对照组	1.34±0.28	0.78±0.22	0.26±0.19	0.58±0.25
高脂组	2.68±0.70++	1.05±0.28	1.33±0.60++	0.98±0.38+
十两茶提取物 50mg/kg	2.41±0.57	0.99±0.28	1.12±0.60	0.64±0.25*
十两茶提取物 100mg/kg	1.70±0.44**	0.92±0.25	0.54±0.25**	0.50±0.22**
十两茶提取物 200mg/kg	1.63±0.32**	0.89±0.25	0.54±0.60**	0.47±0.22**
辛伐他汀阳性对照组 10mg/kg	1.49±0.22**	0.85±0.09	0.39±0.19**	0.38±0.13**

注:与对照组比较,+ $P<0.05$,++ $P<0.01$;与高脂组比较,* $P<0.05$,** $P<0.01$。

(2)十两茶提取物对胸主动脉内皮舒张功能的影响。与正常对照组比较,高胆固醇血症组大鼠的胸主动脉对 ACh 诱导的血管内皮依赖性舒张效应显著降低($P<0.01$)。3个剂量十两茶提取物(50、100、200mg/kg)能显著改善高脂诱导的血管内皮舒张功能障碍,且呈浓度依赖性。阳性药辛伐他汀(10mg/kg)也可显著改善高脂诱导的血管内皮舒张功能障碍(图1)。

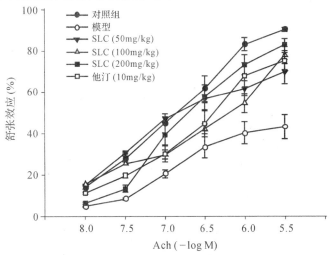

图1 十两茶提取物对高脂血症大鼠胸主动脉内皮依赖性舒张功能的影响)

(3)十两茶提取物对血浆 NO 浓度的影响。与正常对照组比较,高脂饲养大鼠的血浆 NO 水平显著降低($P<0.01$)。3 个剂量十两茶提取物(50、100、200mg/kg)能浓度依赖性增加高脂诱导的血浆 NO 水平降低。阳性药辛伐他汀(10mg/kg)也可显著增加高脂诱导的血浆 NO 水平降低(表2)。

表2　十两茶提取物对高脂血症大鼠血浆 NO、ADMA 和 MDA 的影响

(均数±标准差,$n=10$)

组　别	NO(μmol/L)	ADM(μmol/L)	MDA(μmol/L)
对照组	36.1±12.0	0.43±0.19	7.4±2.2
高脂组	16.4±12.3++	1.23±0.41++	15.5±3.8++
十两茶提取物 50mg/kg	30.3±19.9	1.01±0.32	12.4±4.7
十两茶提取物 100mg/kg	42.6±13.6**	0.64±0.35**	12.2±2.5
十两茶提取物 200mg/kg	42.1±15.8**	0.59±0.38**	9.2±4.7**
辛伐他汀阳性对照组 10mg/kg	45.2±4.2**	0.65±0.44**	7.9±2.8**

注:与对照组比,++ $P<0.01$;与高脂组比,** $P>0.01$。

(4)十两茶提取物对血浆 ADMA 水平的影响。与正常对照组比较,高脂饲养大鼠的血浆 ADMA 水平显著增加($P<0.01$)。3 个剂量十两茶提取物(50、100、200mg/kg)能浓度依赖性降低高脂诱导的血浆 ADMA 水平增加。阳性药辛伐他汀(10mg/kg)也可显著降低高脂诱导的血浆 ADMA 水平升高(表2)。

(5)十两茶提取物对血浆 MDA 含量的影响。与正常对照组比较,高脂饲养大鼠的血浆 MDA 水平显著增加($P<0.01$)。3 个剂量十两茶提取物(50、100、200mg/kg)能浓度依赖性降低高脂诱导的血浆 MDA 水平增加。阳性药辛伐他汀(10mg/kg)也可显著降低高脂诱导的血浆 MDA 水平升高(表2)。

3　讨　论

本研究的主要结果:①十两茶提取物能显著降低高脂饲养诱导的 TC、LDL 和 TG 水平的增加;②十两茶提取物能显著改善高脂血症大鼠的血管内皮依赖性舒张功能;③十两茶提取物能降低高脂血症大鼠血浆 ADMA 和 MDA 水平,增加 NO 水平。

血脂异常(TC、LDL 和 TG 增加,HDL 降低)被认为是 AS 的独立危险因素之一。研究证明,LDL 能够在内皮聚集和氧化及诱导血管炎症反应,进而导致内皮依赖性舒张反应降低和单核-内皮细胞黏附增加,促进 AS 的发生发展。茶叶是由山茶科植物芽叶加工制成的,其中富含多酚类化合物如茶多酚和儿茶

素等。众多研究表明,多种茶类如绿茶和普洱茶的提取物具有抑制 LDL 氧化和降脂及抗 AS 作用[3,7]。十两茶属于湖南黑茶,产于湖南安化。本实验发现,给予十两茶提取物能显著降低高脂饲养所致的大鼠血浆 TC、TG 和 LDL 水平的升高和改善血管内皮依赖性舒张功能,提示十两茶提取物具有明显的调脂和内皮保护作用。

NO 在维持正常血管内皮功能中起关键作用,由 NOS 催化 L-精氨酸转化而来。近期发现,ADMA 能抑制内皮细胞 NOS 活性和减少 NO 生成以及损伤血管内皮依赖性舒张功能[2]。此外,ADMA 还显示了多种直接促 AS 形成作用,包括诱导内皮细胞凋亡、促进单核-内皮细胞黏附和加速泡沫细胞形成等[8-10]。动物和临床实验均发现,高脂血症时血中内源性 ADMA 显著升高,并伴血管内皮舒张功能障碍,表明内源性 ADMA 含量升高是高脂血症血管内皮功能不全发生的关键原因[2]。我们近期的系列研究发现,多种心血管系统药物和天然药物包含茶叶的单体成分 ECGC,均能通过降低 ADMA 水平而发挥保护内皮功能作用[4,11-13]。目前的研究显示,十两茶提取物能呈浓度依赖性降低高脂诱导的大鼠血浆 ADMA 水平和增加 NO 浓度,提示其内皮保护作用可能与调节 ADMA/NO 系统有关。

ADMA 主要被二甲基精氨酸-二甲胺水解酶(DDAH)水解失活。DDAH 活性位点含氧敏感的巯基,因此容易被氧化失活。在培养的内皮细胞,外源性氧化型 LDL 可降低 DDAH 活性和升高细胞内 ADMA 含量[14]。因此,氧化应激是导致 ADMA 水平升高的主要原因。文献报道,在高胆固醇症模型与外源性给予 LDL 损伤血管内皮模型,天然多酚类化合物口山酮均能降低血浆 ADMA 浓度[15-16]。我们前期发现,十两茶中茶多酚含量较之其他茶品种为高(达 52.1%),提示十两茶可能具有较强的抗氧化活性。本实验也显示,十两茶提取物能显著降低高脂血症大鼠血浆 MDA 含量,抑制脂质过氧化水平。

综上所述,十两茶提取物具有明显的降脂和保护血管内皮功能作用,其机制与抑制氧化应激、调节 ADMA/NO 系统有关。

参考文献

[1] Vanhoutte P M. Endothelial dysfunction and atherosclerosis[J]. Eur Heart, 1997, 18 (suppl): E19-E29.

[2] 姜德建,李元建. 非对称二甲基精氨酸——新的心血管疾病危险因子和药物防治靶点[J]. 中国动脉硬化杂志, 2005, 13(3): 379-382.

[3] Stangl V, Lorenz M, Stangl K. The role of tea and tea flavonoids in car-

diovascular health [J]. Mol Nutr Food Res,2006,50(2):218-228.

[4] 唐伟军,陈美芳,江俊麟等. 表没食子儿茶素没食子酸酯对低密度脂蛋白诱导的血管内皮损伤的保护作用[J]. 中国动脉硬化杂志,2006,14(1):21-24.

[5] 刘仲华,黄建安. 儿茶素提制工艺技术研究[J]. 湖南农业大学学报,1997,23(6):537-542.

[6] 钟萝. 茶叶品质理化分析[M]. 上海:上海科学技术出版社,1989. 266-268.

[7] 赵献云,张辉,何丽芬. 绿茶提取物对高脂血症大鼠的降脂和抗氧化作用[J]. 武警医学院学报,2006,15(6):574-575.

[8] Jiang D J, Jia S J, Dai Z, et al. Asymmetric dimethylarginine induces apoptosis via p38 MAPK/caspase-3-dependent signaling pathway in endothelial cells [J]. J Mol Cell Cardiol,2006,40(4):529-539.

[9] Böger R H, Bode-Böger S M, Tsao P S, et al. An endogenous inhibitor of nitric oxide synthase regulates endothelial adhesiveness for monocytes[J]. J Am Coll Cardiol,2000,36(7):2287-2295.

[10] Smirnova I V, Kajstura M, Sawamura T, et al. Asymmetric dimethylarginine upregulates LOX-1 in activated macrophages:Role in foam cell formation [J]. Am J Physiol Heart Circ Physiol,2004,287(2):782-790.

[11] Tan B, Jiang D J, Huang H, et al. Taurine protects against low-density lipoprotein-induced endothelial dysfunction by the DDAH/ADMA pathway [J]. Vascul Pharmacol,2007,46(5):338-345.

[12] 陈美芳,谢秀梅,杨天伦等. 氯沙坦通过降低ADMA水平诱导血管保护作用[J]. 中南药学,2007,5(3):193-198.

[13] Jiang D J, Jiang J L, Zhu H Q, et al. Demethylbellidifolin preserves endothelial function by reduction of the endogenous nitric oxide synthase inhibitor level [J]. J Ethnopharmacol,2004,93(2/3):295-306.

[14] Jiang D J, Hu G Y, Jiang J L, et al. Relationship between protective effect of xanthone on endothelial cells and endogenous nitric oxide synthase inhibitors [J]. Bioorg Med Chem,2003,11(23):5171-5177.

[15] 姜德建,江俊麟,谭桂山等. 川东獐牙菜素A保护溶血性磷脂酰胆碱诱导的内皮细胞损伤[J]. 中南药学,2003,1(2):75-79.

[16] 贾素洁,李年生,王珊等. 口山酮对高胆固醇血症大鼠血脂的影响[J]. 中国动脉硬化杂志,2005,13(3):317-319.

茶皂甙结构修饰型引气剂及其引气混凝土性能

朱伯荣[1]　叶勇[2]　杨杨[3]　瞿佳[1]

1. 杭州中野天然植物科技有限公司,中国 浙江;
2. 华南理工大学,中国 广东;3. 浙江工业大学,中国 浙江

摘　要:本文介绍了茶皂甙结构修饰型混凝土引气剂(下称 ZY-99 引气剂)的性能,测定了该引气剂在混凝土中的各项性能指标,并且研究了其引气混凝土性能。结果表明:ZY-99 引气剂是一种原料来源和性能独特的新型引气剂,其对混凝土各项性能的影响程度可与国外有代表性的 Vinsol 引气剂相媲美,是国内混凝土工程中值得推广应用的一种优质引气剂。

关键词:ZY-99 茶皂甙结构修饰型引气剂;混凝土;和易性;强度;抗冻性

Structural Modified Tea Saponin as an Air Entraining Agent and its Performance in Concrete

ZHU Bo-rong[1]　YE Yong[2]　YANG Yang[3]　QU Jia[1]

1. Hangzhou Zhongye Natural Plant Technology Company, Zhejiang, China;
2. South China University of Technology, Guangdong, China;
3. Zhejiang University of Technology, Zhejiang, China

Abstract: The properties of structural modified tea saponin as a new air entraining agent (named ZY-99) in concrete is introduced in this research. Its performance index in concrete was determined to evaluate its quality. The results showed ZY-99 was distinctive new type of air entraining agent in crude material origin and properties. Its effect on improvement of quality of concrete is similar to the representative Vinsol air entraining agent in abroad. ZY-99 is a high quality air entraining agent worthy spreading its application.

Keywords: ZY-99 tea saponin structural modified air entraining agent; concrete; peaceability; intensity; freezing resistance

混凝土耐久性是当今国内外混凝土领域的研究热点,而混凝土的抗冻性和抗盐冻剥蚀性是混凝土耐久性研究领域中最重要的组成部分之一,也是目前我国北方地区混凝土工程所面临的最严峻的问题之一。

混凝土中通过掺加引气剂,引入大量均匀、稳定而封闭的微小气泡,是大幅度提高混凝土耐久性,特别是抗冻性和抗盐冻剥蚀性的最有效技术措施之一。在北美、北欧和日本等发达国家早已普遍推广引气技术,据报道他们80%以上的混凝土都掺入引气剂,特别在水工、港工、道桥等重要工程更是明确规定了必须掺加引气剂。然而国内长期以来由于传统松香皂、松香热聚物型引气剂产品质量欠佳,存在诸如溶制较困难、与其他外加剂尤其是聚羧酸盐系高性能引气剂复合性欠佳以及引气混凝土强度损失较大等缺点,加上人们对混凝土工程耐久性和引气剂的应用认识不足,制约着引气剂在国内混凝土工程的推广应用,致使混凝土工程耐久性,特别是抗冻和抗盐冻耐久性急剧下降。

研究和开发性能优异、质量稳定、使用方便、技术经济效果显著的新品种引气剂具有极其重要的意义。ZY-99 茶皂甙结构修饰型混凝土引气剂的研制成功,无疑是国内混凝土引气剂品种和性能改善上的一大突破,ZY-99 茶皂甙结构修饰型混凝土引气剂除各项性能指标达到国家引气剂品质指标一等品外,同时具有许多国内传统引气剂所无法相比的性能优势。本文主要介绍 ZY-99 新型引气剂性能,并且研究其引气混凝土性能。

1 ZY-99 引气剂产品化学、物理性能

ZY-99 茶皂甙结构修饰型混凝土引气剂是利用我国特有的可再生资源——茶籽饼粕为主要原料,采用萃取、多效提纯等精深加工技术,从粕料中萃取出高纯度茶皂甙,并经选择性水解与对茶皂甙中的仲羟基进行保护以激活伯羟基,再与二乙氧基甲烷、蔗糖醚化等反应研制出一种两端亲水、中间亲油的双子型表面活性剂,最后应用制剂学原理经优化、复合配伍工艺技术制成的高效混凝土引气剂。由于其带有两端亲水、中间亲油的分子结构,分子极易定向排列在气-液界面上,从而有效降低溶液的表面张力,便于小气泡的形成和稳定。ZY-99 引气剂对酸、碱和硬水有较强的化学稳定性。

产品经浙江省化工研究院检测,其理化指标如下:①外观为棕褐色液体;②表面张力 26.4mN/m;③固含量 35%;④活性物含量(以干基计)77.3%;⑤pH值 8.6;⑥水不溶物 0.2%;⑦发泡高度 183mm;⑧5min 泡沫损失率 7.2%。

产品经浙江省建筑科学研究院检测,其在混凝土中的性能指标如下:①减水率 8.2％;②泌水率比 39.6％;③含气量 3.2％;④凝结时间之差:初凝 5min、终凝 20min;⑤抗压强度比:3d 129％、7d 117％、28d 111％;⑥1h 含气量经时变化量＋0.5;⑦收缩率比 106％;⑧抗冻耐久性(200 次)96％。

2 试 验

(1)原材料。试验所用原材料为:水泥:525♯普通硅酸盐水泥;细骨料:密度 2.62g/cm³,细度模数 2.57 的中砂及耐热玻璃砂(用于引气剂改善混凝土碱集料膨胀破坏效能试验);粗骨料:密度 2.74g/cm³,粒径 5～25mm 的碎石;引气剂:包括所研制的 ZY-99 引气剂、对比用国外有代性的美国产 Vinsol 引气剂及国内工程中常用而且质量较好的松香热聚物类引气剂(简称 SR 引气剂)。瑞典 LUND 大学进行的混凝土抗冻性实验所用原材料为当地硅酸盐水泥和普通砂石。

(2)试验方法。混凝土拌和物和易性用混凝土拌和物坍落度来评定;新拌混凝土含气量用气压式含气量测定仪测定;混凝土强度试件尺寸为 10cm×10cm×10cm,试件按常规方法搅拌振动成型,空气中静养 1d 后拆模,进入标准养护室水中养护至各个龄期,进行抗压强度测试;混凝土碱集料膨胀破坏试验按 JGJ53-92 进行;混凝土抗冻性试验按 SD105-82 进行;瑞典 LUND 大学进行的混凝土抗冻性试验按欧洲通用标准,也是目前 RILEMTC117-FCD 推荐的盐冻方法 CEN/TC51/WG12/TG:6/94 进行。

3 结果与讨论

3.1 ZY-99 引气剂对新拌混凝土性能的影响

如同其他引气剂一样,ZY-99 引气剂能显著改善新拌混凝土性能。引气剂引入的大量球形气泡,在混凝土拌和物中如同滚珠作用,以及大量气泡的存在增加了浆体体积、浆体黏度和屈服应力,因此其新拌混凝土的和易性、塑性和内聚性得到显著提高,离析和泌水现象显著降低。在原材料比例不变的条件下,引气可以提高混凝土的流动性,而在相同坍落度条件下,引气可以降低混凝土拌和用水量。同时,引气还可降低新拌混凝土坍落度损失。

图 1 显示了当基准混凝土配合比按照 GB 8076—87 规定的混凝土外加剂

试验方法进行设计,基准混凝土拌和物坍落度为6cm(含气量为1%)时,在原材料比例不变的条件下,分别掺ZY-99引气剂、Vinsol引气剂及SR引气剂后,含气量对混凝土拌和物坍落度的影响。由结果可见,3种引气剂对混凝土拌和物坍落度的影响规律大致相同,随着含气量的增加,拌和物坍落度逐渐增大。其实,引气对混凝土拌和物工作性的影响要远大于对坍落度的影响。

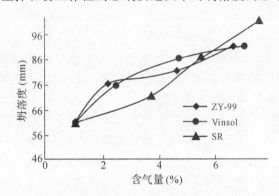

图1 含气量对混凝土拌和物坍落度的影响

随着施工技术和高层建筑的发展需要,混凝土的可泵性显得越来越重要。实际上,可泵性是混凝土工作性良好的一种特殊表现形式。由于引气增加了混凝土的内聚性和物料间的润滑作用,降低了胀流,混凝土泵送时不会过度离析和泌水,因此引气可提高新拌混凝土的可泵性,特别是由于ZY-99引气剂具有极强的水溶性,与其他外加剂具有很强的复合性能,是配制泵送剂等其他复合外加剂的重要组成部分。

3.2 ZY-99引气剂对混凝土强度的影响

混凝土中引入大量微小气泡对混凝土性能带来的负面影响表现在大量气泡的存在引起了混凝土抗压强度的降低,特别当气泡结构较差,存在较多聚合气泡和异形气泡时,其强度损失更大,这对于一些对混凝土强度要求较严的重要工程来说是一个不容忽视的问题。

为了探讨ZY-99引气剂对混凝土抗压强度的影响程度,选择了Vinsol引气剂及SR引气剂作对比,进行了引气混凝土抗压强度对比试验。基准混凝土配合比按照GB 8076—87规定的混凝土外加剂试验方法进行设计、试配和调整后确定为:水泥用量330kg/m³,水灰比0.58,砂率38%。

图2为同水灰比条件下,含气量对混凝土抗压强度的影响;图3为同坍落

度条件下,含气量对混凝土抗压强度的影响。

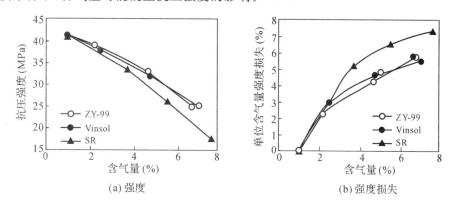

图 2 同水灰比条件下含气量对混凝土抗压强度的影响

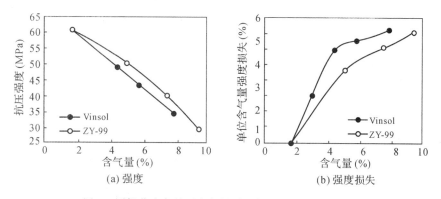

图 3 同坍落度条件下含气量对混凝土抗压强度的影响

由图 2 可见,同水灰比条件下,随着混凝土含气量增加,混凝土抗压强度逐渐下降。比较 3 种引气剂:含气量一定时,ZY-99 引气剂对混凝土抗压强度的降低度与 Vinsol 引气剂大致相同,但明显小于 SR 引气剂,可以推测 ZY-99 引气剂所引入的气泡尺寸较小,气泡结构优于 SR 引气剂,从而对混凝土抗压强度影响较小。同水灰比条件下,ZY-99 引气剂每单位含气量引起的混凝土抗压强度损失率在 3%~6%,比 SR 引气剂低 1~2 个百分点。

3.3 ZY-99 引气剂对混凝土抗冻性和抗盐冻剥蚀性的影响

引气剂能大幅度提高混凝土抗冻性和抗盐冻剥蚀性已为广大混凝土学术界和工程界所一致公认。图 4 是在基准混凝土水泥用量为 $330kg/m^3$,水灰比为 0.54,同坍落度条件下,ZY-99 引气剂(含气量为 5%)对混凝土抗冻性影响。试验结果表明,掺加 ZY-99 引气剂后,其引气混凝土抗冻性显著提高,对于基准非引气混凝土,其 100 次冻融循环后,相对动弹性模量仅为 25.7%;而对于含气量为 5% 的引气混凝土,即使 300 次冻融循环后,相对动弹性模量仍高达 93.8%。

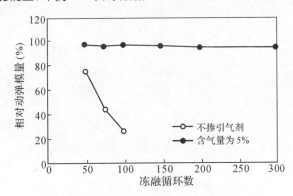

图 4　ZY-99 引气剂对混凝土抗冻性的影响

混凝土拌和物中引入大量微小气泡,改善了拌和物和易性,同坍落度条件下,引气混凝土用水量可降低,从而可不同程度弥补由于引气引起的混凝土抗压强度的损失。从图 3 可见,同坍落度条件下,ZY-99 引气剂和 Vinsol 引气剂效果大致相同,对混凝土抗压强度的降低幅度大大减小,当混凝土含气量低于 3% 时,混凝土抗压强度不但没有下降,还略有提高;含气量在满足抗冻混凝土要求的 5%~6% 时,每单位含气量引起的混凝土抗压强度损失率不大于 3%。

图 5 为在瑞典 LUND 大学,国际上混凝土抗冻性研究权威学者 Fagerlund

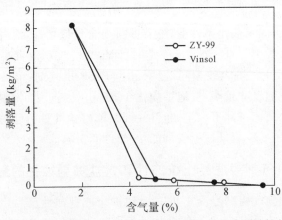

图 5　ZY-99 引气剂和 Vinsol 引气剂对混凝土抗盐冻剥蚀性的影响

G 教授主持的材料研究机构测试的当混凝土试件尺寸为 10cm×10cm×10cm，基准混凝土水泥用量为 380kg/m³，水灰比为 0.54，同水灰比条件下，ZY-99 引气剂和 Vinsol 引气剂对混凝土抗盐冻剥蚀性影响对比试验结果。结果同样表明，ZY-99 引气剂和 Vinsol 引气剂可显著改善混凝土抗盐冻剥蚀性，两种引气剂对混凝土抗盐冻剥蚀性的改善效果大致相同。

3.4 ZY-99 引气剂对混凝土其他耐久性的影响

引气剂除了对混凝土抗冻性和抗盐冻剥蚀性具有显著的改善效果外，试验研究表明对混凝土其他耐久性也有不可忽视的改善效果。由于混凝土中引气产生的大量微小气泡难以被水充满，使得混凝土结构中存在许多均布的微小空间，这些微小空间可以作为体积膨胀的"缓冲阀"，降低和延缓其他物理膨胀(如盐晶体结晶压等)和化学反应膨胀(如碱集料反应和硫酸盐反应等)引起的混凝土破坏。表 1 试验结果证实了 ZY-99 引气剂在这方面产生的显著效果，当混凝土中含气量为 4.5% 时，在 20% NaCl 溶液中经 15 次干湿循环后，盐结晶压产生的膨胀率比非引气混凝土降低约 41.3%；按 JGJ53-92 标准测得的 2 个月后碱集料反应产生的膨胀率比非引气混凝土降低约 57.0%，尽管由于这类膨胀产物的不断富集，不能从根本上消除其膨胀破坏，但是有效起到了延缓这类膨胀破坏的作用。

表 1　ZY-99 引气剂对混凝土抗盐结晶压和碱集料反应膨胀破坏的影响

混凝土品种	20%NaCl 溶液中干湿循环后的膨胀率 ($×10^4$)/膨胀降低率(%)		碱集料反应后的膨胀率 ($×10^4$)/膨胀降低率(%)	
	10 次	15 次	0.5 个月	2 个月
基准混凝土	3.2/0	9.2/0	2.65/0	2.98/0
引气混凝土	1.8/43.8	5.4/41.3	1.00/62.3	1.28/57.0

4　结　论

(1)ZY-99 茶皂甙结构修饰型混凝土引气剂是利用我国特有的可再生资源——茶籽饼粕为主要原料，采用萃取、多效提纯等精深加工技术，从粕料中萃取出高纯度茶皂甙，并经分子修饰与产品优化等综合技术措施得到的新一代混凝土引气剂，它具有水溶性极强，施工方便，可与任何其他外加剂按任何比例复合使用等性能优势,是一种原料和性能独特的新型引气剂。

(2)ZY-99引气剂能显著改善新拌混凝土性能,也是配制泵送剂等其他复合外加剂的重要组成部分。

(3)ZY-99引气剂对混凝土抗压强度的影响程度与国外有代表性的Vinsol引气剂大致相同,但明显优于国内属质量较好的SR引气剂。同和易性条件下,当含气量在满足抗冻混凝土要求的5%~6%时,ZY-99引气剂每单位含气量引起的混凝土抗压强度损失率不大于3%。

(4)ZY-99引气剂能显著提高混凝土抗冻性,它对混凝土抗盐冻剥蚀性(或抗冻性)的改善效果与Vinsol引气剂大致相同。此外,ZY-99引气剂对混凝土的综合耐久性也有不可忽视的改善效果,如可明显延缓盐结晶压和碱骨料反应等膨胀引起的混凝土破坏。

关于茶树修剪枝再利用的探讨

郑生宏[1]　柴红玲[2]　李　阳[1]　何卫中[1]

1. 浙江省丽水市农业科学研究院，中国 浙江；
2. 浙江省丽水职业技术学院，中国 浙江

摘　要：本文分类总结了目前生产上对茶树修剪枝的再利用现状，分析了茶树修剪枝再利用过程中存在的一些问题，同时就今后茶树修剪枝再利用的对策和注意事项进行了探讨。

关键词：茶树修剪枝；再利用现状；问题分析；对策探讨

Study on Reutilization for Tea Branches Produced by Pruning

ZHENG Sheng-hong[1]　　CHAI Hong-ling[2]
LI Yang[1]　　HE Wei-zhong[1]

1. Lishui Academy of Agricultural Sciences，Zhejiang，china；
2. Lishui Vocational and Technical College，Zhejiang，China

Abstract：This paper summarized the present situation on the reutilization for tea branches produced by pruning, then analysis on the problems during the process of tea branches reutilization was conducted. Meanwhile, in order to make full use of such tea branches, discussion on countermeasures and precautions for future reutilization of tea branches by pruning was also given in the paper.

Keywords：tea branches by pruning；present situation on reutilization；analysis on the problems；countermeasure discussion

　　修剪不仅可以解除顶端优势，刺激腋芽萌发，获得更多的新梢，同时还能带走一部分弱生长势和有病虫害的枝条，从而保证茶树更好地生长。然而，长久以来，人们只关注修剪本身，对于修剪下来的枝条却很少问津，大量的茶树修剪枝只能弃于茶园中。研究表明废弃的修剪枝中亦含有丰富的内含物，可作为茶

系列产品和部分蛋白质提取的原料[1-2]。此外,梁月荣等[3]研究发现,将修剪叶施用于茶园,能显著增加土壤有机质含量,有利于促进根系的生长,同时还能降低土壤酸度和活性铝含量。因此茶树修剪枝具有较大的可再利用价值,若能进一步分类再利用,必将为茶叶种植带来可观的经济附加值。本文总结了目前生产上对茶树修剪枝的再利用途径与现状,分析了修剪枝再利用过程中存在的一些问题与不足。在此基础上,对强化茶树修剪枝再利用的对策与注意事项进行了探讨。

1 茶树修剪枝的再利用现状

1.1 健壮修剪枝作为插穗育苗

短穗扦插作为茶树无性繁殖中技术性最强的繁殖方法之一,现已成为国内各茶区进行茶树种苗繁育的常规方法。部分种苗繁育单位为此专门建设采穗母本园,大大增加了建设成本和管理费用。在实际生产中,部分健壮茶树修剪枝已能够满足插穗条件。如浙江诸暨市创办的茶树良种繁育基地通过利用栽后第2、第3年定型修剪下来的枝条进行短穗扦插,同时从第4年起在春茶采摘名优茶后进行修剪养穗,于9月上旬至11月上旬剪穗扦插,探索出了一条快速有效的无性系良种繁育途径。类似的修剪管理还有来自浙江安吉白茶区,幼龄茶树采用以采代剪的办法,通过春茶采摘后,以养树为主,直到夏、秋季再进行修剪,剪下的枝条进行扦插育苗。因此,若能够对茶树健壮修剪枝进行及时收集、分拣和简单处理将其作为扦插育苗材料,既可以节约成本,还可带来额外的经济效益。

1.2 修剪枝还田以提高土壤肥力

茶树修剪枝有机质含量高,养分丰富,是茶园很好的有机肥源,可直接将修剪下的枝条作肥料埋入土中;修剪枝也可作土壤覆盖物铺于行间,防止杂草滋生;此外,干旱季节茶行铺修剪枝还可起到茶园保水作用。近些年来浙江部分茶区,尤其是山区茶园,针对市场求早求新的需要,全年只采摘春茶一季,并探索出一条与之相配套的促进春茶早发优质高效的修剪技术,即"三年两头刈"技术;成龄投产茶园每年春茶采摘结束后立即进行重修剪或台刈,然后将修剪枝铺于茶行中间,既能保持水土,又能减少养分流失和抑制杂草滋生,防止杂草与茶树争水、争光、争肥。铺园的修剪枝腐烂后能增加土壤有机质和其他矿物质养分,改善土壤的理化性质和生物性状,提高土壤的保肥和供肥能力。试验表明,永嘉"乌牛早"用这种方法修剪(台刈)后,茶树次年开采期较常规茶园提早3~5d,而且芽头粗壮,起到了早采摘、早上市的效果,春茶经济效益显著。而金华箬阳乡茶农根据当地山区地形、气候特点,采用上述修剪技术,茶枝还田,可使茶园地温提高2~3℃,空间湿度提高5%~10%,对增强茶树抗旱、防冻能力和增加茶叶产量效果显著。

1.3 修剪枝作为再加工利用原料

近些年,随着生态茶园建设力度不断加大以及茶园标准化进程不断推进,茶树修剪成为茶园管理中必不可少的农艺措施,由此而产生的大量茶树修剪枝除了作为育苗插穗以及还田处理外,还可以回收作为再加工原料进行利用,其资源综合利用已初步取得良好效果。

1.3.1 利用茶树修剪枝生产畜牧饲料添加剂

茶叶中含有多糖、咖啡因、氨基酸、叶绿素、儿茶素等600多种有效成分,茶树修剪枝叶在化学成分上类似于茶树叶片,因此为了扩大产业链,提高茶农收入,可以将废弃的茶树修剪枝通过回收加工后应用于家禽饲料中,从而实现变废为宝。目前,在畜牧养殖中,有人将茶叶、茶渣或茶叶提取物按一定比例加入日粮中以提高饲料的转化率[4]。日本佐野满昭认为在鸡饲料中添加茶粉,可提高鸡肉鲜嫩度和鸡蛋的品质、产量和耐贮藏性[5];赖建辉研究认为奶牛食用一定比例的乌龙茶粉其产奶量可提高10%左右[6]。张广强等人用茶树修剪后的枝叶做成茶汤拌料饲喂土鸡,结果表明散养土鸡的生产性能、免疫力和蛋品质都得到了明显提高[7],该研究为茶树修剪枝应用于家禽饲料提供了依据。在实际应用中,作为家禽饲料的茶树修剪枝必须是来自于无公害茶园或者是农药用量少、无污染的茶树,从而确保饲料的安全性。

1.3.2 利用茶树修剪枝作为生产食用菌的培养基质

在食用菌生产过程中,菌种的培养基质主要是一些内含物丰富的工农业副产品及下脚料,如棉籽壳、木屑、玉米秸秆等。它们富含纤维素、半纤维素和木质素等有机物,是食用菌生长的主要营养源。茶树修剪枝在化学组成上主要也是由纤维素、半纤维素和木质素组成,同时叶片部分还富含多糖和氨基酸等营养成分,是食用菌生产培养基的良好原料来源。近年来,食用菌栽培中木材消耗对当地生态的影响已逐渐凸显,部分地方已逐渐转型升级,寻找新的培养料替代品。据报道,已有一些茶区利用茶树修剪枝进行食用菌生产,并取得了良好的经济效益。福建安溪县、浙江的淳安县等地利用废弃茶树枝条加工成的茶树屑栽培香菇、杏鲍菇、黑木耳等食用菌。据了解利用茶树枝丫材料生产食用菌,每筒菌种成本仅 0.5 元,而生菇市场价每千克 20 元,经济效益显著。该技术的成功突破不仅推动了食用菌生产的发展,而且充分利用了茶叶生产过程中所产生的废弃枝叶和枝丫,提高了茶园经济附加值。

1.3.3 利用茶树修剪枝加工生产茶片、茶末

将内含物丰富的茶树修剪枝回收后通过一定的加工工艺做成茶片、茶末是近年来对茶树修剪枝再利用的一个新途径。浙江省淳安县临岐镇叶家畈村茶厂从 2003 年开始利用茶树修剪枝生产茶片、茶末,取得了很好的经济效益。据研究,修剪枝大概经过如下的摊青、杀青、揉捻、初烘、辉炒、切片、拣梗和包装 8 个加工步骤,最后成为符合要求的茶片和茶末,进而可以加工成袋泡茶或者作为天然食品添加剂加以利用。茶片和茶末的开发使得昔日废弃的茶树修剪枝发挥了最佳的经济效益[8]。

此外,近年来,食品加工企业充分利用茶树枝中内含物丰富且安全可食用的特性,将茶树枝运用到食品加工中去。松阳诚天和食品公司结合当地茶叶生产实际,利用茶农修剪下来的茶叶老枝条,熏制出含带茶叶清香的茶叶熏火腿,变传统的"喝茶"为"吃茶",从而提高产品的附加值,同时延伸了茶叶产业链。类似的还可利用茶叶枝条熏鸡、烤鸭等,形成具有茶香特色风味的卤制品。

2 茶树修剪枝再利用存在的问题

尽管目前生产上对茶树修剪枝的再利用随着茶园面积的不断扩大和修剪枝量的不断增加而相继展开,但总体说来,对茶树修剪枝的再利用还处于起步阶段,生产上还未真正重视起来,导致其利用规模小、数量有限,茶树修剪枝再利用上存在一些问题。

2.1 可用作插穗的修剪枝量小,且扦插易受时间影响

茶叶生产中对无性系良种穗条的量有着较大需求,采穗母本园里的茶树由于生长条件一致、管理方法具有针对性,所剪插穗质量较好且均匀整齐,从而使得扦插育苗省时省力、节约成本。然而利用修剪枝作为育苗扦插枝条,因修剪枝生长各异、性状不一、质量参差不齐,只有部分健壮修剪枝可以满足插穗枝条

要求,因此利用修剪枝作为育苗插穗,其数量有限,只能部分替代母本园所剪插穗;加之拣剔这些满足插穗要求的修剪枝势必会带来额外的人工成本和经济成本,从而使得修剪枝作为插穗利用率较低。此外,实际生产中,由于穗条扦插对于天气、插穗、场地等条件有着较高要求,修剪下来的枝条本身受天气影响大,容易失水,又没有较好的收集保存办法;用作插穗的修剪枝还要进行药剂处理等工序后才可有效扦插,因而使得作为插穗的修剪枝难以及时扦插,其成活率难以得到保证。

2.2 修剪枝还田方式单一,还田效果不显著

作为茶树修剪枝最为广泛的利用途径,修剪枝还田可以起到改善土壤理化性质和生物性状,提高土壤的保肥和供肥能力。然而,长久以来,人们对于修剪枝的还田利用方式单一,操作简单,通常只是将修剪下来的茶树枝条从茶丛中取下,然后铺于茶行间。这样操作固然能够发挥其肥效作用,但由于修剪枝只是简单的覆盖于行间道上,其表层枝叶的腐烂程度不如里层,加上行间土壤只有表层与修剪枝接触,从而使得修剪枝不能完全发挥其增加土壤肥力的作用;此外,对于那些茶园农药使用严重或者病虫害发生较严重的茶树,其修剪枝叶往往农药残留严重或者病虫枝叶较多,如果不将其清理出园,简单的放置于茶行间,就会使农残严重的枝叶随着腐烂程度的加剧,致使水溶性农药渗入土壤乃至被根系所吸收,从而对茶树造成毒害;同时,那些带病虫害的枝条也会因修剪枝的转移而使病虫危害范围进一步扩大,然而由于这些细节很容易被忽视,使得修剪枝还田缺乏精细作业,今后应加强对这种还田方式的研究,从而使得茶枝还田更加合理有效。另一方面,修剪枝还田处理缺乏系统研究,没有具体的实施方案和统计数字,还田效果不显著。也正是由于缺乏相关数据佐证,致使部分人误以为修剪枝只是茶叶生产废弃物,可利用价值不高,导致茶树修剪枝再利用进一步被边缘化。

2.3 修剪枝作为再加工原料,开发应用力度小

目前,尽管一些企业利用茶树修剪枝进行再加工利用,但均处于试研制阶段,他们或者按照茶鲜叶加工工艺,将修剪的嫩枝加工成茶片、茶末,作为食品添加剂和畜牧饲料添加剂使用;或者参考其他作物枝条利用途径(如桑叶枝条),将部分较粗老的修剪枝加工成食用菌培养基辅料。这些修剪枝加工所得到的成品和半成品因程序复杂、辅助性强、用量小等特点而容易被人们所忽视,从而使得其开发应用力度小,这也是阻碍茶树修剪枝进一步推广利用的障碍之

一。深加工提取方面,虽然修剪枝内含物丰富,可以作为茶多酚、氨基酸、咖啡因等有效成分提取原料,但由于其主要利用的是修剪枝上面的叶子部分,由此带来的收集处理,尤其是去枝工序较繁琐,加之一些加工设备、酶制剂等价格昂贵,综合人工成本和加工成本考虑,从而使得这部分开发利用目前还停留在实验阶段。以茶树修剪叶为原料,用水或乙醇萃取,经离心过滤,再喷雾干燥或冷冻干燥后可以得到茶叶天然色素[9]。刘菁等利用外源酶对废弃老叶进行处理以制取其风味物质[1]。李言等人利用生物酶技术对茶树修剪叶中的蛋白质进行提取,同时对提取出的蛋白质进行了氨基酸分析和简单的功能研究[2]。由此可见,鲜有的相关研究报道也多侧重于对修剪枝上面叶子部分的再利用。

3 对策与注意事项

针对茶树修剪枝在生产利用中出现的以上问题,有必要制定相应的对策措施,从而确保茶树修剪枝能够得到最大化的开发利用。

3.1 加强对茶树修剪枝的分类再利用

所谓茶树修剪枝的分类再利用是指依据不同时期的茶树修剪枝特点或者茶叶生产需要,对茶树修剪枝进行分类再利用。茶树各修剪枝中的健壮枝条,因其具有插穗的特点,可以对这部分枝条进行简单拣剔处理,用作插穗进行无性繁殖育苗用,尤其是对一些珍稀名贵品种相当可取,其具有一定的经济效益。名优茶热下的茶叶生产往往只采一季春茶,并且要求春茶早上市,这种背景下所进行的茶树修剪通常采用重修剪和台刈,修剪枝叶多数较长且较粗老,对这部分修剪枝进行还田处理不仅省时省力,而且因修剪枝经过长时间的腐烂,能增加土壤有机质和其他矿物质养分,改善土壤的理化性质和生物性状提高土壤的保肥和供肥能力,有利于提高下季的春茶产量和品质。而对于多季生产茶园,由于每季茶采摘结束后基本都要进行一定程度的轻、深修剪,由此而产生的茶树修剪枝一方面很难满足作为插穗的条件进行无性繁殖,另一方面如果还田以作肥用,短期内很难达到预期效果,因此,对这部分茶树修剪枝进行回收以作再加工再利用原料,将会达到很好的资源综合利用效果。

3.2 对满足插穗要求的修剪枝实行就近就地剪插

生产上,由于茶树修剪枝条量大且质量参差不齐,使得修剪枝中可用于育苗扦插材料的枝条,受到天气、园地、人工等因素的影响,不能及时、有效的插到园地中,从而难以保证其成活率。针对这种因时间差异带来的不便,如果能够有意识地将准备新建的茶园园地开辟在这些修剪茶园附近,待到茶树进行修剪时,其修剪枝就能够就近、就地进行插穗。同时,为了进一步减小气候条件对插穗的影响,还可以建立遮荫设施,它可以有效防止作为插穗的修剪枝和土壤水分过快蒸发,保证修剪枝具有较高成活率和正常生长发育。

3.3 对茶枝还田实施精细作业和系统研究

茶枝还田能够起到一定的肥料作用,今后应加强修剪枝还田精细作业。首先要对各个茶树生长势、茶树病虫害情况、茶园用药情况等进行系统地了解,在修剪茶树的时候,有意识的观察茶树病虫害情况,对于发病较严重的枝条或者病虫危害中心的枝叶单独进行剪除,最好一并带出茶园,焚烧后再还田。其次,为了使修剪枝与土壤能够很好地相互接触,从而更加充分的发挥其增加土壤肥力的效果,在将修剪枝从茶丛上取下铺于茶行间前,对土壤进行一次清草、施肥的同时,对行间土壤进行深翻显得很有必要,根据需要,也可将土壤与修剪枝进行适时翻拌互混,以利于其肥效发挥。同时,对于使用化学农药比较重的茶园,建议将修剪枝平铺于茶行间以作覆盖物防寒保湿用为主。除此之外,还应加强对茶树修剪枝还田处理的系统研究,在制定具体实施方案基础上,做好相关观察、采集、统计、分析等工作,为实际生产提供可靠的参考依据。

3.4 加强对修剪枝的开发应用

与茶树鲜叶比较,茶树修剪枝具有质次、处理繁琐、加工不便等劣势特点,因此如果完全按照鲜叶加工方式进行加工,不仅存在加工成本较高的问题,更主要的是其加工成品亦没有丝毫竞争优势。但鉴于茶树修剪枝量大、嫩度尚可、茶风味物质丰富等特点,应在参照或改进茶鲜叶加工工艺的基础上,对修剪枝加工预期成品重新进行产品定位,根据目前生产上对这部分枝条的利用现状,一些辅料产品,如前面提到的家禽饲料添加剂、食用菌培养基质添加剂、茶片茶末以及其他一些添加剂等都是修剪枝再加工预期成品的良好选择。这些辅料产品也是目前茶树修剪枝再利用的主流方向之一,应加强开发利用,使其尽早实现商品化。与此同时,为了加强对这部分茶树修剪枝叶内含风味物质的再利用,鉴于其去枝工序比较繁杂,可以在原料拣剔过程中尽量选取枝叶较嫩的枝条进行混合加工,这样既可以保证原料茶风味物质的较高含量,同时还能略去去枝工序,最大限度地降低成本。再者,生产过程中,还应充分利用"公司＋合作社＋基地＋农户"的合作模式,由公司组织专业技术人员,从茶叶种植、修剪、采摘、喷药等方面对茶农田间管理进行科学系统的指导,确保修剪枝安全无公害收集再利用。此外,对于新近发展起来的利用茶树修剪枝熏制特色食品,因其具有操作简单、优势明显等鲜明特点,今后应加强宣传,推而广之。

参考文献

[1] 刘菁,陶文沂. 酶法制取茶树废弃老叶中的茶风味物质[J]. 食品与发酵工业,2007,33(3):157－160.

[2] 李言,章海燕,胡敏等. 酶法提取茶树修剪叶中的蛋白及其性质研究[J]. 2011,32(3):127－130.

[3] 梁月荣,赵启泉,陆建良等. 茶树修剪叶和不同氮肥对土壤 pH 和活性铝含量的影响[J]. 茶叶,2000,26(4):205－208.

[4] 李荣林,周维仁,申爱华. 茶叶及其提取物在畜牧和饲料中的应用[J]. 粮食与饲料工业,2002(7):30－31.

[5] 佐野满绍,佐佐木清隆. 茶叶粉末喂鸡对鸡肉鲜嫩度的影响[J]. Journal of the Food Hygienic Society of Japan,1996,52(7):823.

[6] 赖建辉,刘中秋,田超. 乌龙茶用作奶牛饲料添加剂的初步效果[J]. 茶叶科学,1994,14(1):258.

[7] 张广强,张希可. 茶树修剪后的枝叶在散养土鸡中的应用[J]. 当代畜牧,2011(7):28－29.

[8] 郑平汉. 茶树修剪枝叶的利用生产技术初探[J]. 茶业通报,2007,29(4):171－172.

[9] 张利平. 茶叶副产物的开发利用综述[J]. 茶叶,1999(1):5－6.

[10] 童启庆. 茶树栽培学[M]. 3版. 北京:中国农业出版社,1999.

钙离子对绿茶茶汤沉淀形成的影响

许勇泉　钟小玉　陈根生　邓余良　袁海波　尹军峰

国家茶产业工程技术研究中心，中国农业科学院茶叶研究所，中国 浙江

摘 要：研究了钙离子对绿茶茶汤主要理化成分浸出及沉淀形成的影响。研究结果表明，随着钙离子质量浓度的提高，绿茶浸提液茶多酚、蛋白质、有机酸等主要化学成分含量略有下降；钙离子质量浓度40mg/L，在一定程度降低茶叶化学成分浸出的前提下能够明显促进茶汤沉淀的形成，钙离子明显促进了茶多酚、咖啡因、蛋白质及草酸等参与绿茶茶汤沉淀的形成；可逆沉淀与不可逆沉淀化学组分具有显著差异，钙等大量金属元素参与不可逆沉淀的形成。

关键词：钙离子；沉淀形成；绿茶茶汤

Effect of Calcium on Tea Sediment Formation in Green Tea Infusion

XU Yong-quan　ZHONG Xiao-yu　CHEN Gen-sheng
DENG Yu-liang　YUAN Hai-bo　YIN Jun-feng

Engineering Research Center for Tea Processing, Tea Research Institute, Chinese Academy of Agricultural Sciences, Zhejiang, China

Abstract: The effect of calcium on the main chemical components content extracting and tea sediment formation of green tea infusion was investigated. The results showed that, with the increasing of calcium concentration, the content of main chemical components such as polyphenols, protein and organic acids decrease. Under the treatment of 40mg/L calcium, the contents of main chemical components extracted from tea leaves decreased and the tea sediment increased. Calcium promoted the formation of tea sediment in green tea infusion. There was great difference of chemical constituents in reversible and irreversible tea sediment. Large amount of minerals including calcium participate the formation of irreversible tea sediment.

Keywords: calcium; tea sediment formation; green tea infusion

茶饮料是近年来国际上发展最快的健康饮料之一,在茶产业中占有举足轻重的地位。2011年我国茶饮料总产量超过1100万t,产值超过500亿元[1]。目前茶类饮料加工中主要面临香低、色变和沉淀等三大关键问题。由于茶汤沉淀形成过程复杂,迄今缺乏明确、系统的形成机理和满意的解决方法,茶汤沉淀不仅大量损失茶叶功能性成分,降低其保健价值,而且会造成茶汤外观品质和内质风味的明显劣变,然而相关研究极少。

冷后浑(又称茶乳酪)是指茶汤在冷却后形成的浑浊现象,是茶汤体系沉淀产生的前期过程。茶汤浓度、浸提温度、pH值及不同化学组成是茶乳酪产生的主要影响因素[2-5]。许勇泉等研究发现咖啡因、蛋白质、茶多酚、黄酮化合物及钙离子等化学组分更容易参与绿茶冷后浑的形成,其中咖啡因与酯型儿茶素是绿茶冷后浑形成的关键化学成分[5]。钙离子是茶叶中含量较高且具有典型代表性的金属离子,是影响和参与茶汤沉淀形成的重要金属离子。郭炳莹等[6]通过研究茶汤组分与金属离子的络合性能发现,Ca^{2+}、Mg^{2+}、Mn^{2+}、Mn^{7+}、Ag^{+}、Fe^{2+}、Fe^{3+}、Pb^{2+}等22种金属离子可与茶汤组分发生络合,络合物的溶解度较低;Ca^{2+}、Ag^{+}、Fe^{2+}、Fe^{3+}、Hg^{2+}等10种金属离子可与茶多酚发生络合。

钙离子是茶叶中含量较高且具有典型代表性的金属离子,对茶叶风味品质及功能成分影响较大。许勇泉等[7]研究表明,不同钙离子浓度对茶叶中的茶多酚、儿茶素、蛋白质及金属离子等品质成分的浸出有一定的影响,而对氨基酸、黄酮化合物及咖啡因等的浸出影响不显著;40mg/L的钙离子对茶叶中的草酸、奎尼酸、苹果酸及柠檬酸等的浸出有显著影响,其浸出量明显下降[8]。方元超等[9]研究表明,钙离子与茶汤组分生成的络合物,其溶解度及稳定性随茶汤pH值升高而下降,钙络合沉淀中的主要茶汤组分是茶多酚,其中以酯型儿茶素含量最高,在pH值为9.5时,酯型儿茶素(EGCG和ECG)几乎全部被钙离子络合而沉淀;在茶汤组分——钙络合沉淀中,除茶多酚外还含有果胶、水溶性蛋白质和咖啡因,此外还有少量的氨基酸、水溶性碳水化合物等[6]。本试验通过研究钙离子对绿茶茶汤沉淀形成的影响来揭示茶汤沉淀的形成机理,以期为抑制茶饮料沉淀形成,提高茶饮料风味、品质、稳定性等提供理论基础。

1 材料与方法

(1)茶叶原料。本实验用茶叶原料分别采购自武义汤记高山茶业有限公司和龙游茗达茶业有限公司。无水$CaCl_2$:分析纯,分子量110.99,兰溪市华盛化工试剂有限公司;水:娃哈哈桶装纯净水。

（2）可逆沉淀与不可逆沉淀。浸提茶汤经低温冷却后产生冷后浑，冷后浑经过一定时间的沉降后形成沉淀，部分沉淀在一定温度（60℃）下处理后可重新溶解形成茶汤，即为可逆沉淀，而不能重新溶解的沉淀即为不可逆沉淀。

（3）茶汤浸提及沉淀分离。分别称取1.5g茶叶置于150mL不同钙离子浓度（0、5mg/L、10mg/L、20mg/L、40mg/L、100mg/L）的水中，在75℃条件下浸提30min，过滤，分离茶汤和茶渣，迅速冷却至室温，每个处理3个重复。

茶叶（炒青与烘青）→浸提（茶水比10g：1000mL，时间30min，温度80℃），分别在钙离子浓度为0和40mg/L水中浸提→粗滤（双层，300目滤布）→冷却（水冷，至室温）→精滤（离心4000r/min，10℃，15min）→分析茶汤理化成分→灌装（于50mL离心管中灌装40mL，每个处理灌装6支）→4℃低温冷藏24h→离心分离沉淀（8000r/min，4℃，15min）→倒出上清液，得底部沉淀（茶汤总沉淀）→添加60℃纯水到离心管中至40mL刻度→60℃水浴搅拌30min，溶解可逆沉淀→冷却至室温→离心分离沉淀（8000r/min，4℃，15min）→倒出上清液，定容至40mL，获得可逆沉淀溶液，底部沉淀物为不可逆沉淀

以上每个处理制备6支离心管茶汤样品（重复），其中3支离心管茶汤样品用于分析茶汤总沉淀量，3支离心管茶汤样品用于分析可逆沉淀理化成分含量和不可逆沉淀量。

（4）茶汤沉淀分离及测定。茶汤经离心后，倒出上清液，得底部沉淀，用纯水将沉淀洗出至蒸发皿中，先蒸干，然后于105℃下烘干，冷却，称量，计算得沉淀量。

（5）茶汤与沉淀中化学成分分析。茶汤固形物、茶多酚含量测定[10]。固形物含量测定采用蒸发皿蒸干、105℃烘干测定，茶多酚含量采用酒石酸亚铁比色法。

茶汤黄酮化合物、蛋白质含量测定。黄酮化合物总量测定采用三氯化铝比色法[10]。蛋白质测定采用Bradford蛋白浓度测定试剂盒方法。

儿茶素及咖啡因含量测定。用0.22μm微孔滤膜（有机膜）过滤，滤液待检测。Waters E2695高效液相色谱仪（Waters公司生产），VWD检测器；色谱柱：ZORBAX SB-C18 ODS，5μm，4.6mm×250mm；A为0.5%甲酸，流动相B为100%乙腈，流速1mL/min，柱温40℃，检测波长280nm；进样量：10μL，梯度洗脱，流动相B在16min内由6.5%线性梯度变化到25%，25min回到初始状态，平衡10min。

有机酸含量测定。采用高效液相色谱法，色谱柱为Sun Fire™ C18（15μm，4.6mm×250mm）柱；流动相为磷酸二氢钾（用磷酸调pH值至2.8），流速为

0.6mL/min,共 30min,UV 检测波长为 210nm,柱温为 30℃。

金属元素及 C、N 元素含量测定。金属离子含量检测采用 ICP-OES 检测,分析条件如下:检测器:CID;低波长最大间隔时间:15s;高波长最大间隔时间:5s;喷雾器压力:193kPa;泵速率:100r/min;辅助气体:中速(1L/min);RF 功率:1150W。并在每种金属有其特定的分析波长下,测定各标准液、试剂空白及样品,计算样品中所测元素的含量。C、N 元素含量测定采用德国的碳氮元素分析仪 Vario MAX 测定。

(6)数据分析。样品均设 3 次重复。表或图中数据为平均值±标准偏差,方差分析采用 SPSS13.0 软件进行运算,处理间平均值的比较用最小显著差数法(LSD)。

2 结果与分析

2.1 不同钙离子浓度对绿茶理化成分浸出及浊度的影响

茶叶经不同质量浓度钙离子水浸提后,茶汤理化指标发生明显变化(表1)。随着钙离子质量浓度的增高,茶汤浊度呈逐渐增加趋势(图1),钙离子质量浓度 0~40mg/L 区间,茶汤浊度曲线斜率较大,增加迅速,而当钙离子质量浓度大于 40mg/L 后,茶汤浊度曲线斜率较小,增加变缓。茶汤浊度增加,可能与钙离子促进冷后浑形成有关。

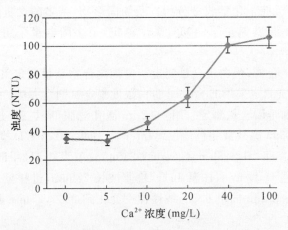

图 1 不同钙离子浓度绿茶浸提液浊度变化曲线

茶叶经不同质量浓度钙离子水浸提后,茶汤主要风味化学成分含量发生明显变化(表1)。随着钙离子质量浓度的增高,茶汤中茶多酚、蛋白质及草酸、苹果酸、柠檬酸等有机酸含量略有下降,而黄酮类化合物与咖啡因含量变化差异不显著。其中,蛋白质、苹果酸和柠檬酸含量在钙离子质量浓度达到5mg/L时就明显下降,而钙离子含量的进一步增加对其影响不显著;茶多酚与草酸含量是在钙离子质量浓度达到40mg/L时才开始出现显著下降,而在低浓度时差异不显著。钙离子主要通过两种途径影响绿茶浸出液品质化学成分含量,一种途径是直接影响茶品质化学成分的浸出,另一种途径是通过促进浸出液中冷后浑或沉淀的产生从而降低品质化学成分含量。

表1　钙离子质量浓度对茶汤主要化学成分的影响　　　（单位:mg/L）

化学成分	Ca^{2+} 质量浓度				
	0	10	20	40	100
茶多酚	2.10 ± 0.02^a	2.08 ± 0.02^a	2.08 ± 0.02^a	2.01 ± 0.01^b	1.97 ± 0.02^c
蛋白质	0.29 ± 0.03^a	0.23 ± 0.02^b	0.23 ± 0.02^b	0.23 ± 0.01^b	0.22 ± 0.01^b
黄酮化合物	0.05 ± 0.00^a	0.050 ± 0.00^a	0.05 ± 0.00^a	0.05 ± 0.00^a	0.05 ± 0.00^a
咖啡因	0.30 ± 0.00^a	0.30 ± 0.00^a	0.31 ± 0.01^a	0.30 ± 0.00^a	0.30 ± 0.00^a
草酸	0.058 ± 0.001^a	0.058 ± 0.001^a	0.056 ± 0.001^a	0.049 ± 0.004^b	0.038 ± 0.003^c
苹果酸	0.082 ± 0.001^a	0.074 ± 0.003^b	0.076 ± 0.001^b	0.074 ± 0.002^b	0.075 ± 0.001^b
柠檬酸	0.304 ± 0.08^a	0.262 ± 0.005^b	0.274 ± 0.006^{bc}	0.273 ± 0.006^{bc}	0.279 ± 0.005^c

注:同一列不同字母表示经LSD法检验在0.05水平上差异显著。

2.2　钙离子对绿茶茶汤沉淀产生的影响

钙离子(40mg/L)处理后,绿茶茶汤沉淀含量显著增加(表2)。以烘青和炒青为原料,选择40mg/L作为钙离子处理浓度,经浸提后,茶汤固形物含量及茶多酚、咖啡因、蛋白质及儿茶素等主要品质化学成分含量略有下降(表3),但大部分差异不显著,而茶汤经冷藏后产生的沉淀量有明显增加。茶汤中草酸含量受钙离子浓度影响较大,经钙离子处理后都有明显下降,钙离子可能容易与草酸等有机酸结合形成沉淀。茶汤沉淀包括可逆与不可逆沉淀都有明显增加。茶汤沉淀量主要决定于茶汤中的化学成分含量[5],这些化学成分包括茶多酚、咖啡因、儿茶素、蛋白质等主要参与茶汤沉淀形成的化学成分。该试验中烘青浸提液中主要化学成分含量明显高于炒青浸提液,其产生的沉淀量也明显高于炒青浸提液,但是两者经钙离子处理后沉淀量都明显增加,可见钙离子具有促进茶汤沉淀产生的作用。

表2 钙离子(40mg/L)对茶汤沉淀含量的影响

处理	原料	钙离子(mg/mL)	茶汤pH	总沉淀(mg/mL)	不可逆沉淀(mg/mL)	可逆沉淀(mg/mL)
1	烘青	0	5.75±0.01^a	4.64±0.22^a	0.18±0.01^a	4.46±0.18^a
2	烘青	40	5.73±0.01^a	5.11±0.08^b	0.27±0.04^b	4.84±0.07^b
3	炒青	0	5.61±0.01^b	3.00±0.11^c	0.14±0.01^c	2.86±0.09^c
4	炒青	40	5.56±0.01^b	3.15±0.02^d	0.18±0.01^d	2.97±0.01^d

注:数据为3个重复平均值;同一列数据中标注相同字母的差异不显著($P \geqslant 0.05$)。

表3 钙离子(40mg/L)对浸出茶汤主要化学成分含量的影响　　mg/mL

处理	固形物	茶多酚	蛋白质	咖啡因	EGC	EC	EGCG	ECG	草酸
1	29.17^a	14.33^a	1.63^a	1.61^a	1.40^a	0.95^a	1.99^a	0.64^a	0.22^a
2	25.33^b	10.42^b	1.52^b	1.38^b	0.79^b	0.77^b	1.40^b	0.37^b	0.18^b
3	28.96^c	14.22^a	1.33^c	1.62^a	1.36^c	0.91^a	1.95^c	0.63^a	0.23^a
4	25.68^b	10.60^b	1.10^d	1.34^c	0.74^d	0.76^b	1.33^b	0.36^b	0.15^c

注:数据为3个重复平均值;同一列数据中标注相同字母的差异不显著($P \geqslant 0.05$)。

2.3 绿茶茶汤沉淀主要化学成分

通过沉淀产生前后上清液中化学成分含量的比较,可以发现钙离子明显促进了茶多酚等主要化学成分参与沉淀的形成(表4)。茶多酚、黄酮化合物、咖啡因、蛋白质、儿茶素及草酸等相对对照参与沉淀形成的量都有差异,可见钙离子确实可以促进茶汤中主要化学成分参与沉淀的形成。但是,由于不可逆沉淀不能够复溶,因此不能明确钙离子在可逆沉淀及不可逆沉淀形成中的作用。通过元素分析,我们发现大量金属离子参与到不可逆沉淀形成中,包括Al、Ca、Cu、Fe、Ga、Mg、Mn、Zn等,在不可逆沉淀中的含量大大高于上清液及可逆沉淀,其中特别是Ca、Ga、Mg与Mn元素的含量分别超高1%,即10000mg/kg(表5)。而可逆沉淀中金属元素的含量与上清液中的含量大都差异不显著,只有Cu、Ga、Mg、Mn、Na、Zn元素含量显著高于上清液中的含量。不可逆沉淀中金属元素含量超过11%,而上清液及可逆沉淀中的含量都只有6%左右(表5),但是上清液及可逆沉淀中的C和N元素含量都大大高于不可逆沉淀。可见,茶汤中可逆沉淀与不可逆沉淀的化学组成具有很大差异。

表 4 参与茶汤沉淀的化学组分含量　　　　　　　　　　　　　mg/mL

处理	茶多酚	黄酮化合物	咖啡因	蛋白质	EGCG	GCG	ECG	草酸
1	4.18[a]	0.05[a]	0.39[a]	0.09[a]	0.83[a]	0.09[a]	0.23[a]	0.05[a]
2	4.63[b]	0.07[b]	0.51[b]	0.11[b]	0.89[b]	0.10[b]	0.24[b]	0.08[b]
3	2.26[c]	0.04[a]	0.18[c]	0.07[c]	0.34[c]	0.05[c]	0.09[c]	0.04[c]
4	2.71[d]	0.07[b]	0.23[d]	0.08[c]	0.39[c]	0.06[c]	0.09[c]	0.10[d]

注:数据为3个重复平均值;同一列数据中标注相同字母的差异不显著($P \geqslant 0.05$).

表 5 绿茶茶汤沉淀组分差异分析

序号	元素	上清液	可逆沉淀	不可逆沉淀
1	Al(mg/kg)	1139±22[a]	1132±29[a]	2541±16[b]
2	Ca(mg/kg)	495±43[a]	595±173[a]	11625±177[b]
3	Cu(mg/kg)	7±0[a]	45±4[b]	646±168[c]
4	Fe(mg/kg)	102±49[a]	164±27[a]	495±18[b]
5	Ga(mg/kg)	267±10[a]	457±4[b]	14420±240[c]
6	K(mg/kg)	54325±431[a]	49970±1711[b]	16735±375[c]
7	Mg(mg/kg)	2138±16[a]	2783±168[b]	43585±1068[c]
8	Mn(mg/kg)	468±2[a]	810±36[b]	20635±544[c]
9	Na(mg/kg)	1013±209[a]	1671±28[b]	2009±506[c]
10	Zn(mg/kg)	74±39[a]	155±43[a]	3263±768[b]
11	P(mg/kg)	4383±127[a]	4739±209[a]	3026±508[b]
12	S(mg/kg)	10375±248[a]	8954±356[b]	3672±129[c]
13	C(%)	43.51±0.03[a]	44.45±0.04[b]	31.87±0.06[c]
14	N(%)	3.41±0.01[a]	3.83±0.05[b]	2.78±0.02[c]

注:数据为3个重复平均值;同一行数据中标注相同字母的差异不显著($P \geqslant 0.05$).

3 讨 论

钙离子主要通过两种途径影响茶叶浸出液品质化学成分含量,一种途径是直接影响品质化学成分的浸出,另一种途径是通过促进浸出液冷后浑或沉淀的产生从而降低品质化学成分含量。Spiro 等[11]研究发现,高钙水质对茶黄素和咖啡因的浸出率显著低于超纯水；Mossion 等[12]报道高矿物质水会影响茶叶中Al、总有机碳和茶多酚等的浸出，他们认为这可能是由于钙被茶叶吸收，钙与茶叶细胞壁上的果胶质结合从而影响茶叶中有效物质的浸出。另外，钙离子具有促进茶乳酪产生的能力，主要是与多酚类、咖啡因或蛋白质等的结合，Jobstl 等[13]研究认为钙是参与茶乳酪形成的重要因素，茶叶中的草酸容易与钙结合形成沉淀。钟小玉[8]研究认为，钙离子与茶叶中草酸、酒石酸等有机酸结合产生颗粒状沉淀，是导致茶汤沉淀形成的重要原因。本实验明确钙离子影响绿茶品质化学成分的浸出，促进茶汤沉淀的形成。尽管如此，针对钙离子如何参与茶汤沉淀形成还没有明确的报道。

冷后浑主要是由茶多酚－蛋白质、茶多酚－咖啡因、蛋白质－咖啡因、茶多酚－蛋白质－咖啡因等通过氢键、盐键、酯键及疏水作用等共同作用形成的[14]，冷后浑经沉积产生沉淀，其中大部分茶汤沉淀是可逆的，只有少部分沉淀是不可逆的，不可逆沉淀可能主要是通过共价键形成的，而部分共价键难以通过加热重新解开。本研究团队前期试验表明[15]，不可逆沉淀与可逆沉淀化学组成具有显著差异，且首次发现大量金属元素（超过 11%）参与不可逆沉淀的形成。以钙离子为代表的金属离子是形成不可逆沉淀的关键因素，因此研究并明确金属离子参与茶汤沉淀，特别是不可逆沉淀的形成是探明不可逆沉淀形成机理及其调控基础的关键点。

参考文献

[1] 赵亚利. 中国茶与咖啡饮料发展趋势探讨[J]. 中国饮料, 2012, 6: 100－102.

[2] Liang Y R, Lu J L, Zhang L Y. Comparative study of cream in infusions of black tea and green tea [Camellia sinensis (L.) O. Kuntze][J]. Journal of Food Science and Technology, 2002: 37, 627－634.

[3] Liang Y R, Xu Y R. Effect of extraction temperature on cream and extractability of black tea [Camellia sinensis (L.) O. Kuntze][J]. International Journal of Food Science and Technology, 2003, 38: 37－45.

[4] Liang Y R, Xu Y R. Effect of pH on cream particle formation and solids extraction yield of black tea [J]. Food Chemistry, 2001, 74: 155－160.

[5] Yin J F, Xu Y Q, Yuan H B, et al. Cream formation and main chemical components of green tea infusions processed from different parts of new shoots [J]. Food Chemistry, 2009, 114: 665－670.

[6] 郭炳莹, 程启坤. 茶汤组分与金属离子的络合性能[J]. 茶叶科学, 1991,

11(2): 139—144.

[7] 许勇泉,陈根生,钟小玉等. 钙离子对绿茶浸提茶汤理化与感官品质的影响[J]. 茶叶科学, 2011, 31(3): 230—236.

[8] 钟小玉. 钙离子对茶汤品质成分及混浊产生影响的研究[D]. 北京: 中国农业科学院, 2012.

[9] 方元超,赵晋府. 茶饮料生产技术[M]. 北京: 中国轻工业出版社, 2001. 32—34.

[10] 钟萝. 茶叶品质理化分析[M]. 上海: 上海科学技术出版社, 1989. 250—350.

[11] Spiro M, Jaganyi D. What causes scum on tea? [J]. Nature, 1993, 364: 581.

[12] Mossion A, Potin-Gautier M, Delerue S, et al. Effect of water composition on aluminium, calcium and organic carbon extraction in tea infusions [J]. Food Chemistry, 2008, 106: 1467—1475.

[13] Jobstl E, Fairclough J P A, Davies A P, et al. Creaming in black tea [J]. Journal of Agriculture and Food Chemistry, 2005, 53: 7997—8002.

[14] Charlton A J, Davis A L, Jones D P, et al. The self-association of the black tea polyphenol theaflavin and its complexation with caffeine [J]. Journal of the Chemical Society-Perkin Transactions, 2000, 2: 317—322.

[15] Xu Y Q, Chen G S, Wang Q S, et al. Irreversible sediment formation in green tea infusions [J]. Journal of Food Science, 2012, 3: 298—302.

附: 基金项目: 浙江省自然科学基金(R3090394)及国家基金(31070615)。

基于响应曲面法的茶黄素发酵工艺优化

孔俊豪 杨秀芳 张士康 涂云飞

中华全国供销合作总社杭州茶叶研究院,中国 浙江

摘 要：采用响应面分析法对影响茶黄素合成累积量的过程变量进行了系统筛选和优化。首先通过Plackett-Burman方法对相关影响因素的效应进行了评价,并确定酶量、反应温度、体系pH为显著性因素;其次以最陡爬坡路径逼近最大累积浓度区域;最后由中心组合实验及响应面分析确定了主要影响因素的最佳催化条件。得到了茶黄素合成的优化工艺条件：温度＝29.5℃,pH＝4.9,加酶量＝14400U,在优化反应条件下茶黄素合成浓度可达15.1mg/mL。

关键词：茶黄素；发酵；响应面优化

Fermentation Technique Optimization of Theaflavins Based on Response Surface Methodology

KONG Jun-hao YANG Xiu-fang ZHANG Shi-kang TU Yun-fei

Hangzhou Tea Research Institute, CHINA COOP, Zhejiang, China

Abstract: The process variables that had effect on the accumulation of theaflavins were systematically selected and optimized by response surface analysis. The effects of related factors were evaluated firstly by the Plackett-Burman method, and the amount of enzyme, reaction temperature, pH have been identified as significant factors; the maximum cumulative concentration regions of theaflavins were obtained through steepest ascent optimization algorithm, and the main factors of the optimal catalytic conditions were established by the central composite and response surface analysis. The optimal conditions were as follows: temperature=29.5℃, pH=4.9, enzyme concentration=14400U, under conditions of that, the accumulation of theaflavins concentration could reach to 15.1mg/mL.

Keywords: theaflavins; fermentation; response optimization

茶黄素是从红茶中发现的以大芳香环为主的高活性酚类化合物[1]，医疗功效较强，同时具有抗氧化、抑制肿瘤、抗病毒等多种功能[2-7]，被誉为有益人体健康的茶叶"软黄金"，在国外已被用于保健产品和功能食品的开发，市场前景可期。

红茶中的茶黄素主要是儿茶素类物质在多酚氧化酶（PPO）、过氧化物酶（POD）等氧化酶类的连续催化下转化形成[8-10]。受原料和季节限制，目前茶黄素的生产主要依靠体外模拟氧化法[11-14]。儿茶素体外酶促氧化是以PPO催化为主导因子的复杂动态发酵过程，其涉及影响因子较多（物理变量、化学变量以及物理化学变量等）[15-16]，这些宏观变量由于酶促反应的耦合性和时变性，对茶黄素合成的影响在统计学上无法忽视。

本研究通过Plackett-Burman试验、最速上升法、Central-Composite组合设计等响应分析，对影响茶黄素累积量的诸因素进行筛选评价和重点考察，验证了非均相催化-过程分离工艺的可行性，为工艺放大及中试生产提供技术依据。

1 材料与方法

(1) 材料与试剂。材料：茶多酚（TP90，购自浙江东方茶叶科技有限公司）；鲜叶：龙井43，杭州翠峰茶叶基地提供；

试剂：乙酸乙酯AR（杭州双林化工试剂厂）、柠檬酸AR（如皋市金陵试剂厂）、磷酸氢二钠AR（湖州湖试化学试剂有限公司）、甲醇AR（衢州巨化试剂有限公司）、蒸馏水。

(2) 仪器与设备。MJ-250BP02A植物组织捣碎机（广州美的生活电器制造有限公司）、3K15低温冷冻离心机（SIGMA，德国）、MDF-U338低温冰箱（大连三洋冷链有限公司）、UV-2102 PC型分光光度计［尤尼柯（上海）仪器有限公司］、H.H.S.11-2K电热恒温水浴锅（上海医疗器械五厂）。

1.3 方　法

(1) 酶液的提取及活性测定[17]。足量称取预处理鲜叶，加入适量柠檬酸-磷酸氢二钠缓冲液，匀浆，离心，收集上清液备用。酶活性以单位体积酶液每分钟OD_{460}增加0.001为一个活性单位（U）。

(2) 茶黄素酶促制备工艺。用柠檬酸-磷酸氢二钠缓冲体系溶解茶多酚，加入鲜叶PPO酶液和乙酸乙酯，通入氧气，设定搅拌转速，于恒温水浴中进行酶促化反应,反应结束后收集酯相层,转溶浓缩干燥,得到茶黄素粗制品。

(3)耦合发酵工艺优化设计。以酯相层茶黄素浓度作为响应目标,通过3个步骤进行优化:①利用Plackett-Burman试验设计方案筛选出对酯相层茶黄素富集量影响较大的因素,各因素水平设置见表1;②用最速上升路径逼近最大响应区域;③利用响应面分析法中的Central-Composite试验方案进行设计,通过试验数据拟合得到二阶响应面模型,最终确定最优试验条件,并进行验证。

表1 Plackett-Burman试验因素设计

水平	温度(℃)	酶量(U)	C虚拟因子	通氧(L/min)	酯/水	F虚拟因子	pH	时间(min)	I虚拟因子	TP(%)	K虚拟因子
低(−)	25	2000		0.5	0.5		4	30		2.5	
高(+)	35	8000		1	1		6	60		5	

(4)茶黄素的HPLC测定方法。参照文献[18]方法进行测定。

2 结果与分析

2.1 显著性因素筛选试验

前期研究均表明影响茶黄素酶促合成的可能因素包括:溶氧量、加酶量、反应温度、pH值、酯/水、反应时间、底物浓度等。本研究对上述7个因素进行同步考察,选用$N=12$的Plackett-Burman设计,并余留4个空项,作误差分析,每个因素取两个水平,试验因素设计如表1。

由表2结果及表3的效应评价可知主要因素为加酶量、pH值、温度和通氧量,酶量和通氧量为正效应,pH值和温度为负效应。但作响应面实验时,考察因素超过3个会使实验次数显著增加,注意到通氧量的效应在这4个因素中最小,因此仍保持通氧量不变。

表2 $N=12$的Plackett-Burman试验结果

序号	反应温度(℃)	酶量(U)	通氧量(L/min)	体系(pH)	酯/水	反应时间(min)	TP(%)	TFs浓度(mg/mL)
1	25	2000	0.5	4	0.5	30	2.5	3.38
2	25	8000	0.5	6	1	30	5	4.39

续表

序号	反应温度(℃)	酶量(U)	通氧量(L/min)	体系(pH)	酯/水	反应时间(min)	TP(%)	TFs浓度(mg/mL)
3	35	8000	1	4	0.5	30	5	5.99
4	35	2000	0.5	4	1	30	5	1.29
5	35	8000	0.5	6	1	60	2.5	3.87
6	25	8000	1	4	1	60	5	11.28
7	25	8000	1	6	0.5	30	2.5	4.07
8	35	2000	1	6	1	30	2.5	2.17
9	35	8000	0.5	4	0.5	60	2.5	5.98
10	35	2000	1	6	0.5	60	5	1.73
11	25	2000	1	4	1	60	5	4.58
12	25	2000	0.5	6	0.5	60	5	1.94

注：试验结果测定3次，取平均值，下同。

表3 各因素效应及重要性评价

因素		水平		效应	t检验
代码	名称	低(−)	高(+)	$|\Sigma(+)-\Sigma(-)|/6$	
A	温度	25	35	−1.43	−1.708
B	酶量	2000	8000	3.42	4.067
(C)	虚拟	—	—	−1.10	−1.310
D	通氧	0.5	1	1.49	1.779
E	相比	0.5	1	0.84	1.004
(F)	虚拟	—	—	−0.09	−0.109
G	pH	4	6	−2.39	−2.843
H	时间	30	60	1.35	1.603
(I)	虚拟	—	—	−1.20	−1.425
J	底物	2.5	5	0.43	0.509
(K)	虚拟	—	—	0.47	0.555

2.2 最速上升试验设计

最速上升法以试验值变化的梯度方向为上升方向,根据各因素效应值的大小确定变化步长,快速逼近最佳值区域[19]。根据因素效应评价结果,3个主因素效应大小的比例以及实际试验结果设计其变化方向及步长进行最速上升路径设计,△A=0.5,△B=1,△G=0.6,试验结果见表4。由表可知,茶黄素转化浓度的最优条件可能在处理4和处理5之间,优化试验以处理4的条件为中心组合试验的中心点。

表4 最速上升试验设计及结果

试验编号	参数变量							TFs浓度 (mg/mL)
	温度(℃)	酶量(U)	pH	时间(min)	通氧量(L/min)	TP(%)	酯水比	
1	35	4000	6.2	60	1.0	5	1/1	2.0
2	33	7000	5.6					6.2
3	31	10000	5.2					9.4
4	29	13000	4.6					11.0
5	27	16000	4					11.5
6	25	19000	3.4					6.3

2.3 耦合发酵条件的优化

根据最速上升试验,响应变量Y值逼近最大合成浓度区域,以处理4为中心点进行中心组合试验设计。为使拟合响应方程具有旋转性和通用性[20],选择中心点试验数为6,星号臂长γ=1.682,各自变量水平见表5,试验设计及结果见表6。

表5 中心组合试验因素变量

因素	水平				
	-1.682	-1	0	1	1.682
A:酶量/U	7960	10000	13000	16000	18040
B:pH	3.6	4	4.6	5	5.4
C:反应温度/℃	25.6	27	29	31	32.4

表6 中心组合试验设计及其结果

试验序号	因素			浓度(g/mL)响应值(Y)
	A:酶量(U)	B:pH	C:反应温度(℃)	
1	0	0	0	11.5
2	1	1	−1	11.8
3	0	0	0	11.5
4	1	1	1	10.88
5	0	0	0	11.6
6	0	0	0	11.5
7	0	0	−1.682	9.5
8	−1.682	0	0	9.9
9	0	1.682	0	9.2
10	1	−1	−1	7.91
11	−1	−1	1	7.7
12	0	−1.682	0	5.7
13	0	0	1.682	11.5
14	−1	1	−1	10.3
15	1.682	0	0	12.0
16	0	0	0	11.5
17	−1	1	1	10.6
18	0	0	0	11.5
19	1	−1	1	6.0
20	−1	−1	−1	9.0

由表7可见,该模型的 P 值为 0.0006($P<0.001$),说明该模型回归显著。

表7 二次多项式模型及其方差分析

项目	变异来源	平方和	自由度	均方	F值	P值
模型		0.651792	9	0.072421	10.10214	0.0006
线性	A	0.004733	1	0.004733	0.660188	0.4354
	B	0.260028	1	0.260028	36.27163	0.0001
	C	0.000314	1	0.000314	0.043777	0.8385
交互项	AB	0.02624	1	0.02624	3.660216	0.0848
	AC	0.004083	1	0.004083	0.569533	0.4678
	BC	0.008485	1	0.008485	1.183515	0.3022
二次项	A^2	0.011485	1	0.011485	1.602093	0.2343
	B^2	0.334182	1	0.334182	46.61553	<0.0001
	C^2	0.027854	1	0.027854	3.885398	0.0770
残差	残差	0.071689	10	0.007169		
失拟	失拟	0.071689	5	0.014338		
纯误差	纯误差	0	5	0		
总和	总和	0.723481	19			

利用 Design Expert 分析软件对实验结果进行分析得到酶量(A)、pH 值(B)、反应温度(C)与茶黄素转化浓度(Y)之间的数学模型回归方程:

$$Y = -17.1348 + 0.005028A + 4.317248B + 0.537474C + 0.007636AB - 0.00075AC + 0.032566BC - 0.00013A^2 - 0.60912B^2 - 0.01099C^2$$

茶黄素转化浓度与酶量、pH 值、反应温度的相关系数 $R^2 = 0.90$,表明该数学模型3个因素对茶黄素转化浓度的影响占 90%,其他因素的影响和误差占 10%。

求解方程得到响应 Y(茶黄素浓度)的极值点参数为:温度=29.5℃,pH=4.9,加酶量=14400U,此条件下预测值为 14.8mg/mL。

2.4 优化反应条件的验证

在优化后的耦合发酵工艺条件下进行验证实验,实测酯相层浓度为 15.1mg/mL,证明该模型能较好的预测反应的实际情况。优化后的茶黄素收集液经浓缩干燥,得到粗品,其溶液的 HPLC 分析如图1所示。

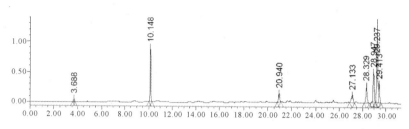

图 1 优化工艺茶黄素溶液的 HPLC 扫描图谱

3 结果与讨论

酶催化反应是工业化生产茶黄素的关键技术,提高发酵反应的定向生物转化率成为研究的难点和重点。因过程反应的时变性和连串反应的发生,影响催化反应平衡及目标产物积累量的因素较为复杂。常规的部分因子试验和常规试验设计对上述过程反应缺乏同步准确的判别和拟合。

Plackett-Burman 设计是一种近饱和的两水平试验设计方法,该法基于非完全平衡原理,能以最少试验次数估计出主要的影响因素,从众多的考察因素中快速、有效地筛选出最为重要的几个因素进行进一步试验研究[21-22]。研究过程组合采用 Plackett-Burman 设计、最速上升法、旋转中心组合试验来准确寻求目标参数的最优区域,通过二阶回归方程得到了优化工艺条件:温度 = 29.5 ℃,pH = 4.9,加酶量 = 14400U,在此反应条件下实际产物浓度为 15.1mg/mL,与预测产物浓度(14.8mg/mL)高度吻合,验证了模型的预测合理性。随着保健医药和天然食品走热,国内外对茶黄素为主体的茶色素需求呈增加趋势,而茶黄素工业化生产仍处于起步阶段,国内总产能仅为 300t 左右。利用自源酶催化合成茶黄素,可同时为夏秋茶资源、茶叶萃取物转化为附加值较高的产品提供成熟的生产技术,二次提升茶叶功能组分的附加值,满足国内外功能产品开发的新需求。

参考文献

[1] Roberts E A H, Myers M. The phenolic substances of manufactured tea. VI. The preparation of theaflavin and of theaflavin gallate [J]. J Sci Food Agric,1959,176-179.

[2] Arghya A, Suchismita M, Lakshmishri L, et al. Theaflavins retard human breast cancer cell migration by inhibiting NF-jB via p53-ROS cross-

talk [J]. FEBS Letters, 2010(584): 7—14.

[3] Trina K, Subhabrata D, Madhumita R, et al. Induction of apoptosis in human leukemia cells by black tea and its polyphenol theaflavin [J]. Cancer Letters, 2005(230): 111—121.

[4] Cheng P, Ho Y E C, Yuk M L, et al. Black tea theaflavins extend the lifespan of fruit flies [J]. Experimental Gerontology, 2009(44): 773—783.

[5] Ya-Lun S, Jin-Ze X, Chi H N, et al. Antioxidant activity of tea theaflavins and methylated catechins in canola oil [J]. JAOCS, 2004, 81(3): 269—275.

[6] 屠幼英. 茶黄素抗癌作用机理[J]. 中国茶叶, 2008, 30(2): 11—13.

[7] 傅冬和, 毛清黎, 郑海涛等. 茶色素药理作用研究进展[J]. 茶叶通讯, 2003(2): 11—15.

[8] Samuel B, Adrienne L, Davis J R, et al. A model oxidation system to study oxidised phenolic compounds present in black tea [J]. Food Chemistry, 2003(83): 485—492.

[9] Yosuke M, Takashi T, Isao. A new mechanism for oxidation of epigallocatechin and production of benzotropolone pigments [J]. Tetrahedron, 2006(62): 4774—4783.

[10] Lopez S J, Thomas J, Pius P K, et al. A reliable technique to identify superior quality clones from tea germplasm [J]. Food Chemistry, 2005(91): 771—778.

[11] 王坤波, 刘仲华, 黄建安. 茶黄素的提取分离与纯化研究进展[J]. 湖南农业大学学报: 自然科学版, 2002, 28(4): 355—358.

[12] 谷记平, 刘仲华, 黄建安等. 单双液相下不同多酚氧化酶源对酶性合成茶黄素的影响[J]. 茶叶科学, 2007, 27(1): 76—82.

[13] Kapil S, Shamsher S B, Harsh P S. Biotransformation of tea catechins into theaflavins with immobilized polyphenol oxidase [J]. Journal of Molecular Catalysis B: Enzymatic, 2009(56): 253—258.

[14] Francis M N, John K W, Symon M M, et al. Catechins depletion patterns in relation to theaflavin and thearubigins formation [J]. Food Chemistry, 2009(115): 8—14.

[15] 李适, 黄建安, 刘仲华. 茶黄素体外氧化制备方法的影响因子研究进展[J]. 茶叶通讯, 2005, 32(2): 36—39.

[16] 毛清黎,朱旗,刘仲华等. 红茶发酵中 pH 调控对多酚氧化酶活性及茶黄素形成的影响[J]. 湖南农业大学学报:自然科学版,2005,31(5):524-526.

[17] 钟萝. 茶叶品质理化分析[M]. 上海:上海科学技术出版社,1989:469-470.

[18] 周卫龙,徐建峰,许凌. 茶叶中茶黄素测定的提取方法探讨[J]. 中国茶叶加工,2007(3):42-44.

[19] 胡升,梅乐和,姚善泾. 响应面法优化纳豆激酶液体发酵[J]. 食品与发酵工业,2002,29(1):13-17.

[20] 张巧艳,钱俊青. 响应面发优化黄杆菌突变株产脂肪酶摇瓶发酵条件[J]. 浙江工业大学学报,2009,37(2):156-160.

[21] 袁帅,胡承,曹海鹏等. 响应面法优化细菌纤维素发酵合成工艺[J]. 时珍国医国药,2010,21(2):407-410.

[22] 冯培勇,赵彦宏,张丽. 响应面法优化黑曲霉产纤维素酶发酵条件[J]. 食品科学,2009,30(23):335-339.

光照对 PET 瓶装绿茶饮料品质稳定性影响初探

刘 平　许勇泉　汪 芳　袁海波　刘盼盼　尹军峰

国家茶产业工程技术研究中心,中国农业科学院茶叶研究所,中国 浙江

摘 要:本实验研究光照对 PET 瓶装绿茶饮料理化成分及感官品质的影响。研究结果表明,相比对照,光照处理的绿茶饮料色差 L 值下降,a、b 值上升,pH 值下降,茶多酚及主要儿茶素组分含量下降,黄酮化合物含量上升,过氧化氢含量上升,茶汤外观颜色变深,香气、滋味变熟变陈;而添加 VcNa 对光照引起的茶汤劣变有很好的抑制作用,可以较好地抑制绿茶饮料在光照条件下的风味品质劣变。

关键词:光照;绿茶饮料;PET 瓶;稳定性

Effect of Illumination on the Quality Stability of PET-bottled Green Tea Beverage

LIU Ping　XU Yong-quan　WANG Fang
YUAN Hai-bo　LIU Pan-pan　YIN Jun-feng

Engineering Research Center for Tea Processing, Tea Research Institute, Chinese Academy of Agricultural Sciences, Zhejiang, China

Abstract: The effect of illumination on chemical compositions and flavor of green tea beverage during storage was investigated. The results showed, compared to the control, illumination made color L value decreased, color a and b values increased, pH value decreased, the contents of polyphenols and main catechins decreased, the contents of flavones and H_2O_2 increased, the color darkened and the flavor turned inferior. The addition of VcNa helped to inhibit the change of PET-bottled green tea beverage.

Keywords: illumination; green tea beverage; PET bottle; stability

茶饮料因其具有天然、方便、健康、快捷等特点，越来越受到消费者的青睐，成为当今软饮料发展最快、最有发展前景的饮料之一，在我国茶产业中占有举足轻重的地位。2011年我国茶饮料总产量超过1100万t，约占整个饮料行业总量的10%[1]。目前，我国茶饮料以果汁果味调味茶饮料、奶茶饮料和复合茶饮料为主[1]，而随着社会经济的发展和人民生活水平的不断提高，纯茶饮料和保健型茶饮料将是未来茶饮料发展的主要方向。目前诸多因素限制茶饮料的发展，例如茶饮料的色泽难以长时间保持，储藏过程中产生沉淀，香气滋味劣变问题等等都是货架期内品质稳定的问题。另外，由于PET瓶包装具有透明、美观、造型灵活、隔氧性好、质轻方便等优点，目前90%以上的茶饮料产品采用PET瓶包装[2]，但是PET瓶透光特性给绿茶饮料货架期内保持品质稳定带来难题。光照作为一种热能，影响绿茶干茶的储藏保鲜，会促使茶叶叶绿素及酯类物质氧化，加速茶叶的陈化和变质[3-5]。茶饮料中含有的茶绿色素类物质同样受光照影响[6]，赵先明等[7]推测光照使紫芽红茶浓缩液多酚分子中的酚羟基氧化成为醌而使茶汤红变；钱奕等[8]认为，光照会促使绿茶饮料中儿茶素类降解。尽管如此，针对光照对绿茶饮料货架期内的风味品质稳定影响的相关研究较少。本试验以绿茶饮料为研究对象，设置对照、光照、添加VcNa及光照基础上添加VcNa四个处理，研究光照对绿茶饮料风味品质的影响以及添加VcNa的防护效果，以期从理论层面上研究绿茶饮料的品质稳定性及其调控技术。

1 材料与方法

1.1 材料与仪器

茶叶原料，福鼎大白茶鲜叶原料（一芽三四叶）于2010年5月在武义汤记高山茶业有限公司经摊放、杀青、揉捻、烘干，加工成传统烘青绿茶，4℃低温冷藏待用；VcNa，购于Sigma上海公司；PET瓶（330mL），购于杭州百思生物技术有限公司。Waters 2465 series高效液相色谱，Waters公司；UV-2550紫外-可见分光光度计，日本岛津；MODELPHS-3C pH测定仪、雷磁E-201-C电极，上海精密科学仪器有限公司；CM-3500d分光测色仪，日本柯尼卡美能达公司；Model 680酶联免疫测定仪，美国Bio-Rad公司；LRH-250生化培养箱，上海一恒科学仪器有限公司；ZDS-10型照度计，上海市嘉定学联仪表厂。

1.2 方法

(1)茶饮料的制备。取茶样 100g,按 1∶40 茶水比、(70±2)℃浸提 20min,1μm 滤袋过滤后冷却,测量茶多酚(Tp=A×3.913×1000,mg/L),计算稀释至 800mg/L,稀释液分成两部分,一半按体积添加 0.05％的 VcNa,一半不添加,分别采用 UHT 灭菌(135℃,5s),灭菌过的绿茶饮料按如下 4 种处理分别放置于 38℃贮藏观察,每个处理 12 瓶,分别在 0、14d 和 21d 取出进行理化和感官分析,每个处理 3 个重复。培养箱内光照强度经测定为(695±5)lx,不同处理方式见表 1。

表 1　饮料处理方式

处　理	VcNa	光　照
空白	0	No
光照	0	Yes
添加 VcNa	0.05％	No
光照＋VcNa	0.05％	Yes

(2)pH 的测定。茶饮料 pH 采用便携式 pH 计(MODELPHS-3C)测定。

(3)色差的测定。茶饮料色差采用 L^*a^*b 色差系统进行测定,其中 L 值代表明度;a 代表红绿色度,正值代表红色程度,负值代表绿色程度;b 代表黄蓝色度,正值代表黄色程度,负值代表蓝色程度。

(4)生化成分含量测定。茶多酚含量:酒石酸亚铁比色法(GB 8313—2002);氨基酸总量:茚三酮比色法(GB 8314—2002);儿茶素、咖啡因含量检测:高效液相色谱法(茶饮料用 0.22μm 微孔滤膜过滤,Waters 高效液相色谱仪,VWD 检测器;色谱柱:ZORBAX SB－C18 ODS,5μm,4.6mm×150mm;流动相:A 为 0.5％甲酸,B 为乙腈,流速 1mL/min,柱温 40℃,检测波长 280nm;进样量:10μL,梯度洗脱,流动相 B 在 16min 内由 6.5％线性梯度变化到 25％,25min 回到初始状态,平衡 10min);H_2O_2 含量:S0038 过氧化氢检测试剂盒,碧云天生物技术研究所。

(5)茶饮料感官审评。审评小组由 3 个具有中级茶叶审评资格以上的专业审评员组成,采用密码审评方法,感官品质主要包括汤色、香气和滋味,感官审评评价标准:总分＝汤色×30％＋香气×30％＋滋味×40％。

1.3 数据分析

每个样品均设 3 次重复。表和图中数据均为平均值,方差分析采用 SPSS13.0 软件进行运算,处理间平均数的比较用最小显著差数法(LSD)。

2 结果与分析

2.1 不同处理茶饮料汤色稳定性

贮藏过程中绿茶饮料色差 L、a、b 值发生明显变化,不同处理差异较大(图 1~3)。贮藏过程中色差 L 值呈下降趋势,其中空白和光照处理 L 值都有明显下降,光照处理下降比空白更明显,而添加 VcNa 对茶饮料的明亮度有较

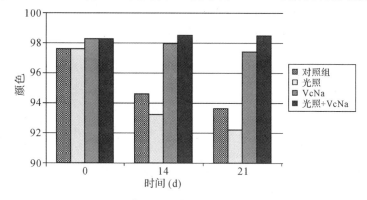

图 1 贮藏过程中茶饮料色差 L 值变化

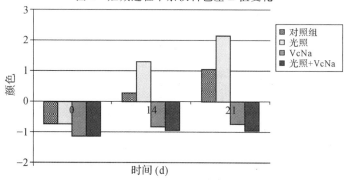

图 2 贮藏过程中茶饮料色差 a 值变化

好的保护作用。贮藏过程中色差 a 值呈上升趋势,其中空白和光照处理 a 值都

有明显上升,a值从负值升为正值,而添加VcNa对茶饮料的绿色程度有较好的保护作用。贮藏过程中色差b值呈上升趋势,其中空白和光照处理b值都有较大幅度上升,而添加VcNa可以有效防止绿茶饮料颜色变深。

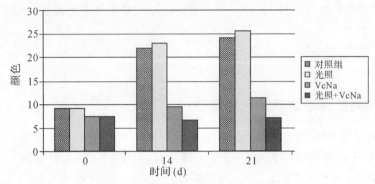

图3 贮藏过程中茶饮料色差b值变化

2.2 不同处理茶饮料理化成分稳定性

不同处理对茶饮料pH值影响差异较大,贮藏过程中绿茶饮料pH值呈下降趋势(图4)。贮藏过程中,空白和光照处理茶饮料pH值都呈大幅度下降,而添加VcNa的茶饮料pH值下降较缓。光照处理相比无光照处理茶饮料pH值下降较快。茶饮料pH值下降主要是由于多酚类物质如儿茶素发生水解产生的没食子酸等引起的,是茶饮料品质劣变的一个标志,可见光照处理可以促使茶饮料劣变,而添加VcNa可以有效减缓茶饮料的品质劣变。

不同处理对茶饮料多酚含量有一定影响,贮藏过程中绿茶饮料多酚含量呈

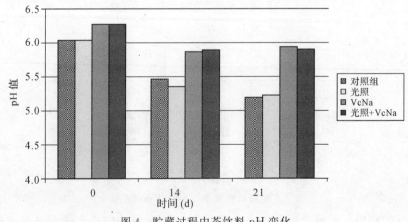

图4 贮藏过程中茶饮料pH变化

下降趋势(图5)。空白和光照处理茶饮料多酚含量值都呈大幅度下降,而添加VcNa的茶饮料多酚含量基本保持不变。多酚含量下降也是茶饮料品质劣变的重要标志[9],茶多酚在高温环境条件下发生氧化降解,光照促进茶多酚氧化降解,VcNa对茶饮料中的多酚物质起到较好的保护作用。

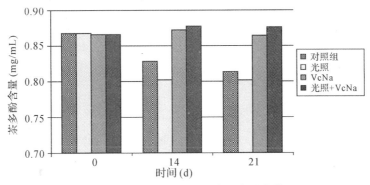

图5 贮藏过程中茶饮料多酚含量变化

不同处理对茶饮料黄酮化合物含量有一定影响,贮藏过程中绿茶饮料黄酮化合物含量呈上升趋势(图6)。绿茶饮料经过储藏后,各处理黄酮化合物含量都呈现不同程度的增加,未添加VcNa的茶饮料增加更明显,光照处理也加剧了黄酮化合物含量的增加趋势,而添加VcNa可以有效抑制黄酮化合物含量的上升。许勇泉等[9]也报道过,绿茶饮料贮藏过程中黄酮化合物含量呈上升趋势,这可能与多酚物质降解有关。

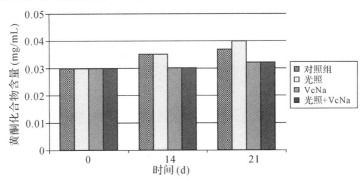

图6 贮藏过程中茶饮料黄酮化合物含量变化

不同处理茶饮料主要儿茶素含量变化有一定差异,贮藏过程中绿茶饮料主要儿茶素含量呈下降趋势,没食子酸含量呈上升趋势(表2)。EGCG及ECG等酯型儿茶素在贮藏过程中降解产生简单儿茶素和没食子酸。光照处理加剧了酯型儿茶素的降解,而添加 VcNa 可以有效抑制酯型儿茶素的降解。

表 2　贮藏过程中茶饮料中儿茶素组分含量变化　　　　mg/mL

儿茶素	0 d		14 d			
	对照组	VcNa	对照组	光　照	VcNa	光照＋VcNa
EGCG	0.118	0.129	0.095	0.089	0.121	0.122
ECG	0.061	0.062	0.048	0.048	0.049	0.049
GA	0.006	0.005	0.014	0.011	0.009	0.008

不同处理对茶饮料中过氧化氢含量有一定影响,贮藏过程中绿茶饮料过氧化氢含量呈上升趋势(图 7)。绿茶饮料经过储藏后,各处理过氧化氢含量都呈现不同程度的增加,未添加 VcNa 的茶饮料增加更明显,光照处理也加剧了过氧化氢含量的增加趋势,而添加 VcNa 可以有效减缓过氧化氢含量的上升。有研究证实,绿茶、红茶等多酚类饮料在贮藏过程中都会产生过氧化氢,其中儿茶素 EGCG 和 EGC 是产生过氧化氢最强的化合物,且过氧化氢的产生量与茶饮料中多酚类物质的浓度呈正相关[10]。窦宏亮等[11]也研究证实了绿茶饮料在储藏过程中会产生过氧化氢,且其生成量呈动态变化,先升后降,在贮藏 1 个月时达到最高峰。

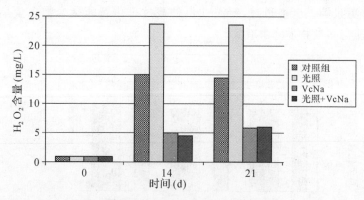

图 7　贮藏过程中茶饮料过氧化氢含量变化

2.3　不同处理茶饮料感官品质变化

不同处理绿茶饮料感官品质差异较大(表 3)。贮藏过程中绿茶饮料风味品质呈下降趋势,茶饮料外观颜色变黄,有明显熟味产生,光照处理加剧了绿茶饮料的风味品质劣变,而添加 VcNa 可以较好减缓绿茶饮料的风味品质下降,尽管如此,在添加 VcNa 和光照处理同时存在的情况下绿茶饮料风味品质仍然快速下降,该变化趋势还有待进一步验证和明确。

表3 贮藏过程中茶饮料感官品质变化

时间(d)	处理	汤色	得分	香气	得分	滋味	得分	总分
0	对照组	浅绿黄	84	稍有熟味	82	略有陈熟	81	82.7
	VcNa	浅黄绿	86	尚纯	84	入口微咸带陈熟	78	83.9
14	对照组	黄亮	79	有熟味	81	有陈味稍带熟	80	80.0
	光照	黄尚明	76	熟汤味	78	有陈熟味	79	77.3
	VcNa	浅黄绿	84	稍有熟	82	尚醇,稍带熟	82	82.8
	光照+VcNa	黄绿	85	熟味重	75	熟味重	74	79.0
21	对照组	黄	77	有陈熟	79	有陈熟	79	78.2
	光照	黄	76	陈熟味	77	陈熟味重	76	76.4
	VcNa	浅黄绿	82	略有陈熟	81	略带陈熟	81	81.4
	光照+VcNa	黄绿	84	熟带酸	74	熟味重	76	78.5

3 讨 论

绿茶饮料在高温(38℃)贮藏过程中,风味品质劣变,外观颜色变深、变黄,茶多酚、儿茶素含量显著下降,黄酮化合物、没食子酸含量显著上升,咖啡因、氨基酸含量略有下降,感官风味品质明显下降,熟汤味明显;该变化趋势与许勇泉等[9]的研究报道一致,该研究认为,随着茶多酚浓度升高,绿茶饮料的感官理化品质变化趋势加剧。光照明显加剧了绿茶饮料的品质劣变,钱奕等[8]研究认为,光照促使茶饮料品质劣变,饮料中茶多酚和儿茶素含量与光照的强度呈负相关,并推测茶饮料在无氧条件下,光照促进了茶多酚(主要是儿茶素)的降解。另外有研究显示光照可使诸多色素类物质降解和异构化,Lee 等[12]研究发现光照可使番茄红素降解和异构化,Chen 等[13-14]研究发现光照可以引起叶绿素、α-胡萝卜素和β-胡萝卜素降解和异构化,但这也仅仅是光照对各单体物质的反应作用,在食品或饮料中的相关作用还有待进一步验证。

对绿茶饮料品质劣变的机理性的研究也有较多报道。Wang 等[15]研究认为,绿茶饮料灭菌后储藏过程中的感官品质变化主要是由 EGCG、EGC 的变化引起,儿茶素特别是酯型儿茶素是产生异构化或氧化反应的主体。Kim 等[16]研究报道茶汤在受热处理时,EGCG、EGC、EC、ECG 部分差向异构化,造成总儿茶素含量下降。Akagawa 等[10]研究发现多酚类饮料在储藏过程中会产生 H_2O_2,其中以 EGCG 和 EGC 产生 H_2O_2 的能力最强,且 H_2O_2 的生成量与多酚类化合物的浓度成正比。也有学者研究发现,添加苹果汁和柠檬汁等物质可以减少瓶装绿茶饮料 H_2O_2 的产生,并且也认为其产生量与多酚的浓度有关[17]。本文试验中,光照加剧了 H_2O_2 的产生,其产生量明显大于对照,而添加 VcNa 可以有效抑制 H_2O_2 的产生,该结果与 Ayabe 等[17]研究报道一致。窦宏亮等[11]研究发现,绿茶饮料在贮藏过程中 H_2O_2 含量变化与主要儿茶素 EGCG 含量的变化相关系数达 0.892,并认为 H_2O_2 的氧化是导致儿茶素 EGCG 含量减少的重要原因,但是通过添加柠檬酸、苹果酸、琥珀酸及抗坏血酸等降低茶汤的 pH 值可以大大减少 H_2O_2 的产生量[18]。

绿茶饮料货架期内的品质稳定性是茶饮料企业开发绿茶纯茶饮料的主要障碍,本文实验研究证实添加 0.5% VcNa 可以有效减缓绿茶饮料的主要品质成分劣变,保持绿茶饮料的外观颜色,然而对绿茶饮料在高温环境下产生熟汤味等并没有很好的减缓效果,因此针对绿茶饮料货架期内的风味品质稳定问题还应进一步研究和探索。

参考文献

[1] 中国饮料工业协会. 2012 中国茶与咖啡饮料发展研讨会[M]. 2012.

[2] 尹军峰,许勇泉,袁海波. 中国大陆液态茶饮料发展趋势及主要技术需求分析[J]. 茶叶科学,2010,30(增刊1):588-592.

[3] 汪毅,龚正礼,骆耀平. 茶叶保鲜技术及质变成因的比较研究[J]. 中国食品添加剂,2005(5):19-22,64.

[4] 顾谦,陆锦时,叶宝存. 茶叶化学[M]. 北京:中国科学技术大学出版社,2002.

[5] 袁建兴,宓晓黎,成恒嵩等. 典型相关分析在茶叶贮藏保鲜中的应用[J]. 江苏林业科技,1993,1:30-33.

[6] Martinus A B. Kinetic modeling in food science:A case study on chlorophyll degradation in olives [J]. Journal of the Science of Food and Agricultural,2000,80:3-9.

[7] 赵先明,汪艳霞,王孝仕等. 紫芽茶制备红茶饮料浓缩液及其呈色稳定性研究[J]. 西南大学学报:自然科学版,2011,33(6):165-172.

[8] 钱奕,卢立新. 贮藏条件对茶饮料品质影响的试验研究[J]. 包装工程,2009,30(11):34-35,60.

[9] 许勇泉,刘平,陈根生等. 茶多酚浓度对绿茶饮料稳定性的影响研究[J]. 茶叶科学,2011,31(6):525-531.

[10] Akagawa M, Shigemitsu T, Suyama K. Production of hydrogen peroxide by polyphenols and polyphenols and polyphenols-rich beverages under quasi-physiological conditions[J]. Bioscience, Biotechnology and Biochemistry, 2003, 12:2632-2640.

[11] 窦宏亮,李春美,郝菊芳等. 绿茶饮料贮藏期间主要生化成分和典型香气成分的变化及相关性研究[J]. 茶叶科学,2008,28(3):181-188.

[12] Lee M T, Chen B H. Stability of lycopene during heating and illumination in a model system[J]. Food Chemistry, 2002, 78:425-432.

[13] Chen B H, Chen T M, Chie J T N. Kinetic model for studying the isomerization of α- and β-carotene during heating and illumination[J]. Journal of Agricultural and Food Chemistry, 1994, 42:2391-2397.

[14] Chen B H, Huang J H. Degradation and isomerization of chlorophy II α and β-carotene as affected by various heating and illumination treatments[J]. Food Chemistry, 1998, 62:299-307.

[15] Wang L F, Kim D M, Lee C Y. Effect of heat processing and storage on flavanols and sensory qualities of green tea beverage[J]. Journal of Agricultural and Food Chemistry, 2000, 48:4227-4232.

[16] Kim E S, Liang Y R, Jin J, et al. Impact of heating on chemical composition of green tea liquor[J]. Food Chemistry, 2007, 103:1263-1267.

[17] Ayabe S, Aoshima H. Aqueous extract of citrus peel reduces production of hydrogen peroxide in catechin-enriched green tea[J]. Food Chemistry, 2007, 104:1594-1598.

[18] Aoshima H, Ayabe S. Prevention of the deterioration of polyphenol-rich beverages[J]. Food Chemistry, 2007,100:350-355.

附:基金项目:十二五科技支撑(2011BAD01B03-5-5)及国家基金(31070615)。

茶籽多糖的表征和生物活性研究

王元凤　姜慧仙　毛芳芳　魏新林

上海师范大学工程食品研究所,中国 上海

摘　要:应用 DEAE-52 纤维素阴离子交换色谱法和 Sephadex G-200 凝胶渗透色谱法从茶树籽水溶性粗提物中获得三种纯化多糖成分(NTSPS、ATSPS1-1 和 ATSPS2)。离子色谱法测定纯品的单糖组成,结果显示 NTSPS 含有阿拉伯糖、半乳糖和葡萄糖;而 ATSPS1-1 含有蛋白质(2.83%),是由岩藻糖、鼠李糖、阿拉伯糖、半乳糖、葡萄糖、木糖、甘露糖和葡萄糖组成的杂多糖;ATSPS2 是杂多糖,蛋白质含量为 2.12%,单糖组成为半乳糖、葡萄糖、木糖和核糖。高效凝胶渗透色谱分析表明,三种纯化多糖的平均分子量分别为 4.588? kD、500? kD 和 100? kD。NTSPS、ATSPS1-1 和 ATSPS2 在高浓度时具有很高的脾淋巴细胞增殖活性和 K562 细胞体外抑制活性。

Characterization and Biological Activities of Purified Polysaccharides from Crude Extract of Tea Seeds

WANG Yuan-feng　JIANG Hui-xian　MAO Fang-fang　WEI Xin-lin

Institute of Food Engineering, Shanghai Normal University, Shanghai, China

Abstract: Three water-soluble polysaccharide fractions (NTSPS, ATSPS1-1 and ATSPS2) were extracted from the tea seeds (*Camellia sinensis*), and obtained by DEAE-52 cellulose anion-exchange chromatography, and Sephadex G-200 gel-permeation chromatography. The monosaccharide components of three polysaccharides were characterized by ion chromatography (IC), results indicated that NTSPS was composed of arabinose, galactose and glucose, while ATSPS1-1 was a hetero-polysaccharide which bounded with protein (2.83%) and comprised of fucose, rhamnose, arabinose, galactose, glucose, xylose, mannose and glucose, ATSPS2 was a hetero-polysaccharide bounded with protein (2.12%), and composed of galactose glucose, xylose and ribose. The high-performance gel-permeation chromatography (HPGPC) analysis showed that the average molecular weight of three polysaccharides were approximately 4.588, 500 and 100 KD, respectively. NTSPS, ATSPS1-1 and ATSPS2 with high concentrations had a high proliferation effect on spleen lymphocytes and inhibition activity of K562 cells in vitro.

松阳茶叶精深加工现状及今后发展方向

叶火香

松阳县农业局,中国 浙江

摘　要：近年来,松阳的茶叶产业迅猛地发展,形成了较为完善的茶叶产业发展模型,产业链完整,为农村的经济发展起到了积极的推进作用。尤其是近两年,茶叶精深加工、茶产品开发得到发展,为松阳茶产业持续、快速、健康发展,实现松阳茶叶从传统向现代转型,延伸产业链,提高附加值,以工业产品的理念来生产、加工农业产品作出贡献。实现茶叶由初级农产品向精深加工产品转变,提高了茶产品的科技含量和经济效益,为提升茶产品市场竞争力,有效地规避市场风险打下基础。

关键词：茶叶;精深加工;茶多酚

Multipurpose Utilization of Tea in Songyang and its Developing Trend

YE Huo-xiang

Department of Agriculture, Songyang People's Government, Zhejiang, China

Abstract：In recent years, tea industry of Songyang county has developed quickly. Tea industry has formed its own chains and plays a vital important role on income of tea farmers and pushes forward the pace to build new countryside. In the last two years, industry of multipurpose utilization of tea in Songyang started to develop and gained achievements to expend chain of tea industry and increase additional value of tea. This paper introduces the experience to develop multipurpose utilization of tea and future plan of Songyang.

Keywords：tea; refined processing; tea polyphenol

1 茶叶精深加工是当前茶叶生产的重要内容和发展方向

茶叶精深加工是当前茶叶生产领域的一项重要内容和发展方向。近年来饮料业发展迅速,新产品不断涌现,竞争激烈。茶叶作为一种传统的饮料,受到其他饮料的挑战,固定不变的产品结构已不能满足市场需求,茶叶行业必须采用新技术开发新产品,才能求得生存和发展。茶叶市场供求目前基本平衡,如何进一步开发利用大量的中低档茶是未来茶产业发展要解决的重要问题。据统计,我国现有茶园面积 220 万 hm^2,至 2015 年约有几百万亩新开辟茶园投入生产,到时茶叶市场供过于求的现象可能会出现。茶叶精深加工将能有效地解决中低档茶出路。茶多酚等茶叶功能性有效成分开发越来越受到关注。由茶叶中提取的茶多酚具有天然的抗氧化活性,备受人们的重视,从而推动茶多酚等茶叶功能性有效成分的开发和利用。四是茶叶保健药效研究成果喜人。茶叶最早是作为药物被人们利用的,现代科学研究也证实了茶叶的药用功能,长期研究表明,茶叶具有降血糖、降血脂、降血压、抗衰老、防辐射等功效,因此茶叶药用成分的提取利用也成为国内外天然药物开发关注的热点。

2 松阳茶叶精深加工的现状

茶叶精深加工包括两个方面:①将传统工艺加工的成品进行更深层次的加工,形成新型茶饮料品种;②提取和利用茶叶中功能性成分,并将这些产品应用于医药、食品、化工等行业。松阳从 2008 年开始涉足茶叶精深加工领域,到 2010 年,茶叶精深加工有了一定的规模,目前至少有 5 家企业专门从事茶叶深加工产品开发和生产,取得显著的经济效益。

2.1 建设完成了年产 2500t 的速溶茶加工生产线

近几年,在国家政策的扶持下,全国的茶叶生产规模迅速扩张,茶叶的市场竞争越来越激烈,根据国内大中型饮料生产企业的生产需要,从 2008 年开始,松阳开始有茶叶企业着手建设速溶茶加工生产线,并发了红茶、绿茶、白茶等速溶茶,并与福建、广东、湖南等省茶叶企业和进出口公司建立了良好的技术和销售合作关系。目前该公司已建立了年产 2500t 的速溶茶加工生产线,通过对中低档茶、茶末、茶片、茶梗以及茶树修剪叶的综合开发利用,将有效地解决低档茶市场销路问题,最大限度地利用茶叶资源,丰富松阳县茶叶产品结构,增加农民收入。

2.2 合作开发、提取利用茶叶中的有效成分,开发新产品

2011年松阳新成立了一家茶叶生物技术开发公司,主要通过利用茶叶中的有效成分茶多酚、咖啡因等,加工生产茶爽、茶宁片、茶多酚片等茶叶深加工产品,这些产品具有很好的保健药效,能辅助降低血糖、血压、血脂等。茶叶深加工产品的开发开启了松阳茶产业向食品、药品行业拓展,延伸茶叶产业链,增加茶叶生产附加值。

2.3 茶餐饮、茶食品的开发悄然兴起

目前,松阳还有3家专门生产茶叶食品、糕点、菜肴的公司,产品包括茶叶香肠、茶多酚酱油肉、茶叶年糕、茶叶青糕、茶叶熏腿等。火腿是我国传承千年的传统腌肉制品,在江、浙、沪等地区有妇女产后、病人疗伤、养生食用火腿之风俗。茶叶熏腿按照传统工艺加工腌制,并结合现代工艺,使火腿经过修剪茶树枝叶的熏制,吸收了茶叶中挥发的有效成分,可延长火腿保质期,降低亚硝酸盐合成,还形成了风味独特、醇香味美、营养丰富的特异品质。茶叶熏腿2012年荣获浙江省第三届农民创富大赛创新项目类好点子奖。

3 今后的发展思路

速溶茶等茶叶深加工、茶叶提取物的应用,茶餐饮、茶食品等茶叶衍生品的开发利用,极大地丰富了茶叶产品,延伸了茶产业链,提高了茶叶附加值,实现了茶叶从数量扩张型向规模效益型转变,增强了市场竞争能力,对促进松阳茶产业升级有着极为重要的意义。为此,我们将努力抓好以下方面。

3.1 建设茶叶精深加工园区,培育茶叶精深加工企业,实现茶产业升级

强化政府服务,解决茶产业发展及茶叶精深加工中的土地制约问题。目前已启动了26.7hm^2的茶叶精深加工园区建设项目,通过集中供地、完善和配套园区相关设施、招商引资和鼓励有实力的茶叶企业从事茶叶精深加工,吸引茶叶深加工企业向加工园区集聚,加快培育茶叶精深加工产业,实现松阳茶叶的加工升级,增强松阳茶叶抵御市场风险的能力。

3.2 依托科研院所,拓展我县茶叶精深加工领域

今后,我们在稳定以"松阳银猴"和"松阳香茶"为主的名优茶生产的基础上,重视低档茶资源的开发利用,利用丰富的茶叶原料资源,加大精深加工终端产品开发,开发多元化的茶产品。依托国内茶叶科研单位,在"十二五"期间实施茶产业升级转化工程,开发含茶降血脂与抗氧化系列健康食品以及功能性成分在食品加工中的产业化应用技术与示范;计划建立茶叶深加工核心示范基地1～2个;安装1～2套速溶茶提取设备和1套双效浓缩设备;批量生产茶多酚口含片、茶多酚片、茶黄素胶囊、茶爽茶叶籽油等茶食品、茶保健品;建立1条"茶—烘焙食品"产业化示范线,拟开发茶末花生、茶桔饼、茶叶蜜饯等新产品,拓展茶资源利用广度和深度,带动茶业与当地传统食品业的联动发展。

茶氨酸的保健功能研究进展

沈晓玲　龚正礼

西南大学食品科学学院,中国 重庆

摘　要:茶氨酸是茶叶和部分山茶科植物特有的酰胺类物质,也是重要功能成分。随着人们对茶叶健康作用的重视,科学家对茶氨酸的药理学和生理功能研究不断深入,茶氨酸新的生理功能也不断被发现。本文综述了茶氨酸的理化性质、生理功能、应用及前景。

关键词:茶氨酸;理化性质;保健功能

Research Development of Health function of Theanine

SHEN Xiaoling　GONG Zheng-li

College of Food Science, Southwest University, Chongqing, China

Abstract: Theanine is a kind of amide that only can be extracted from tea and part of Camellia secco plants, and is the important functional component. As people pay attention to the health effects of tea, scientists keep studying on pharmacological and physiological function of theanine, and find its new physiological function. This paper reviews the physical and chemical properties, physiological function, application and prospect of theanine.

Keywords: theanine; physical and chemical properties; health function

茶叶中氨基酸是其品质的重要组分,目前人们在茶叶中已发现 25 种氨基酸,茶氨酸(Theanine,γ-glutamylethylamide)被认为是茶叶的特征氨基酸,属酰胺类化合物,是绿茶中有效的呈味物质。茶氨酸的含量直接决定茶叶的品质,一般占茶叶中游离氨基酸的 40% 以上,占干茶重的 1%～2%[1]。除了茶叶,只在茶梅、山茶、油茶、蘑菇四种天然植物中检测出其微量存在,其他植物中尚未发现[2]。茶氨酸是 1950 年日本学者酒户弥二郎首次从茶叶中提取、精制并确定其化学结构,此后国内外茶叶学者对它进行了较系统的研究。

1 茶氨酸的理化性质

茶氨酸(N-乙基-r-谷氨酰胺)又名谷氨酰乙胺,分子结构为:

$$CH_3-CH_2-NH-\overset{\overset{O}{\|}}{C}-CH_2-CH_2-\underset{\underset{NH_2}{|}}{CH}-CHOON$$

是谷氨酸的乙胺诱导体,在茶树根中由谷氨酸和乙胺合成,通过树干运送到叶中贮藏,在日光照射下可转化为儿茶素[3]。自然存在的茶氨酸均为 L 型,纯品为白色针状结晶,熔点 217~218℃,比旋光度 0.7°。茶氨酸极易溶于水,在茶汤中的泡出率可达 80%,其水溶液呈微酸性,不溶于无水乙醇和无水乙醚。茶氨酸的性质较稳定,如将茶氨酸溶液煮沸 5min 或将茶氨酸溶于 pH 值 3.0 的溶液并在 25℃下储放 1 年,其中的茶氨酸含量不会改变[4]。茶氨酸有焦糖香及类似味精的鲜爽味,能缓解苦涩味,增加茶汤的鲜甜度,其在日本已作为食品添加剂[5]。研究证明茶氨酸的含量与绿茶的品质密切相关,其相关系数为 0.787~0.876。经安全性实验表明茶氨酸安全无毒,在摄取量上没有任何限制[3]。茶氨酸可经 25%硫酸或 6mol/L 的盐酸水解为 L-谷氨酸和乙胺;用茚三酮显色呈紫色,其发色强度与谷氨酸相似;可用醋酸汞和碳酸钠沉淀,易与碱或铜生成淡紫色柱状铜盐[6]。茶氨酸的定性、定量检测可用碱式碳酸铜沉淀法、强酸性阴离子交换树脂层析法、茚三酮显色法、纸层析法、薄层色谱法、气相色谱法、氨基酸自动分析仪、高效液相及胶束电动气毛细管色谱法等[7]。

2 茶氨酸的保健功能

2.1 增强记忆力

林雪玲[6]等进行的茶氨酸对小鼠学习记忆能力影响的研究证明,无论是复杂的水迷宫实验还是跳台实验,不同剂量的茶氨酸均显示出增强小鼠学习记忆能力的作用。茶氨酸能缩短正常小鼠游完水迷宫的时间,减少进入盲端的次数;对记忆获得障碍小鼠可延长首次错误的出现时间,减少触电次数。而且茶氨酸对学习记忆力的增强作用随其剂量的增加而增大。

以 D-半乳糖致亚急性衰老小鼠为模型,采用 Y 电迷宫法检测各组小鼠学习记忆能力,结果显示,茶氨酸可明显提高衰老小鼠的学习记忆能力。脑组织匀浆酶法测定表明,茶氨酸显著提高超氧化物歧化酶(SOD)和胆碱酯酶(AchE)活性,降低丙二醛(MDA)含量。推断茶氨酸提高记忆能力的作用机理可能与其清除自由基、促进中枢胆碱等功能有关[7]。茶氨酸(0.3 和 1mg/kg)可以抑制反复脑缺血引起的海马体神经元细胞死亡,防止脑缺血导致的空间记忆损害,有助于预防脑血管疾病[8]。

茶氨酸通过刷毛缘膜上的 Na+偶合协同转运蛋白被吸收,然后由亮氨酸优先运输系统通过血脑屏障结合到脑部;茶氨酸的衍生物谷氨酸是脑的主要神经递质之一,其对血压的调节是通过影响脑和周围神经系统中含儿茶酚胺和含血清素的神经元而起作用。

2.2 抗疲劳

给小鼠口服 L-茶氨酸(4.2～16.6mg/kg)30 天后进行负重游泳实验。以茶氨酸 5.6mg/(kg·D)(低剂量)、8.4mg/(kg·D)(中剂量)和 12.6mg/(kg·D)(高剂量)连续灌胃 30d,并以等量蒸馏水为对照。试验结果表明,不同剂量的 L-茶氨酸能明显延长小鼠负重游泳时间,减少肝糖原的消耗量,降低运动时血清尿素氮(BUN)水平,对小鼠运动后血乳酸升高有明显的抑制作用,能促进运动后血乳酸的消除。有关生化指标测定表明,中、高剂量组的负重游泳时间和肝糖原含量均显著高于对照组($P<0.05$);运动后,中剂量组 BUN 增加量显著低于其他各组;休息 60min 后,各茶氨酸剂量组的 BUN 含量明显低于对照组($P<0.05$);运动前、后各剂量组 5-羟色胺(5-HT)含量低于对照组;运动后中、高剂量组多巴胺(DA)含量高于对照组;运动前各剂量组和运动后中剂量组 DA/5-HT 比值高于对照组,差异显著($P<0.05$)。说明茶氨酸具有延缓运动性疲劳作用,其机制可能与茶氨酸能增加脑组织中 DA 含量,可抑制 5-羟色胺分泌,促进儿茶酚胺分泌有关(5-羟色胺对中枢神经系统具有抑制作用,而儿茶酚胺具有兴奋作用)[9]。

2.3 抗肿瘤作用

茶氨酸抗肿瘤的作用机理如下。①可增强抗癌药物的作用效果,如茶氨酸与抗肿瘤药物阿霉素(DOX)同时使用,还可以增强阿霉素的抗肿瘤活性,但不增加阿霉素的副毒作用。②可减轻某些抗癌药物的副毒作用。如茶氨酸可以降低 DOX 引起的脂质过氧化水平并降低谷胱甘肽过氧化酶活性,使细胞的谷氨酸和谷胱甘肽保持较高水平,进而减轻 DOX 的毒性[10]。③有助于保持细胞的代谢平衡,茶氨酸处理使血液总胆固醇、低密度脂蛋白、致动脉粥样硬化指数、甘油三酸酯、脂质过氧化物等指标显著低于空白对照,但高密度脂蛋白显著高于对照。④茶氨酸可以增强抗癌药物对癌细胞转移的抑制作用,茶氨酸与DOX同时使用,可以抑制小鼠肝癌细胞的转移。总之,茶氨酸对癌症患者临床化学治疗是十分有益的,它不仅促进某些抗肿瘤药物作用,而且还能提高癌症病人的生活质量。

2.4 降血压作用

降压的机理是通过影响末梢神经血管系统来实现的,而不是通过提高脑中5-羟色胺水平来实现的。当给自发性高血压大鼠注射不同剂量茶氨酸时血压下降,尤其在高剂量组可观察到血压下降明显;但对正常血压的大鼠,即便是给予最高剂量茶氨酸,血压也不会改变[11]。咖啡因对自发性高血压大鼠不但有增加血压的作用,而且还增强其自测机敏性和神经过敏性;但谷酰胺甲胺和茶氨酸除了降血压以外,对自测机敏性、神经过敏性没有影响[12]。茶氨酸是否对人体也有降压作用却至今还未见报道。

2.5 抗糖尿病作用

锌的异常代谢与某些生理代谢紊乱有关,如出现糖尿病并发症。锌在保护糖尿病心肌病人心脏免受各种氧化应激方面起着关键作用。补充锌是预防心脏氧化损伤的重要措施;锌补充剂可以预防或延缓糖尿病心肌病。虽然茶氨酸单独使用的降糖效果并不明显,但茶氨酸-锌复合物 Zn(gln-e)2 可以明显降低小鼠 KK-Ay 血糖和糖化血红蛋白(HbA1c,与糖尿病视网膜病变有关)水平[13]。茶氨酸-锌复合物可以作为锌补充剂在临床上用于预防糖尿病综合征有积极的意义。

2.6 对脑神经细胞起保护作用

环境中部分神经毒素和氧化胁迫具有选择性影响帕金森综合征(PD)多巴胺激导的神经细胞易受伤的部位,从而对神经系统产生伤害。Cho[14]等研究表明,L-茶氨酸被认为具有保护神经免受 PD 相关神经毒素伤害的作用,在临床上可用于预防帕金森综合征。

大鼠母鼠分娩期后随机饲喂茶氨酸试验表明,处理幼鼠体重与对照无区别,但处理幼鼠大脑中部分神经递质,如多巴胺、5-羟色胺(血管收缩素)、甘氨酸和 GABA(γ-胺基丁酸)浓度提高,而且大脑皮质和海马体的神经生长因子

(NGF)mRNA 表达水平上升[15]。说明在神经成熟期间，茶氨酸能增强 NGF 和神经递质合成，促进中枢神经成熟，有利于脑功能健康发育。

2.7 保护心脑血管

血清胆固醇是导致冠心病(CHD)的危险因素之一，降低胆固醇有助于减少心脑血管疾病风险。在饲料中添加 0.028% 茶氨酸喂养小鼠 16 周后，小鼠腹腔脂肪减少到对照组的 58%，血清中性脂肪及胆固醇含量比对照组分别减少 32%和15%，肝脏胆固醇含量比对照减少 28%[16]。以脑缺血损伤大鼠试验表明，茶氨酸组大鼠神经症状评分和脑组织含水量分较脑缺血组显著降低；血清、脑组织 SOD 活力较脑缺血组显著增高；脑组织 MDA 含量较脑缺血组明显降低。说明茶氨酸对大鼠脑缺血损伤的保护作用可能与调节脑缺血损伤引起的自由基代谢失调有关[17]。以脑缺血损伤家兔模型试验显示，茶氨酸组脑超微结构改变以及血清白细胞介素-8(IL-8)和神经元特异性烯醇化酶(NSE)含量变化均明显轻于脑缺血组[18]。流行病学调查认为，饮茶可能对降低脑中风发生的概率有积极作用[19]。

2.8 增强免疫力

在"非典"流行期间，国内外专家都提到饮茶可增强抵抗力，预防传染病。据美国研究显示，茶叶的这种作用可能与茶氨酸有关。用感染流行性感冒病毒的小鼠实验表明，同时口服 L-胱氨酸和 L-茶氨酸可以提高特异性抗原免疫球蛋白的产生[20]。将疗养院的老年受试者分成 2 组，在免疫接种之前分别口服 L-胱氨酸和 L-茶氨酸，口服 14 天后接种流行性感冒病毒疫苗，接种后 4 星期检测，3 种病毒都能造成血凝集抑制(HI)滴度提高，两组之间没有显著差异；但层状分析显示，血清总蛋白和血红蛋白低的受试者中，实验组的血清转化率显著高于对照组[20]。表明疫苗接种前同时口服 L-胱氨酸和 L-茶氨酸有助于增强血清总蛋白和血红蛋白低的老年人对流行性感冒病毒疫苗的免疫响应。

2.9 减轻酒精对肝脏的伤害

过量饮酒导致肝脏自由基大量产生，谷胱甘肽酶(GSH)活力水平降低，脂质过氧化物浓度提高，进而损伤肝脏。以"酒精"(对照组)和"酒精+茶氨酸"(茶氨酸组)分别腹腔注射小鼠(Male CDF1)，1h 后与对照组相比，茶氨酸组血液酒精浓度明显低于对照组，乙醇脱氢酶和乙醛脱氢酶活力显著高于对照组，细胞色素 P450(CYP)2E1 得到控制[21]。说明茶氨酸可以有效减轻酒精引起的肝损伤。

2.10 其他功能

除上述保健功能外,茶氨酸还具有降脂、抗衰老、缓解抑郁症[22]、改善经期综合征、松弛神经紧张和放松等作用。

3 应用及前景

茶氨酸作为一种食品添加剂,不仅易溶于水,还有抑制苦味及改善食品风味的特点,可广泛应用于点心类、糖果及果冻、饮料、口香糖等几乎所有食品中,并且添加于食品后,不会产生好食用、不好饮用、口感下降的问题。另外茶氨酸具有良好的安全性、稳定性,急性、亚急性毒性试验和致突变试验均未发现毒性,细菌恢复突变试验也证明未导致基因变异,无诱变性。因此,茶氨酸的摄入量是不受限制的。

除上述作用外,茶氨酸还可以作为镇静药、抗腐蚀剂、消臭剂、抗酸化剂和预防牙周病的牙膏的添加剂,化妆品中作保湿剂等[23]。

随着经济发展、人们生活水平的提高,功能食品已成为人们生活中的一种追求。茶氨酸是继茶多酚之后的又一大功能性物质,安全、无毒、具有多种生理功能。在今后的发展中,还应加强这些方面的研究以提高茶氨酸的转化率、产率及生产效率。并应致力于拓宽其在食品领域的应用,利用茶氨酸的特殊生理功能,开发功能性药物等等。

参考文献

[1] 宛小春.茶叶生物化学第三版[M].北京:中国农业出版社,2008.34-35
[2] 张健,荣绍丰,龚钢明等.不同茶叶中茶氨酸含量的测定比较[J].食品科学,2008,29(4).
[3] 帅玉英,张涛,江波等.茶氨酸的研究进展[J].食品与发酵工业,2008,34(11):117-123.
[4] 李靓.茶氨酸的保健功效研究及其保健食品开发[J].茶叶.2009.
[5] GAO X-H,YUAN H,YU Z-R. Research development of theanine [J]. Chemical and biological engineering,2004,21(1).
[6] 林雪玲,程朝辉,黄才欢等.茶氨酸对小鼠学习记忆能力的影响[J].食品科学,2004,25(5):171-173.
[7] 刘显明,李月芬,李国平.茶氨酸对D-半乳糖衰老模型小鼠抗衰老作用的实

验研究[J].创伤外科杂志,2008,3:257—259.

[8] Egashira N, Ishigami N, Pu F, et al. Theanine prevents memory impairment induced by repeated cerebral ischemia in Rats [J]. Phytother Res, 2008,22:65—68.

[9] 李敏,沈新南,姚国英.茶氨酸延缓运动性疲劳及其作用机制研究[J].营养学报,2005,27(4):326—329.

[10] Sugiyama T, Sadzukab Y. Theanine, a specific glutamate derivative in green tea, reduces the adverse reactions of doxorubicin by changing the glutathione level [J]. Cancer Letters, 2004, 212: 177—184.

[11] 刘媛,王健.茶氨酸的最新研究进展[J].中国食物与营养,2007,10:23—25.

[12] Rogers P J, Smith J E, Heatherley S V, et al. Time for tea: Mood, blood pressure and cognitive performance effects of caffeine and theanine administered alone and together [J]. Psychopharmacology, 2008, 195: 569—577.

[13] Matsumoto K, Yamamoto S, et al. Antidiabetic activity of Zn(II) complexes with a derivative of L-glutamine [J]. Bull Chem Soc Jpn, 2005, 78: 1077—1081.

[14] Cho H S, Kim S, Lee S Y, et al. Protective effect of the green tea component, L-theanine on environmental toxins-induced neuronal cell death [J]. Neuro Toxicology, 2008, 29: 656—662.

[15] Yamada T, Terashima T, et al. Theanine, γ-glutamylethylamide, increases eurotransmission concentrations and neurotrophin mRNA levels in the brain during lactation [J]. Life Sciences, 2007, 81:1247—1255.

[16] 林智,杨勇等.茶氨酸提取纯化工艺研究[J].天然产物研究与开发,2004,16(5):442—447.

[17] 崔湘兴,龚雨顺等.717阴离子交换树脂吸附茶氨酸的热力学研究[J].湖南农业大学学报(自然科学版),2008,34(5):601—603.

[18] 唐仕荣,宋慧等.茶氨酸的HPLC快速检测[J].中国食品添加剂分析检测,2006,(4):160—163.

[19] Du Y Y, Liang Y R, et al. A study on the chemical composition of albino tea cultivars [J]. J. Hort. Sci. Biotechnol, 2006, 81(5): 809—812.

[20] Desai M J, Armstrong D W. Analysis of derivatized and underivatized theanine enantiomers by high-performance liquid chromatography/atmos-

pheric pressure ionization-mass spectrometry[J]. Rapid Commun. Mass Spectrom,2004,18:251－256.

[21] 杨秋生,徐平湘,李宇航等.茶氨酸和厚朴提取物对7日龄小鸡分离应激过程的影响[J].中国中药杂志,2007,32(19):2040－2043.

[22] 吴春兰,王娟,黄亚辉.茶氨酸的研究进展[J].茶叶通讯,2010,37(9):15－20.

[23] Matsumoto K,Yamamoto S,et al. Antidiabetic Activity of Zn(II) Complexes with a Derivative of L-Glutamine[J]. Bull Chem SocJpn, 2005, 78:1077－1081.